AGENT-BASED MODELING OF TAX EVASION

Wiley Series in Computational and Quantitative Social Science

Computational Social Science is an interdisciplinary field undergoing rapid growth due to the availability of ever increasing computational power leading to new areas of research. Embracing a spectrum from theoretical foundations to realworld applications, theWiley Series in Computational and Quantitative Social Science is a series of titles ranging from high level student texts, explanation and dissemination of technology and good practice, through to interesting and important research that is immediately relevant to social / scientific development or practice. Books within the series will be of interest to senior undergraduate and graduate students, researchers and practitioners within statistics and social science.

Behavioral Computational Social Science

Riccardo Boero

Tipping Points: Modeling Social Problems and Health

John Bissell (Editor), Camila Caiado (Editor), Sarah Curtis (Editor), Michael Goldstein (Editor), Brian Straughan (Editor)

Understanding Large Temporal Networks and Spatial Networks: Exploration, Pattern Searching, Visualization and Network Evolution

Vladimir Batagelj, Patrick Doreian, Anuska Ferligoj, Natasa Kejzar

Analytical Sociology: Actions and Networks

Gianluca Manzo (Editor)

Computational Approaches to Studying the Co-evolution of Networks and Behavior in Social Dilemmas

Rense Corten

The Visualisation of Spatial Social Structure

Danny Dorling

AGENT-BASED MODELING OF TAX EVASION
THEORETICAL ASPECTS AND COMPUTATIONAL SIMULATIONS

Edited by

Sascha Hokamp
Universität Hamburg

László Gulyás
Eötvös Loránd University

Matthew Koehler
The MITRE Corporation

Sanith Wijesinghe
The MITRE Corporation

WILEY

Registered Offices
John Wiley & Sons, Inc., 111 River Street, Hoboken, NJ 07030, USA
John Wiley & Sons Ltd, The Atrium, Southern Gate, Chichester, West Sussex, PO19 8SQ, UK

Editorial Office
9600 Garsington Road, Oxford, OX4 2DQ, UK

For details of our global editorial offices, customer services, and more information about Wiley products visit us at www.wiley.com.

Wiley also publishes its books in a variety of electronic formats and by print-on-demand. Some content that appears in standard print versions of this book may not be available in other formats.

Library of Congress Cataloging-in-Publication Data applied for

Hardback ISBN: 9781119155683

Cover image: © KrulUA/Gettyimages

Set in 10/12pt TimesLTStd by SPi Global, Chennai, India

Printed in Singapore by C.O.S. Printers Pte Ltd

10 9 8 7 6 5 4 3 2 1

To our families

Contents

Notes on Contributors

Amanda Andrei is a social scientist and senior artificial intelligence engineer at The MITRE Corporation specializing in social analytics, development of innovative processes and spaces, and application of mixed methods to sociotechnical problems. She received her BA in Anthropology from the College of William & Mary, her Certificate in Computational Social Science from George Mason University, and her MA in Communication, Culture, and Technology from Georgetown University.

Kim M. Bloomquist is an operations research analyst with the U.S. Internal Revenue Service's Taxpayer Advocate Service. He has authored and coauthored numerous papers and book chapters on the economics of taxpayer compliance and agent-based modeling. His research has been cited in Congressional testimony, the *Washington Post*, *Tax Notes*, and *Tax Notes International*. He has received several awards for his research on tax compliance including the Organization for Economic Cooperation and Development's (OECD) Jan Francke Tax Research Award, the Cedric Sandford medal for the best paper at the seventh International Conference on Tax Administration in Sydney, Australia, and the IRS Research Community Award for Research Technical Expertise. Kim received his PhD (Computational Social Science) from George Mason University, Fairfax, Virginia in 2012.

Kevin Comer is a Modeling and Simulation Engineer at the MITRE Corporation, specializing in agent-based modeling design, development and validation. His primary focus is in simulating the dynamics of the individual health insurance market. He also works on modeling processes in cybersecurity and homeland security. He received his B.S. in Systems Engineering and Economics from the University of Virginia, his M.S. in Operations Research from George Mason University Volgenau School of Engineering, and his Ph.D. in Computational Social Science from George Mason University's Department of Computational Social Science. His email address is ktcomer@mitre.org.

Andrés M. Cuervo Díaz completed his BSc in Physics from Universidad de los Andes, Bogota, Colombia, in 2014 in the field of quantum optics and is a master's student in the study program "Integrated Climate System Sciences" at the Cluster of Excellence (DFG EXC 177 CliSAP), participating in the research group Climate Change and Security (CLISEC). He is mainly interested in agent-based and integrated assessment modeling for the evaluation of policies' performance and possible strategies to enhance environmental protection and climate change action.

László Gulyás holds a PhD in Computer Science from Eötvös Loránd University, Hungary. He is an assistant professor at Eötvös Loránd University, Budapest, and held the position of head of division at AITIA International, Inc. He has been doing research on agent-based modeling and multiagent systems since 1996. He has been involved in teaching both graduate and undergraduate level courses in agent-based modeling and simulation at Harvard University, at the Central European University and at the Eötvös Loránd University, Hungary. He has authored several book chapters and journal articles. In particular, László has published on agent-based modeling of tax evasion since 2009.

Nigar Hashimzade obtained her PhD in Economics from Cornell University in 2003. Prior to her current position at Durham she has worked at the University of Exeter and the University of Reading. Her research interests are in applied microeconomic theory and in quantitative methods. Since 2012, she has been involved in research on the behavioral approach to tax evasion funded jointly by the ESRC, HMRC, and HMT. Among other projects, she works on developing behavioral models of tax compliance decisions in social networks and on building agent-based models that can be used to assess and compare tax enforcement policies in a complex environment. Nigar has published in leading international academic journals and contributed to a number of monographs, for some of which she has also been a coeditor. Her previous career was in theoretical physics.

Christine Harvey is a high-performance and analytic computing engineer at The MITRE Corporation in Washington, D.C. She specializes in data analysis and high-performance computing in simulations. In addition, her research interests include agent-based modeling and big data analysis, particularly in the field of Healthcare. She completed her masters in Computational Science from Stockton University in 2013 and is expected to finish her PhD in Computational Science and Informatics at George Mason University in 2018.

Erik Hemberg is a postdoctoral researcher in the ALFA group in CSAIL at the Massachusetts Institute of Technology. He performs research regarding scaleable machine learning. He is currently involved in research regarding tax evasion and physiological time series prediction. He received his PhD in Computer Science from the University College Dublin, Ireland. At ICAIL 2015 he co-authored the Peter Jackson Best Innovative Application paper.

Sascha Hokamp obtained his PhD in Public Economics from the Brandenburg University of Technology Cottbus, Germany, in 2013. He was awarded a stipend from the Deutsche Bundesbank via the "Verein für Socialpolitik" in 2011 and 2012. He is involved in organizing the biannual "Shadow Economy" conference series, founded at the Westfälische Wilhelms-Universität Münster, Germany, in 2009. He served in 2015 as a guest editor in *Economics of Governance* on "The Shadow Economy, Tax Evasion, and Governance." Sascha is a member of the Research Unit for Sustainability and Global Change (FNU) and of the Center for Earth System Research and Sustainability (CEN), Universität Hamburg, Germany, and he is participating researcher at the Cluster of Excellence "Integrated Climate System Analysis and Prediction" (DFG EXC 177 CliSAP), on the project "Societal Use of Climate Information." His research topics are integrated assessment modeling of climate change, agent-based modeling of environmental challenges, and climate policy as well as illicit activities (tax evasion and doping in elite-sports) and the shadow economy.

Matthew Koehler is the Applied Complexity Sciences Area Lead for the U.S. Treasury/Internal Revenue Service, U.S. Commerce, and Social Security Administration Program Division at The MITRE Corporation. He has concentrated on decision support using agent-based models and simulations, and the analysis and visualization of large datasets coming from complex systems. He received his AB in Anthropology from Kenyon College, his MPA from Indiana University's School of Public and Environmental Affairs, his JD from George Washington University's Law School, and his PhD in Computational Social Science from George Mason University's Krasnow Institute for Advanced Study, Department of Computational Social Science.

Toni Llacer holds a PhD in Sociology from the Universitat Autónoma de Barcelona (UAB), Spain. His research field is the interdisciplinary study of tax evasion. He obtained a bachelor's degree in Economics from Universitat Pompeu Fabra, a bachelor's degree in Philosophy from Universitat de Barcelona (Academic Excellence Award), and a master of science in Applied Social Research from UAB.

Tamás Máhr has a PhD in Computer Sciences from the Delft University of Technology, The Netherlands, and project manager of the simulation group of AITIA International Inc. He has been the senior software architect of the CRISIS project on agent-based modeling of the macro-financial system, responsible for the CRISIS Game Architecture, as well as for the CRISIS Integrated Simulator. His past research involved agent-based transportation planning and robustness of multiagent logistical planning, but he also published on auction algorithms, which are commonly used mechanisms in multiagent systems. Before the PhD track, Tamás got his MSc from the Budapest University of Technology and Economics, Hungary.

Shaun Michel is a computational sociologist in The MITRE Corporation's Department of Artificial Intelligence and Cognitive Science, specializing in simulation modeling, social behavior, and globalization. He began working on a PhD in Sociology at George Mason University in 2012 and he holds a master's degree in Sociology from East Tennessee State University.

Francisco J. Miguel Quesada is an associate professor at the Universitat Autónoma de Barcelona (UAB) teaching Methodology for the Social Sciences, Sociology of Consumption, and Applied Statistics for Marketing Analysis. He holds a PhD in Sociology from the UAB and a university specialist degree in Sociology of Consumption from the Universidad Complutense de Madrid. He has conducted research in sociology of consumption, social indicators, and school-to-work transitions. At present, he mainly works in the domain of computational sociology as GSADI group member. As Head of the "Laboratory for Socio-Historical Dynamics Simulation" (LSDS), he has been involved in several projects on the use of agent-based social simulation for modeling social networks dynamics and evolution of social behavior.

Luigi Mittone is a full professor of Economics at the University of Trento, Italy. At the University of Trento he is the Director of the Doctoral School of Social Sciences, Director of the Cognitive and Experimental Economics Laboratory, and coordinator of the International Master in Economics (MEC). He is also the coordinator of the research project in Experimental Economics and Nudging at the Bruno Kessler Research Center. He attended his master's degree in Economics at the Bocconi University of Milan (Italy), his master's degree in Social Sciences at the University of Birmingham (UK), and his PhD in Economics at the University of Bristol (UK). Luigi's main research interests and publications are in the field of Experimental and Behavioral Economics: fiscal evasion theory, consumer behavior, mental modeling of uncertain events, intertemporal choices, and cooperation among agents; Computational Economics: fiscal system dynamics with heterogeneous agents; and Public and Health Economics.

Gareth Myles is Professor of Economics and a Research Fellow at the Institute for Fiscal Studies. He is an associate editor of the *Journal of Public Economic Theory* and was managing editor of *Fiscal Studies* from 1998 to 2013. He obtained his BA from Warwick in 1983, his MSc from the London School of Economics in 1984, and his DPhil in 1987 from Oxford. His major research interest is in

Public Economics and his publications include *Public Economics* (1995), *Intermediate Public Economics* (2006) and numerous papers in *International Tax and Public Finance*, the *Journal of Public Economic Theory*, and the *Journal of Public Economics*. Gareth is an academic adviser to HM Treasury and HM Revenue and Customs. He has also provided economic advice to international bodies including the European Commission and the OECD.

José A. Noguera is an associate professor (2002) in the Department of Sociology at the Universitat Autónoma de Barcelona, and Director of the Analytical Sociology and Institutional Design Group (GSADI). He holds a PhD in Sociology (1998) from the Universitat Autónoma de Barcelona and has been visiting researcher at the University of California, Berkeley, and at the London School of Economics and Political Science. He has been the principal investigator of several GSADI projects funded by the R&D Spanish National Plan since 2006. His empirical research is currently focused on tax compliance, social influence dynamics, prosocial motivations, and the feasibility of universal welfare policies such as basic income. José's publications also cover sociological theory, philosophy of social science, social policy, and normative social theory. His work in sociological theory aims to demonstrate the explanatory power of analytical sociology and the social mechanisms approach in sociology, with a strong focus on methodological individualism, the theory of rationality, social ontology, and the philosophy of social science. He is editor of *PAPERS: Revista de Sociologia*, an editorial board member of Basic Income Studies, and has served on the editorial board of *Revista Española de Investigaciones Sociológicas*. He is a member of the International Network of Analytical Sociologists (INAS) and serves on the Board of the Spanish Basic Income Network (RRB) and on the International Advisory Board of the Basic Income Earth Network (BIEN).

Una-May O'Reilly is the founder of the AnyScale Learning For All (ALFA) group at the Massachusetts Institute of Technology Computer Science and Artificial Intelligence Laboratory. She has expertise in big data, scalable machine learning, evolutionary algorithms, and frameworks for large-scale, automated knowledge mining, prediction, and analytics. She holds a PhD from Carleton University, Ottawa, Canada. She is an elected Fellow of the International Society for Genetic and Evolutionary Computation, holds the EvoStar Award for Outstanding Achievements in Evolutionary Computation, is Vice Chair of ACM SIGEVO, area editor for Data Analytics and Knowledge Discovery for Genetic Programming and Evolvable Machines (Kluwer), editor for Evolutionary Computation (MIT Press), and action editor for the Journal of Machine Learning Research. At ICAIL 2015 she co-authored the Peter Jackson Best Innovative Application paper.

Aloys L. Prinz is a full professor of Economics at Westfälische Wilhelms-Universität Münster, Institute of Public Economics. His work focuses primarily on public economics and taxation. He is the co-organizer of the biannual conferences on the shadow economy and tax evasion. In addition to publishing intensively in scientific journals, he has also coauthored books for general readers on social policy, public debt, monetary policy and taxation. Moreover, Aloys is also an active researcher in economic ethics.

Jacob Rosen is a chief technology officer at Deftr. His research focuses on the abstract representation of financial transactions and the use of artificial intelligence techniques to characterize human behaviors. Jacob received an MSc in Technology and Policy from the Massachusetts Institute of Technology.

Viola L. Saredi is a junior research fellow at the Cognitive and Experimental Economics Laboratory at the University of Trento, Italy. As a junior researcher, she has recently visited the Department of Applied Psychology at the University of Vienna. She holds a PhD degree in Behavioral and Experimental Economics from the University of Trento, and a master's degree in Economics from the Catholic University of the Sacred Heart, Milan, Italy. In keeping with the interdisciplinary nature of her field of interest, Viola's research mainly focuses on the application of innovative experimental methods to the study of psychological drivers and biases affecting consumers' behavior and clandestine activities (such as tax evasion, bribing, and agency dilemmas).

Götz Seibold is a professor at the Brandenburg University of Technology Cottbus-Senftenberg, Germany, where he holds the Chair for Computational Physics. He received his PhD in Physics from the University of Stuttgart, Germany, in 1995. He works in the general area of solid-state theory with a focus on superconductivity and strongly correlated electron systems. He is internationally recognized for his work on electronic inhomogeneities in high-temperature superconductors and the time-dependent Gutzwiller approximation. Besides, Götz is interested in the field of econophysics where he works on agent-based models related to tax evasion and the shadow economy.

David Slater is a lead operations research analyst within The MITRE Corporation's Operations Research Department. He has spent the last 5 years at MITRE applying Applied Mathematics to some of the United States' most critical and challenging Systems Engineering problems. His research interests include complex systems analysis, network science, and complexity science. He received his BS from Michigan Technological University in 2005 and an MS and PhD in Applied Mathematics from Cornell University in 2009 and 2011, respectively.

Eduardo Tapia Tejada holds a PhD in Sociology from the Universitat Autònoma de Barcelona and is a postdoctoral researcher at The Institute for Analytical Sociology (Sweden). He is a graduate in Sociology from the Universidad Federico

Villarreal of Peru and he holds a master's degree in Sociological Research from the Universitat Autònoma de Barcelona. He also holds a degree in Design and Evaluation of Social Projects from the Pontificia Universidad Católica del Perú. He was a postdoctoral research fellow at GSADI (2013–2014) in the project Social and environmental transitions: SIMULPAST. His research interests involve two main areas: diffusion dynamics, particularly how certain beliefs spread through social groups, and beliefs evaluation mechanisms. In order to tackle these issues, Eduardo uses tools such as network analysis, statistical modeling, and agent-based simulation. The main theoretical question that motivates his research is to analyze how actions at the micro level generate patterns at the macro level.

István János Tóth is senior research fellow at the Institute of Economics of the Hungarian Academy of Sciences in Budapest and managing director of Corruption Research Center Budapest (CRCB). He graduated at the Corvinus (former Karl Marx) University of Budapest in 1984, and studied at École des Hautes Études en Sciences Sociales (EHESS) in Paris in 1990–1991. He holds a master's degree in Economics and Sociology. István János obtained his PhD at the Hungarian Academy of Sciences in 1998. His main research interests are the problems of corruption and economic institutions, hidden economy, and tax evasion.

Geoffrey Warner is a lead information systems engineer at The MITRE Corporation. He received his PhD in theoretical condensed matter physics from the University of Illinois at Urbana-Champaign. His research involves the application of mathematical and computational modeling to a diverse range of problems, from tax evasion to the physics of remote sensing devices.

Sanith Wijesinghe is an Innovation Area Leader at The MITRE Corporation where he oversees research and development efforts to support multiple federal government agencies. His most recent work on detecting tax evasion schemes was awarded the Peter Jackson award by the International Association of Artificial Intelligence and Law and was featured in the New York Times. Prior to joining MITRE, he was Vice President for Project Deployment at MillenniumIT, a software development firm serving the capital markets industry. He received his master's and PhD degrees in Aeronautics/Astronautics from the Massachusetts Institute of Technology.

Foreword

We model social systems to better explain, predict, design, and act. Macroeconomic models explain growth rates and patterns. Financial models predict stock prices. Models of correlated assets help the Federal Communications Commission design their spectrum auctions. And industrial organization models of firm competition inform decisions to intervene by the Justice Department.

The aforementioned macroeconomic models differ in their assumptions and methodologies. Macroeconomists gloss over firm level strategic choices that lie at the core of the industrial organization models. Auction models assume strategically rational actors, whereas some financial models assume actors who buy and hold.

Models include some variables and leave out others. They simplify the world, and as a result, err. We can correct for the biases and faults in any one model by constructing complementary models. To be of use, these other models must capture or include different features or relationships. They can then fill in the gaps missed by the original model. By offering more coverage and encompassing a larger number of processes, an ensemble of models provides deeper and broader knowledge and more refined insights than does a single model.

This book exemplifies this many-model approach. The chapters within explore a single policy domain – tax evasion – through the lenses of a variety of agent-based models (ABMs). While each model makes a substantial stand-alone contribution, the volume in its entirety makes a much larger one.

ABMs consist of situated agents endowed with attributes who follow rules instantiated in computer programs. The modeler defines the agents, their attributes, their physical or social placement, the rules they follow, and the processes by which their behaviors aggregate. She then hits the return key and watches phenomena emerge. Hence, many refer to ABMs as a bottom-up methodology.

An ABM of friendship formation might position people in a space, assign them physical features and rules for making and breaking friendships based on those features, and then allow networks to form. The modeler builds in the assumptions. The friendships emerge.

ABMs offer finer granularity and more flexibility than game theory models. Game theory models often consist of two, three, or an infinity of types – as these make deriving equilibria possible. Game theory models also assume, for the most part, rational actors. An ABM can include any number of types as well as any behavioral rule that can be encoded as an algorithm.

ABMs can also include heterogeneity within types, network structures, learning and adaptation, and (almost) any size population. That flexibility forms an Achilles heel large enough to be hit with an arrow from about 10,000 feet. Almost anyone can program an ABM. Not many people can solve for a Bayesian Perfect Subgame Equilibrium. Thus, there exists an abundance of not so useful, baked in the results, ABMs. It might even be that as time goes to infinity, so too does the ratio of bad to useful ABMs.

Books such as this can reverse that trend. The editors have selected a collection of sound exemplars. And, as if to rein in over-eager newbies, the editors lead with a treatise on the art and practice of writing scientific ABMs. They take the stand that ABMs can assist, aid, and improve, and do not claim that ABMs are the be all or bee's knees. This modesty stems from an awareness that ABMs have been touted (overhyped) as a breakthrough methodology for social science for more than a quarter-century.

ABMs have had successes to date, particularly in domains such as epidemiology, traffic, crowd dispersion, racial segregation, and meme spreading. These are all domains in which the network of interactions or geography plays a large role and behavior is low-dimensional or rule-based. This foray into tax evasion offers an opportunity to push the frontier of impact into economic policy.

At the moment, ABMs occupy an awkward position within economic analysis. On the one hand, they have become ubiquitous and ridiculously broad in scope: The set of ABMs range from elaborate government-funded, million agent models with household level granularity, to laptop models of supply chain dynamics, to models by physicists of spinning magnets. On the other hand, they remain a (somewhat) fringe methodology among the academic elite. Not many ABMs make their way into the top journals. That said, journal editors do serve up prodigious quantities of the near beer of numerical estimations and calibration exercises.

Tax evasion – the singular focus of this work – provides a near ideal context by which to demonstrate the potential of ABMs. That potential can be best seen juxtaposed with neoclassical economic models.

The straw man neoclassical economic model of tax evasion assumes a single representative agent who decides how much to shade reported income. The model also makes assumptions about the tax progressivity, risk-aversion, and auditing policies – do tax authorities look harder at rich people or do they monitor according to an equilibrium strategy?

The amount of evasion predicted by neoclassical models depends on the assumptions. Increasing risk-aversion drives up evasion and progressive taxation biases the income of evaders upward. These models operate and succeed within a small box. They explain why people cheat, they predict who will cheat, they point to better designs for tax policy, and they tell the Internal Revenue Service where to look for evaders.

Though useful as a tool for organizing our thinking about the implications of tax progressivity, changing levels of risk-aversion, and auditing practice, they prove a rather blunt instrument. They provide deeper insights than say "bigger houses sell for more money" and "studying harder improves grades" but not by much.

ABMs allow for a much richer set of assumptions. They also embrace the complexity of the worlds in which evasion takes place. Evasion does not take the form of lying about 5% of your income. Evasion comes in the form of equity swaps, shell companies, misreporting of expenses and gains, and even the now famous Double-Dutch Irish Sandwich used by corporations.

The models in this book emphasize broadening assumptions within four categories: behavior, occupations, networks, and social influence. Each of these four extensions adds value on its own. As an ensemble, they add much more.

By behavioral diversity, I mean differences in how people make decisions about paying taxes. The neoclassical model assumes that each person makes a calculated decision whether or not to cheat. If that model produces boundary decisions, then either people do not cheat, or they all cheat. If it produces interior solutions, everyone cheats a little.

Some of the models discussed in this book include three other types of people: honest people who always pay their taxes (that's me by the way!), socially influenced people who mimic the behaviors of others, and random, idiosyncratic people who evade taxes either on a whim or because they accidentally tossed out a 1099 form thinking it was another American Association of Retired Persons application.

The honest types lower the number of evaders. That alone is a contribution because the economic models suggest far more evasion than can be identified empirically. The socially influenced types create patterns in tax evasion. If evaders avoid the tax authorities, then social influence creates a positive feedback for evasion. The random, idiosyncratic people introduce noise.

Second, the ABMs include occupational variation. This allows for more realistic assumptions. A school teacher or line worker who earns all her income as salary and receives an employer wage reporting form W2 has less room for evasion than a New York City financial consultant pulling in $10 million per year who works part time out of her home in Connecticut. The consultant can claim any number of business-related expenses (a Rolling Stones concert) that the teacher or line worker cannot. The consultant can also claim to live more than half the year in Connecticut and avoid the nearly 4% New York City income tax. Although, if she reports 183 days in CT and 182 in NYC, she will trigger an audit automatically.

The key observation that evasion requires the opportunity to evade is obvious yet ignored by the economic model. To construct an ABM to capture how opportunity does not spread equally across professions requires categorizing occupations by evasion potential and calibrating the numbers to economic data. For the ABM to be the cat's meow and not the dog's breakfast, the categorizations must be germane, the assumed behaviors within a category must be appropriate, and the calibration must be precise. All are doable. None are easy.

Third, ABMs naturally embed social networks and geography. In the real world, people reside in physical space, and they mix with their friends rather than randomly. The same assumptions can be built into an ABM. Adding the network has no effect unless information or influence spreads over the network. Hence, my earlier comment about the whole exceeding the parts. The neoclassical evaders can be put in any network you like. Nothing will change.

However, when we add social influence (the final category), networks do matter. ABMs typically add social influence in one of two ways, both can be found in this book. As already mentioned, an ABM can include some agents who are socially influenced. Alternatively, an ABM can assume that everyone has some level of susceptibility to social pressure. Much like someone drives 68 mph in a 65 mph limit zone because "everyone else does," so too might someone claim a home office even though their father-in-law now sleeps in that room.

To be fair, game theoretic models can also include social influence and they can even capture social influence over networks. However, when we add in behavioral types and occupational types, we soon find that our model world contains too many types and cases to perform all those calculations. Computation proves the only practical path.

Computation does more than overcome the combinatorial explosion of cases. It also reveals the class of outcome that obtains. Some systems go to equilibria quickly. Others produce simple patterns. Others walk randomly within a set. And last, others produce complex, often novel patterns.

The potential for patterns, randomness, and complexity in addition to equilibria challenges the standard comparative static evaluative approach, that is, if we turn this policy knob to the right, the equilibrium moves in this direction. We might also turn a knob and transform a well-behaved system into one that is chaotic.

This book should speak, if not sing, to two kinds of audiences. For those interested in the methodology of ABMs, what follows includes a mixture of novel applications of workhorse models along with new, higher fidelity models. For those interested in tax evasion, and the potential efficacy of interventions, it provides a wealth of ideas and insights. For both audiences, the book will expand their thinking about tax evasion.

In sum, although Ben Franklin claimed that, "In this world nothing can be said to be certain, except death and taxes," in fact, many people do avoid and evade taxes. They have their Double-Dutch Irish Sandwich and eat it too. To explain and predict evasion within the complex financial world, to design systems that reduce it, and to take appropriate actions, we need many models, and we need models that embrace the complexity of the world. This collection of ABMs gets us started on that path.

Scott E. Page
University of Michigan
Santa Fe Institute

Preface

Tax evasion or tax noncompliance is an age-old behavior of humans. It is probably as old as taxation itself. Taxation is an essential tool of any institutionalized community, a means by which resources are collected for the common causes. However, free-riding also appears to be an age-old, ever-existing phenomenon, leading to actors seeking safe ways to avoid the burdens of taxation. Various rulers and governments developed various taxes and tax systems, adding several layers of interpretations for avoiding paying taxes. Subjects of unjust rulers may feel disassociated with the projects that are financed by their taxes and may also feel that they do not possess the power to influence the decisions on the allocation of the taxes collected. This leads to a very complex net of possible reasons, causes, and effects for tax noncompliance that culminate in the repeated decisions by millions of individuals. Such decisions lead to the numbers individuals put in their tax reporting forms.

Agent-based modeling, on the other hand, is a novel methodology to study complex social systems, made possible by the recent advances in computer technology. Its main tenet is to model the individual decision maker, with all its idiosyncrasies and interactions, with specific goals and concerns. The multi-level, networked interactions of these individually modeled decisions are often too complex to follow with the analytical means provided by traditional mathematics. Therefore, the consequences of the micro-level rules are simulated in a computer instead, and the macro-level patterns that emerge are studied.

The mission of our book is to apply the emerging method of agent-based modeling to address the age-old problem of tax evasion.

While the sparking idea for the book emerged in 2014 in Barcelona, at a dinner during the Social Simulation Conference (The 10th Conference of the European Social Simulation Association (ESSA)), the origins can be traced back to the series of Shadow Conferences on the shadow economy, co-organized by *Sascha Hokamp*, where all the other editors of this volume had participated. At the dinner in Barcelona, *Sascha Hokamp* and *László Gulyás* were discussing the increasing number of agent-based contributions to the field of tax noncompliance. They decided that the field would benefit from a collection of the best works in this domain, especially when presented from a unified perspective. After the book project was accepted by John Wiley & Sons Limited, they decided to strengthen the editorial team by inviting two colleagues from the United States: *Matthew Koehler* and *Sanith Wijesinghe*.

The current volume is the result of three-year-long efforts of several people. The editors would like to thank all the authors who accepted our invitations and contributed the best of their work to this volume. We are also indebted to John Wiley & Sons Limited for supporting the production of this book since the very beginning. In particular, we are grateful for the cooperation by *Viktoria Hartl-Vida*, *Debbie Jupe*, *Heather Kay*, *Alison Oliver*, *Blesy Regulas*, *Jo Taylor*, and *Liz Wingett*. Furthermore, we would like to extend our thanks to the numerous additional people at John Wiley & Sons Ltd who were involved in the production of this book but are not named here explicitly. Special thanks are also due to three distinguished scholars, *James Alm*, *Scott E. Page*, and *Klaus G. Troitzsch* for their help with advertising the book. The valuable comments and contributions of 14 anonymous external reviewers are also gratefully acknowledged. *Sascha Hokamp* would like to thank *Silvija Glodaite* and *Vanessa Hesping* for their assistance in Warsaw and Barcelona, respectively. We, the editors, dedicate this volume to our families.

Sascha Hokamp
Universität Hamburg

László Gulyás
Eötvös Loránd University

Matthew Koehler
The MITRE Corporation

Sanith Wijesinghe
The MITRE Corporation
September
2017

Part I
Introduction

1

Agent-Based Modeling and Tax Evasion: Theory and Application

Sascha Hokamp, László Gulyás, Matthew Koehler and Sanith Wijesinghe

1.1 Introduction

While the formal study of tax evasion began with a seminal paper by M.G. Allingham and A. Sandmo titled "Income Tax Evasion: A Theoretical Analysis" in 1972, scholars and practitioners continue to be challenged with designing and implementing policies and incentives to mitigate tax evasion. While early theoretical studies provide a baseline from which to evaluate hypotheses, the methodology underlying the classical formulation; that is, the use of utility functions and the assumptions of taxpayer homogeneity and rationality, fall short in characterizing taxpayer behaviors observed in practice. In this book, we seek to advance the state of the art in the study of tax evasion by presenting an alternative computational approach based on simulating individual agents. These so-called *agent-based models (ABM)* aim to take into account individual preferences and can accommodate a larger variety of intrinsic and extrinsic variables to help explore a broader space of compliance outcomes. In this introductory chapter, we present a formal definition of tax evasion in Section 1.2 and outline the case for why its analysis is a priority not only for tax administrators, but also for society at large. The classical theoretical models of tax evasion are then summarized

Agent-Based Modeling of Tax Evasion: Theoretical Aspects and Computational Simulations,
First Edition. Edited by Sascha Hokamp, László Gulyás, Matthew Koehler, and Sanith Wijesinghe.
© 2018 John Wiley & Sons Ltd. Published 2018 by John Wiley & Sons Ltd.

in Section 1.3 to provide historical context as well as an understanding of the assumptions and abstractions that have helped formalize the preliminary studies in this field. The ABM paradigm is introduced in Section 1.4 with an outline of the Overview, Design Concepts, and Details plus Decision-Making (ODD+D) protocol (Müller *et al.*, 2013) that we propose as a standard to guide both the development and presentation of simulation results. Here we stress the need to calibrate and replicate ABMs to help further advance our collective efforts. A literature review follows in Section 1.6 describing the tax evasion agent-based methods developed to date. In conclusion, Section 1.7 provides an overview of the edited volume by summarizing the remaining chapters that further explore the nuances in this field.

1.2   Tax Evasion, Tax Avoidance and Tax Noncompliance

What constitutes tax evasion? Alm (2012, p. 55) defines tax evasion as the "illegal and intentional actions taken by individuals to reduce their legally due tax obligations." The most common "strategies" for tax evasion involve intentionally underreporting income, claiming fake deductions and credits, exploiting loopholes in the tax regulations, and engineering artificial losses. The US Internal Revenue Service (IRS) links tax evasion with tax fraud, which it defines as "… an intentional wrongdoing, on the part of the taxpayer, with the specific purpose of evading a tax known or believed to be owing" (IRS, 2016b). Note that tax evasion should be distinguished from legal tax avoidance that allows individuals to reduce their tax liability through legitimate means. A grey area exists when individuals exploit tax loopholes as a means for tax avoidance. In certain countries (such as Germany) this may fall within the definition of legal tax avoidance. Notably, Slemrod and Yitzhaki (2002) distinguish tax avoidance and evasion, where the former is the legal usage of tax loopholes to lower the tax burden while the later is illegal. Thus the law draws the line between tax avoidance and evasion.

An additional term "tax noncompliance" also appears in the literature and is sometimes used interchangeably with tax evasion. We consider tax noncompliance as a more general term that includes *both* intentional evasion *and* unintentional errors. However, the tax gap studies by the IRS reports overall tax noncompliance (estimated at $458 billion for 2008–2010, see IRS, 2016a) without attempting to consider this distinction (GAO, 2012, p. 3). Hence, strictly speaking, interchangeable use of tax evasion and tax noncompliance is inconsistent. To be clear, our discussions center on characterizing intentional acts by individuals to reduce their tax liability rather than on factors that may lead to mistakes in tax reporting.

One may ask, why analyze tax evasion? First and foremost, tax is essential to fund public expenditures. Tax evasion is therefore not only a concern for the tax authorities, but also for society at large. Given a nation's investment in healthcare, education, defence, social security, transportation, infrastructure, science and technology are derived substantially from public financing, tax evasion causes misallocations of public goods (Andreoni *et al.*, 1998; Slemrod and Yitzhaki, 2002; Torgler, 2002; Kirchler, 2007; Slemrod, 2007; Alm, 2012, provide surveys, and James and Edwards, 2010, present a bibliography of the literature on tax compliance). Moreover, given the nonnegligible extent of tax evasion as indicated by the tax gap estimates by the IRS (2016a), tax authorities can incur significant expenses to enact deterrence efforts. An analysis of tax evasion can therefore be used to identify factors that influence compliance rates and ultimately help governments reach revenue targets. Secondly, by exploring issues of tax evasion across different tax systems, there is an opportunity to inform and share insight about the effectiveness of different policy initiatives. As an example, even though the tax rate, tax base, and tax-exempt amounts may differ among nations, cross-border comparison of countries with flat income taxes might yield valuable lessons. Now that we have briefly discussed the meaning of terms used to describe tax evasion, tax avoidance, and tax noncompliance, we will now address theories of tax evasion.

1.3 Standard Theories of Tax Evasion

Research on tax evasion started in the 1970s with the seminal works by Allingham and Sandmo (1972) and Srinivasan (1973), who applied the economics-of-crime approach by Becker (1968, 1993) to tax reporting behavior.[1] Comparing these tax reporting models reveals various distinctions and similarities, that can be grouped based on (i) mathematical modeling, (ii) taxpayer's optimal choice, (iii) comparative statics, (iv) framework extensions, and (v) model critique.

To begin with, we briefly describe standard versions of the tax evasion models. Allingham and Sandmo (1972) consider an individual, a taxpayer, who is faced by a decision problem about how much income X of the true income I to declare to the tax authority given an audit probability α, a tax rate θ, and a penalty rate π. Further, the taxpayer is assumed to be risk averse, so that the marginal utility $\mathcal{U}'$ is strictly decreasing;, that is, the resulting utility function $\mathcal{U}$ is concave. To solve the decision problem, the taxpayer maximizes his expected utility

$$\mathcal{E}\mathcal{U}[X] = (1 - \alpha)\mathcal{U}[I - \theta X] + \alpha\mathcal{U}[(1 - \pi)I + (\pi - \theta)X] \quad (1.1)$$

such that the necessary condition for a maximum is

$$(1 - \alpha)(-\theta)\mathcal{U}'[I - \theta X] + \alpha(\pi - \theta)\mathcal{U}'[(1 - \pi)I + (\pi - \theta)X] = 0 \quad (1.2)$$

---

[1] On the one hand, the economics-of-crime approach allows for a rational point of view on individual illegal activities. On the other hand, such an analysis takes into account social welfare maximization and related problems to design optimal policies for combating illegal behavior. At the Nobel Lecture Becker (1993, p. 391) highlighted that the economics-of-crime theory and its analytical tools have allowed the examination of a variety of topics, for instance, tax evasion.

The taxpayer has an incentive toward tax evasion if his marginal expected utility is positive for total evasion, that is, $X = 0$, and negative for full tax compliance, that is, $X = I$; mathematically speaking the first derivative of the expected utility with respect to income declaration X needs to have a sign change, and the second derivative must be negative. The latter condition is fulfilled because of the concavity of the utility function and the former condition leads to

$$\left.\frac{\partial \mathcal{E}\mathcal{U}[X]}{\partial X}\right|_{X=0} = (1 - \alpha)(-\theta)\mathcal{U}'[I] + \alpha(\pi - \theta)\mathcal{U}'[(1 - \pi)I] > 0 \qquad (1.3)$$

and

$$\left.\frac{\partial \mathcal{E}\mathcal{U}[X]}{\partial X}\right|_{X=I} = (1 - \alpha)(-\theta)\mathcal{U}'[(1 - \theta)I] + \alpha(\pi - \theta)\mathcal{U}'[(1 - \theta)I] < 0 \quad (1.4)$$

Rearranging Eqs (1.3) and (1.4), yields the condition which guarantees an interior solution for the income maximization problem

$$\theta > \alpha\pi > \theta\left(\alpha + (1 - \alpha)\frac{\mathcal{U}'[I]}{\mathcal{U}'[(1 - \pi)I]}\right) \qquad (1.5)$$

If the tax rate θ changes, then a fixed penalty rate π on undeclared income $I - X$ might cause a conflict between two effects on the optimal income declared X^*: an income effect and a substitution effect. Yitzhaki (1974) shows that modeling a fine on the evaded tax via a sanction tax rate $\zeta = \pi/\theta > 1$ resolves this conflict.

Implicitly assuming a linear utility function, Srinivasan (1973) investigates the taxpayer's expected income after taxes and penalties

$$\mathcal{E}\mathcal{I}[I] = (1 - \alpha)(I - \mathcal{T}[(1 - \lambda)I]) + \alpha(I - \mathcal{T}[I] - \lambda\mathcal{P}[\lambda]I) \qquad (1.6)$$

where $\mathcal{T}[I]$ denotes the taxes as function of true income, and $\mathcal{P}[\lambda]$ the penalties as function of $\lambda = 1 - X/I$ denoting the fraction of true income not declared to the tax authority. Hence, the first-order condition for a maximum is

$$(1 - \alpha)\mathcal{T}'[(1 - \lambda)I]I - \alpha(\mathcal{P}[\lambda] + \lambda\mathcal{P}'[\lambda])I = 0 \qquad (1.7)$$

Table 1.1 summarizes the mathematical syntax and its meaning in these standard tax reporting theories. The essential distinction in the tax evasion theories is that Allingham and Sandmo (1972) employ expected utility maximization while Srinivasan (1973) maximizes an expected income after tax and penalties without an utility function modeled explicitly. The strict concavity of the utility functions permits Allingham and Sandmo (1972) to explore the tax reporting behavior of risk averse taxpayers. In particular, Allingham and Sandmo (1972) rule out linear utility functions because of the need for a strict decline of marginal utility. To put it differently, Allingham and Sandmo (1972) provide an all-or-nothing solution for the borderline case of a linear utility function, that is, a risk neutral individual. Yet,

Table 1.1 Mathematical syntax for standard theories of tax evasion

Mathematical syntax	Meaning
α	Audit probability
θ	Tax rate
π	Penalty rate
ζ	Sanction tax rate
λ	Fraction of true income not declared to the tax authority
I	True income
X	Income declaration
X^*	Optimal income declaration
$\mathcal{U}$	Utility
$\mathcal{U}'$	Marginal utility
$\mathcal{EU}$	Expected utility
$\mathcal{EI}$	Expected income (after tax and penalties)
$\mathcal{T}$	Tax function
$\mathcal{P}$	Punishment function

risk neutral subjects are implicitly considered by Srinivasan (1973).[2] Furthermore, taxpayers considered in the Allingham and Sandmo (1972) model chose an optimal amount of the income to declare while Srinivasan (1973) assumes the taxpayers to declare an optimal fraction understating the income. Both models suppose taxpayers can report between the true value, that is, full tax compliance, and nothing that is, full tax evasion. A fixed audit probability is the standard assumption in Allingham and Sandmo (1972). Thus, Srinivasan (1973) represents a broader framework in the sense of implementing an audit probability, which varies and may depend on income declaration. Srinivasan (1973) makes use of progressive tax schemes without any kind of tax allowances or deductions while the fixed tax rate in Allingham and Sandmo (1972) reflects a flat tax scenario. Further, a taxpayer faces a fixed penalty rate on nondeclared income if his cheating behavior is uncovered. Srinivasan (1973) varies a penalty rate on nondeclared income. To summarize, the mathematical features in Allingham and Sandmo (1972) and Srinivasan (1973) are quite similar but differ with respect to optimization procedures, perceived taxpayer's risk attitudes, decision variables, audit probabilities, tax tariffs, and penalty functions.

---

[2] Risk neutral taxpayers rather fit to Srinivasan (1973) than suit to Allingham and Sandmo (1972). In line with this Sandmo (2012, p. 6, Note 1) remarks that individual's risk neutrality of Srinivasan (1973) is due to an implicit linear utility function. However, the other way around, Srinivasan (1973, pp. 341–342) emphasizes that applying strictly concave utility functions and, therefore, examining risk averse taxpayers would not necessarily change his theoretical findings.

Second, to find the taxpayer's optimal reporting behavior both theories of tax evasion take into account a first-order condition, which refers either to maximizing the expected utility (Allingham and Sandmo, 1972) or the expected income after tax and penalties (Srinivasan, 1973). Allingham and Sandmo (1972) derive taxpayer's evading conditions (Eq. (1.5)) when to declare no income at all, full income or a fraction of true income, that is, an interior solution. The Allingham and Sandmo (1972) conditions (Eq. (1.5)) depend on the characteristics of taxpayer's utility function and the "tax enforcement variables," audit probability, penalty, and tax rate (e.g., see Cowell, 1992, p. 527). The interior solution vanishes if taxpayer's utility function reflects risk neutrality. Likewise, Srinivasan (1973) proves the existence of an interior solution depending on audit probability and the characteristics of prevailing tax and penalty functions. It is remarkable that this study does not allow for a corner solution, which is quite a converse result to Allingham and Sandmo (1972).

Third, while comparative statics bring up a variety of findings, both tax evasion models show a coinciding result; that is, increasing the audit probability *ceteris paribus* (c.p.) leads to less tax evasion. Allingham and Sandmo (1972) find that higher penalty rates, ceteris paribus, rise the optimal income declared and, thus, reduce the theoretical extent of tax evasion. The remaining comparative static results require additional assumptions. For instance, Srinivasan (1973) considers simultaneously (i) a progressive tax scheme and (ii) an income-independent audit probability. Then, raising the income, ceteris paribus, increases the declared fraction of income understatement and, therefore, tax evasion. To put it differently, richer taxpayers evade more tax – in relative and absolute terms – if a progressive income tax tariff applies. Further, the author examines an increase in income given (i) constant marginal tax rates and (ii) increasing audit probabilities. Here, the opposite effect occurs; that is, a higher income, ceteris paribus, decreases the magnitude of tax evasion. Allingham and Sandmo (1972) consider an increase of income and simultaneously analyze (i) a reduction of absolute risk aversion and (ii) a constant penalty rate, which is equal or greater than one. In this case, a larger income, ceteris paribus, triggers a higher optimal income declared and, therefore, leads to less tax evasion. Next, in addition to optimal income declared, a fraction of income declared is studied. In this context, it is worth noting that fractions of taxpayer's true income (Allingham and Sandmo, 1972) and income understatement (Srinivasan, 1973) are counterparts and add up to unity. The group of authors finds variations of fractions declared to tax authorities depending on taxpayer's relative risk aversion. Allingham and Sandmo (1972) introduce an income and substitution effect to the tax evasion theory, which can be traced back to their modeling of tax and penalty rates. The authors observe negative substitution effects since rising tax rates enhance taxpayer's incentives to evade tax. The sign of the income effect is ambiguous and depends on taxpayer's absolute risk aversion. For instance, decreasing absolute risk aversion yields strictly positive income effects such that the model fails to provide a clear-cut hypothesis. In summary, ceteris paribus, lower extents of tax evasion are caused by separately increasing the two enforcement parameters (i) audit probability and (ii) penalty rate, whereas all other comparative statics results by Allingham and Sandmo (1972) or Srinivasan (1973) require further assumptions.

Fourth, the authors present various extensions of their theoretical frameworks. Allingham and Sandmo (1972) examine three additional features, which are (i) lapse of time effects (or back auditing), (ii) nonpecuniary factors and (iii) variable audit probabilities. Concerning the last feature, comparative statics support the notion that more audit efforts, ceteris paribus, reduce tax evasion if and only if the audit function is monotone decreasing and strictly concave with respect to income. Nonpecuniary factors are added to taxpayer's utility function, for example, to reflect shame and guilt of identified and punished tax evaders. Allingham and Sandmo (1972) show the impact of such a modeling on taxpayer's evading constraints (Eq. (1.5)), even though the signs are not clear. With respect to back auditing, the authors find tipping points when a representative taxpayer starts to declare almost everything honestly.[3] Srinivasan (1973) studies two other extensions, which are (i) progressive versus proportional tax tariffs and (ii) optimization of audit probabilities. Considering various feasible tax tariffs, the author obtains that flat tax, ceteris paribus, yields less tax evasion than progressive tax schemes. With respect to an audit optimization procedure, he deduces a precondition that depends on marginal governmental expenditures due to the costs of raising audit probabilities. To sum up, Allingham and Sandmo (1972) and Srinivasan (1973) find that variable audit probabilities, lapse of time effects and changes of tax tariffs, for example, to implement flat tax, reduce tax evasion while the remaining extensions, for instance nonpecuniary factors, yield no clear-cut conclusions.

Finally, both tax evasion models are subject to notable criticism. One of the outstanding problems is that both theoretical frameworks in their standard version simply over-predict the amount of income tax evasion, which is estimated in reality. This is inconsistent with the real-world magnitudes of tax enforcement variables, which are obviously too low to guarantee deterrence levels that are theoretically necessary for taxpayers to become fully tax compliant. Therefore, the emerging crucial research question is the taxpayer's compliance mystery, why are they so tax compliant or intend to be totally compliant? To put this differently, the challenging issue is not, why tax avoidance, evasion, and the like exist but on the contrary it is the well-known question: "Why do people pay taxes?" (Alm et al., 1992b). Moreover, on the technical side, the Yitzhaki (1974) criticism deals with the problem of ambiguous income and substitution effects in the standard setting by Allingham and Sandmo (1972). Yet, Yitzhaki (1974) himself

---

[3] Hokamp and Pickhardt (2010) consider lapse of time effects via an expected utility approach within their agent-based tax evasion model, which represents the neoclassical (Allingham and Sandmo, 1972) model, if populated with 100% expected utility maximizing a-type agents. Thus, computational simulations might be regarded as numerical approximations for theories.

has suggested a solution; that is, the penalty rate has to depend on evaded tax rather than on nondeclared income. Under these circumstances the assumption of decreasing absolute risk aversion, with increasing income, yields pure income and no substitution effects so that higher tax rates, ceteris paribus, reduce the extent of tax evasion. Obviously, Srinivasan (1973) is not subject to the Yitzhaki (1974) critique because of penalty rates that are based on income understatement.

In addition to the critics of the components of the models, there are also a number of critics of these models based upon missing elements. These critics are not unique to these foundational tax evasion models and relate to issues of the heterogeneity of taxpayer behavior, potential taxpayer learning and adaptation over time, the potential impact of social networks, and the role of intermediaries such as tax preparers.

Despite the aforementioned critics, the standard theoretical model proposed by Allingham and Sandmo (1972) has become the standard theoretical model to analyze income tax evasion. The approach is still a fertile origin for rich work concerning tax evasion and related issues (e.g., for empirical observations see Kleven *et al.*, 2011, for applications to the shadow economy see Buehn and Schneider, 2012; Slemrod and Weber, 2012). Since the early 1970s, considerable extensions have been added to correct, enlarge, and adjust (Allingham and Sandmo, 1972), for instance misperceptions of tax enforcement variables, psychological effects and social norms (e.g., see Myles and Naylor, 1996; Fortin *et al.*, 2007, for a survey on psychological incentives of tax evasion see Kirchler, 2007). Yet, model extensions frequently generate such a complexity that clear-cut hypotheses are not available (e.g., see Alm, 2012, pp. 60–64). Of course, without any doubt, income tax evasion dynamics clearly influence tax revenues and public budgets. The above discussion outlined the canonical economic foundations for the analysis of tax paying behavior; next we will discuss the use of agent-based modeling in this endeavor.

1.4   Agent-Based Models

The rapid increase in computing power over the past several decades has led to a widespread usage of computers for conducting research via simulation. The social sciences in particular have turned to computer simulations as a technique to investigate complex real-world situations. Garson (2009, p. 267) identifies the main research branches of computational social simulation as: (i) ABMs, (ii) network models, (iii) spatial models, and (iv) systems dynamics models. The last mentioned consider digital-programmed systems of equations (or rules) that allow for illuminating complex socio-economic systems, for example, *"The limits of economic growth"* by Meadows *et al.* (1972). Spatial models simulate the interaction between a geographical (or physical) environment and, for instance, human behavior patterns. Network models cover a variety of topics; for example, from neural networks to queuing theory and its application to traffic problems

(Gilbert and Troitzsch, 2005). ABMs, the focus of this book, consist of discrete autonomous agents that behave according to prescribed decision rules. Interaction of agents at the micro-level often leads to complex emergent phenomena at the macro-level.

Individual decisions on tax evasion are embedded in a highly complex environment of behavioral rules, social norms, and tax enforcement (Kirchler, 2007). Our focus on the use of ABMs is driven by their ability to explore the heterogeneous behaviors of large populations that can result from this complexity. This is in rich contrast to the standard neoclassical approach based on a single representative agent (see Alm, 2010; Bloomquist, 2011a). Furthermore, ABMs investigate the interaction of adaptive agents and allow for the use of tools from related branches; for example, systems dynamics models to describe the behavior patterns and the interaction rules of the agents and spatial and network models to describe an agent's physical environment and social neighborhood, respectively (Carley and Maxwell, 2006). Hence, agent-based modeling seems to be a promising tool for research on complex social systems such as tax evasion in societies of behaviorally heterogeneous agents (for overviews on agent-based simulations see Garson, 2009; Heath *et al.*, 2009; Schinckus, 2013, for surveys on agent-based tax evasion models see Bloomquist, 2006; Hokamp, 2013; Pickhardt and Seibold, 2014; Bazart *et al.*, 2016, the insights from tax evasion theory, income declaration experiments and field studies are surveyed by Alm, 2012, the behavioral dynamics of tax compliance are reviewed by Pickhardt and Prinz, 2014). In particular, Garson (2009, p. 271) employs tax evasion as an example to point out that "[...] simulation is a way to explore assumptions, not necessarily to find a "correct" solution or optimal set of parameters [...]." The above discussion focused on the "why" of agent-based modeling, next we discuss how to report ABMs to aid in the creation of a cumulative science.

1.5 Standard Protocols to Describe Agent-Based Models

In this book we focus on model-based science, specifically the use of ABMs to explore more realistic, or higher resolution, representations of taxpayers and the tax payment system. Here, one uses the ABM to generate "sufficiency theorems" (Axtell, 2000a): for example, a population of agents acting under certain behavioral rules and specific conditions are sufficient to produce the observed dynamics and structures. In this work we stress the importance of clearly and thoroughly explaining the simulations used because we feel that this is a science. This being the case, we must ensure that the work can be understood, evaluated, and extended by others; and not seen as a "black box" (Lorscheid *et al.*, 2012). In order to make the claim that we are engaged in science we must create "theories." For our purposes, theories are statements about the world that are sufficiently specific as to be falsifiable, general enough to be comprehensive, and simple enough to be parsimonious (the ratio of predictions to assumptions should be large) (Harte, 2011).

Here, the output of a run of the model serves as a prediction, it is a strict deduction (Epstein, 2006). Ideally, this prediction is sufficiently specific as to be falsifiable. One can then use collections of output (deductions) to inductively build up theories, here, of human behavior.

As with most scientific endeavors, our work should be cumulative and build upon a foundation of previous research. This requires that enough information be reported to allow for replication, and then extension, of previous work. To that end, one may say that all researchers should be required to post their code. Arguably, that could be even better than a well expressed formulation (narrative), and certainly more efficient from a cumulative perspective. However, that would mean that researchers may understand less of the foundation they are using and, more importantly, could perpetuate unexpressed biases in implementation choices and bugs that may go unnoticed. If prior simulation-based work is re-implemented, then these assumptions are more likely to be discovered and previous bugs are less likely to be perpetuated (which is not to imply that potentially new bugs will not be introduced).

A second reason to stress the importance, and use, of a standard protocol to document computational simulations is for validation and verification purposes. These two concepts become increasingly important as the use of the models and simulation move from academic thought experiments to decision support systems used by policy makers. Verification is the process of deciding if the model matches its specification. Validation is the process of deciding how well a model matches the "real-world" phenomena under study. It is through this process that one can develop and document an understanding of how the model relates to the real world and how much "faith" to place in its predictions (a general process typically referred to as accreditation). One process to use in this endeavor is that of Docking (Axtell *et al.*, 1996) and categorizing the ABM's empirical relevance (Axtell, 2005). Docking refers to the process of comparing your model to a referent system (usually another model). There are three degrees of docking: Identity, where the two models produce numerically identical results; Distributional, where the models produce statistically indistinguishable results; and Relational, where the models produce qualitatively similar results that are statistically distinct. The empirical relevance of a model is a way to characterize how well the model represents both the micro-level (individuals) and macro-level (emergent structure) of the real world under study. Level 0 is micro-level qualitative correspondence; Level 1 is macro-level qualitative correspondence; Level 2 is macro-level quantitative correspondence, and finally, Level 3 is micro-level quantitative correspondence. The required level of empirical relevance is a function of how one plans to use the model. This process should be done with the model's intended purpose and audience in mind. For example, if one is creating an ABM simply to help think through a particular generating mechanism, then Level 0, Relational correspondence to the real world is likely to be sufficient. Here, the agents behave plausibly and, as a whole, the system changes in an appropriate manner when inputs are adjusted. However, if one is creating a model to be used by policy makers as they reform a taxation system then a higher standard should be used. Under these circumstances one would likely want to achieve at least Level 2, Empirical Relevance and Distributional Equivalence. Meaning, if income in the real world is Zipf distributed, the income distribution within the simulation is also Zipf distributed and is statistically indistinguishable form that of the real world.

1.5.1   The Overview, Design Concepts, Details, and Decision-Making Protocol

Originally, the Overview, Design Concepts, and Details (ODD) protocol was introduced by Grimm *et al.* (2006) and later refined in Grimm *et al.* (2010). It was created to address the need for a standardized way to adequately document individual-based and ABMs within Ecology. Recently, the ODD protocol has gained increasing acceptance within the computational social sciences as a way to document ABMs focused on human complex systems (e.g., the Computational Social Science Society of the Americas now requires the use of the ODD protocol for papers discussing ABMs). While the ODD protocol is an improvement over ad hoc discussions of ABMs it was not specifically designed around human systems, which meant that important elements of human systems (such as individual decision-making) could be lost or not discussed in adequate detail to allow for complete understanding or replication. Therefore, Müller *et al.* (2013) introduced the ODD+D protocol. In addition to the standard ODD protocol, the authors specifically added Decision-making (+D) to the protocol to ensure this critical component was not lost or under-specified. It is this ODD+D protocol that we use for this book. We would like to note that additional protocols termed the "transparent and comprehensive ecological modeling (TRACE) documentation" (Grimm and Schmolke, 2011) and the "Dahlem ABM documentation guidelines" (Wolf *et al.*, 2013) were proposed as an alternative or add-on to the ODD protocol to better ease description of large-scale ABMs. While the basic structure of the ODD protocol was retained, certain subsections were removed and others added to ensure that the model description could be made more concise. We do not attempt to compare and contrast these three approaches here.

For our purposes, we will adopt the basic definitions used by Müller *et al.* (2013). Therefore, an "agent" is a bundle of "data and behavioral methods representing an entity constituting part of a computationally constructed world" (Tesfatsion, 2006). Moreover, "decision-making" is defined as: "the methods agents use to make decisions about their behavior" (Dibble, 2006). Given that human agents often have dynamic behaviors within a model (Müller *et al.*, 2013) go on to distinguish learning and adaptation. For the authors, adaptation is more "passive" meaning, an agent that changes its behavior under different circumstances while using the same rules is defined as adapting. However, if an agent changes its rules over time, then it is defined as learning. Most other concepts within the ODD+D protocol are adopted from the original ODD protocol. In summary, the protocol is defined using the following taxonomy:

1.5.1.1   Overview of the Model

Purpose

This section discusses why the model was made and for whom. For example, is the model designed as a thought experiment to understand one very specific component of a tax regime; or is it designed to help inform policy makers as they work to reform an existing tax system? It is very important to explain clearly the purpose of the model so that readers will understand the motivation and apply the correct amount of scepticism and rigor to their understanding and evaluation.

Entities, State Variables, and Scales

This section discusses what entities (types of agents) are in the model and what makes each type unique (its state variables), what exogenous forces are included, and the temporal and spatial extent of the model and how it relates to the agents. For example, if there is only one kind of tax paying agent in the model, is heterogeneity introduced via differences by variable values or by different types of agents? Is the taxation authority explicitly represented? Is a time step a year or is a finer resolution of time used? It should be noted that this section should highlight what makes agents unique even if agents use the same submodels but with different parameter values. For example, if all taxpayers have the same submodel for deciding how much income to declare but agents that are expected to declare less income are given a lower risk aversion to induce that behavior; that sort of distinction should be highlighted.

Process Overview and Scheduling

This section discusses what goes on during a model run and in what order. How are agents activated? Are all agents activated in the same order or randomized? Are different types of agents activated together? Will every agent be activated at each time step or is activation probabilistic, or are agents only activated when they have something to do (discrete event)? It is very important to be as explicit as possible in this section. The activation regime can have significant impacts on model results (Axtell, 2000b). Researchers should be very thorough when describing these model elements. Pseudo-code can be particularly useful in this section when describing the structure of the simulation's activation regime.

1.5.1.2 Design Concepts

Theoretical and Empirical Background
This section should cover the theoretical and empirical underpinnings of the model, and the rational and assumptions used in the design choices for the structure of the model. Is the model based on rational choice theory or bounded rationality? How do agents deal with uncertainty, are they based upon behavioral data or psychology? Do agents have an objective or sense of better and worse outcomes?

Individual Decision-Making
Here, one should discuss agent decision-making, how decisions are made; what information is used; how sociocultural norms, uncertainty, spatiality, and time are dealt with; and how decision-making changes over time. What decisions do the agents make? For example, do agents decide how much of their income to report or is that simply a function of their type or class? Does experience over time, or the experience of their friends impact their decisions? What is the influence of social norms, do agents know what the majority of other agents are doing or do they assume (or not care)?

Learning
This section deals with individual learning and, if implemented, collective learning. How do agents incorporate experiences into their decision-making? Do agents have a memory that is used for decision-making?

Individual Sensing
Here, one discusses agent sensing, what variables are updated via sensing, what state variables are exposed for other agents to sense, how is error introduced, the temporal and spatial scale of sensing, and what costs are associated with sensing. How do agents know about their environment? Does the taxing authority make public statements? Do agents exchange information with each other, or is communication indirect?

Individual Prediction
This section covers the methods of prediction used by agents, how they are implemented, what is it the agents are predicting, and how error is introduced. For example, do agents make predictions about changes to the tax code or a likelihood to be caught if they cheat on their taxes?

Interaction
Here one discusses the direct and indirect interactions that take place among the agents, if the interactions depend upon communication, and any coordination

networks that may exist. How do agents interact? Is interaction based upon random connections or do agents have specific networks?

Collective

This section discusses the existence of collectives, how they emerge or are imposed, and how the collective impacts the individual and vice versa. For example, do preparer agents belong to a group that sets standards for itself that are a function of the current knowledge, skills, and abilities of its members? In this case, when an agent becomes part of the collective of tax preparers it must then ensure that it meets the standards of the collective. Over time, with changes in the population of preparers, these standards could change as the make up of the preparers change.

Heterogeneity

Here, one deals explicitly with heterogeneity among agents, what variables it impacts, how it changes, and if it includes agent behavior and decision-making. As mentioned earlier, this section should explicitly and thoroughly discuss what makes an agent (or group of agents) unique. This could come from different sets of variables or different values for the same variables. This heterogeneity could also come from the unique "experiences" that an agent has over the course of the simulation run.

Stochastic

This section specifies where randomness is used within the model, including both runtime and initialization. What stochasticity is in the model? Does it come into the order agents are activated? Do agents also use randomness within their decisions? Does the entire system use a single random number generator or does each agent have their own?

Observation

Here, the modeler discusses how the simulation is observed, what data is collected and when it is collected, and should also include how emergent phenomena are observed and measured. What data is gathered from the simulation? Do agents dump their data or are aggregates measured?

1.5.1.3   Details

Implementation Details

This section specifies how the model was implemented, including simulation framework, hardware needs, and where the source code may be found. What agent-based modeling platform was used to build the simulation, Repast, Net-Logo, MASON, and so on? Did you create your own framework? If so, why? Is the framework that was used available to others? What computational hardware did you use? Is the memory footprint of your simulation large enough that specialized hardware is needed?

Initialization

Here, the design choices made about how the model will start are specified, what values are used and how were they chosen, is there a variation, and what is the overall state in which the model starts. How is the model initialized? Does the model require a "burn-in" period before data is collected? Did you do multiple runs for each given collection of parameter values? Was the random number generator initialized with different values for each run?

Input Data

This sections discusses what input data the model uses, where it came from or how it was derived, and what its form/format is. If the model is not based solely on theory, then it is using some sort of input data. Where did the data that it uses come from? Was data cleaning or transformation necessary? This is an important section as it should also relate to the motivation and audience for your model. Different uses and motivations will call for more or less inclusion of "real world" data.

Submodels

Finally, one details all of the submodels in the simulation, how they were created and parametrized, what variables they act upon, and how they were tested during development. For example, if agents use a specific routine to calculate how much income to declare, one should explain how the agent thinks through this problem, how data is used, and where it comes from. The submodel descriptions should include a narrative of it, the rational for it, and should include equations (if appropriate).

1.5.2   Concluding Remarks on the ODD+D Protocol

The ODD+D protocol provides a comprehensive guide to structure assumptions and to deliberate the level of detail required to address the research question at hand. In the context of developing models to characterize individual tax evasion, this deliberation is key to ensure that outputs from the model are interpretable and can be defended. As will be evident from the discussion of the literature on agent-based tax evasion modeling in the upcoming section, there exists a broad spectrum of prior studies and corresponding findings that cannot be considered in isolation of the underlying modeling methodologies. As we attempt to advance the state of the art, the ODD+D protocol additionally provides a common framework for the basis of extracting findings that might apply under only specific circumstances versus those that might be more broadly applicable. This has implications for tax authorities, taxpayer service providers, and legislative bodies that are potential consumers of this research. This discussion outlined our preferred method for reporting ABMs within the scientific literature, the next section will review the extant tax evasion literature making use of ABMs.

1.6   Literature Review of Agent-Based Tax Evasion Models

In this section, we trace the evolution of agent-based tax compliance modeling and briefly outline recent advances. The agent-based frameworks highlight the need to model tax noncompliance decisions not in isolation, but rather embedded within a complex environment. The characteristics of the different methodologies used are summarized in Tables 1.2–1.4 presenting two highlights (or main areas) per agent-based tax noncompliance methodology and categorizing them with respect to (i) lead authors and research groups, (ii) software, (iii) domain, (iv) population size, (v) public goods, (vi) governmental tasks, (vii) back auditing, (viii) replication and docking studies, (ix) calibration, and (x) model highlights. The literature review is based on six surveys; (i) Bloomquist (2006) on the three early agent-based tax noncompliance frameworks by Mittone and Patelli (2000), Davis *et al.* (2003), and Bloomquist (2004a,b, 2008), (ii) Alm (2012) on tax evasion theories, income declaration experiments, and field studies, (iii) Hokamp (2013) on agent-based tax evasion and noncompliance models (iv) Pickhardt and Prinz (2014) on the behavioral dynamics of tax compliance, (v) Oates (2015) on tax evasion literature, and (vi) Bazart *et al.* (2016) on the calibration of agent-based tax evasion models.

Inspection of Tables 1.2–1.4 permits us to identify more than 30 agent-based settings of tax noncompliance, although Kim (2003), Meacci *et al.* (2012), Bertotti and Modanese (2014a,b, 2016a,b), and Nicolaides (2014) might be considered as borderline cases. The sequential arrangement of methodologies is alphabetic by the lead author of the very first contribution with respect to each research group. Thus, the first letter of the lead author ranges from A to C, in Table 1.2, from D to MA, in Table 1.3, and from MB to W, in Table 1.4. The research groups are not strictly disjoint, for instance L. Antunes, K. M. Bloomquist, S. Hokamp, M. Koehler, L. Mittone, M. Pickhardt, and F. W. S. Lima provide more than one agent-based tax noncompliance methodology. Moreover, seven research groups provide an abbreviation of their ABM: (i) Tax Compliance Simulator (TCS, Bloomquist, 2004a,b, 2008), (ii) Networked Agent-based Compliance Model (NASCM, Korobow *et al.*, 2007), (iii) Agent-based Tax Evasion Simulator (TAXSIM, Szabó *et al.*, 2009, 2010; Gulyás *et al.*, 2015), (iv) Small Business Tax Compliance Simulator (SBTCS, Bloomquist, 2011a), (v) Individual Reporting Compliance Model (IRCM Bloomquist, 2011b, 2013; Bloomquist and Koehler, 2015), (vi) Model C (Méder *et al.*, 2012) and (vii) SIMULFIS (Llacer *et al.*, 2013; Noguera *et al.*, 2014).

Table 1.2 Overview of agent-based tax evasion settings, research groups A–C

Research group	Domain	Software	Population size	Public goods	Governmental tasks	Back auditing	Replication/ docking study	Calibration	Highlights
Andrei et al. (2014) Chapter 8	Economics	NetLogo	441	–	–	–	Korobow et al. (2007) Hokamp and Pickhardt (2010)	–	• Docking study • Network structures
Antunes et al. (2006, 2007a,b) Balsa et al. (2006)	Economics	NetLogo	500	X	–	X	–	–	• Back auditing • Social interactions
Arsian and İcan (2013a,b)	Economics	NetLogo	10,000	–	–	–	Bloomquist (2011a)	Turkish Revenue Administration	• Audit perception • Social networks
Bertotti and Modanese (2014a,b, 2016a,b)	Economics	–	25^a	–	X	–	–	–	• Income classes • Trade
Bloomquist (2004a,b, 2008)	Economics	NetLogo	(i) 300 (ii) 1,000	–	–	–	–	–	• Life span • Social networks
Bloomquist (2011a)	Economics	NetLogo	10,000	–	–	–	–	Experiments / Internal Revenue Service	• Bankruptcy • Calibration
Bloomquist (2011b, 2013) Bloomquist and Koehler (2015) Chapter 7	Economics	RePast	85,000	–	–	–	–	Internal Revenue Service	• Artificial county • Third-party reporting
Carley and Maxwell (2006)	Economics	Construct	–	–	–	–	–	–	• Advertising campaign • Synthetic city
Cline et al. (2014) Chapter 8	Economics	RePast HPC	12,000,000	–	–	–	Bloomquist (2011b) Andrei et al. (2014)	U.S. synthesized population dataset	• Massive-scale • Social networks
Crokidakis (2014)	Econophysics	–	10,000	–	–	–	Zaklan et al. (2009)	–	• Opinion exchange dynamics • Three states of tax declaration

"Research Group" allows for assigning the agent-based tax evasion frameworks to teams of scholars in alphabetic order that refers to the leading author regarding the first contribution. "Software" shows computational tools. "–" denotes no information. "Domain" classifies the settings with respect to the economics or the econophysics modeling branch. "Population Size" lists the number of agents for scenarios in order of appearance, denoted as (i), (ii), and (iii). ^{a} The income classes in Bertotti and Modanese (2014a,b, 2016a,b). "X" represents an agent-based setting of tax noncompliance that incorporates "Public Goods," "Governmental Tasks," and "Back Auditing," respectively. "Replication/Docking Study" permits to figure out the replicated frameworks. "Calibration" presents the data source used. "Highlights" provides two key features per setting.

Table 1.3 Overview of agent-based tax evasion settings, research groups D–MA

Research group	Software	Domain	Population size	Public goods	Governmental tasks	Back auditing	Replication/ docking study	Calibration	Highlights
Davis et al. (2003)	Mathematica	Economics	500	–	–	–	–	–	• Compliance periods • Social norms
Garrido and Mittone (2013) Chapter 3	–	Economics	180	–	×	–	–	Experiments	• Audit frequency • Optimal audit programs
Hashimzade et al. (2014, 2015, 2016) Hashimzade and Myles (2017) Chapter 4	MATLAB	Economics	(i) 1,000 (ii) 6,000	–	×	–	–	–	• Audit strategy • Occupational choice
Hokamp (2014) Chapter 9	RePast	Economics	2,000	×	×	–	Hokamp and Pickhardt (2010)	–	• Pareto-optimality • Social norm updating
Hokamp and Pickhardt (2010) Chapter 9	MATLAB	Economics	150,000	–	×	×	–	–	• Back auditing • Political cycle
Kim (2003)	–	Economics	–	–	×	–	–	–	• Audit policy • Social coordination
Korobow et al. (2007)	–	Economics	1,600	–	–	–	–	–	• Heat maps • Social neighborhoods
Lima (2010, 2012a,b)	–	Econophysics	(i) 1,000,000 (ii) 400 (iii) 4,000	–	×	–	–	–	• Majority voting • Replication study
Lima and Zaklan (2008) Zaklan et al. (2008, 2009)	Fortran	Econophysics	(i) 400 (ii) 1,000,000	–	×	–	Zaklan et al. (2009)	–	• Ising-model • Network structures
Llacer et al. (2013) Noguera et al. (2014) Chapter 5	NetLogo	Economics	–	–	×	–	–	Spanish empirical data	• Learning mechanisms • Social benefits
Magessi and Antunes (2013a,b, 2015)	NetLogo	Economics	(i) 20 (ii) 1,000	–	–	–	–	Portuguese survey data	• Greed • Risk perception
Manhire (2015)	NetLogo	Economics	(i) 19[a];1,986[b] (ii) 0[a]–125[a];1,986[b]	–	×	×	–	–	• Audit effects • Tax compliance puzzle

[a] The number of tax examiners Manhire (2015).

[b] The taxpayers in Manhire. For a brief description of the syntax see Table 1.2.

Table 1.4 Overview of agent-based tax evasion settings, research groups MB–Z. See Table 1.2 for a brief description

Research group	Software	Domain	Population size	Public goods	Governmental tasks	Back auditing	Replication/ docking study	Calibration	Highlights
Meacci et al. (2012)	–	Economics	1,600	–	×	–	–	–	• Cellular automata • Cross-border policy
Méder et al. (2012)	NetLogo	Economics	1,000	×	×	–	–	–	• Laffer-curves • No enforcement
Mittone and Patelli (2000) Chapter 3	SWARM	Economics	300	×	×	–	–	–	• Public goods • Strategic auditing
Nicolaides (2014)	–	Economics	–	×	×	–	–	–	• Institutional quality • Social norms
Nordblom and Žamac (2012)	Java	Economics	10,000	–	–	–	–	Swedisch Tax Agency Survey	• Black market • Life cycle
Pellizzari and Rizzi (2014)	–	Economics	1,000	×	×	–	–	–	• Individual preferences • Slippery-slope framework
Seibold and Pickhardt (2013) Hokamp and Seibold (2014b) Pickhardt and Seibold (2014) Bazart et al. (2016) Chapter 11	Fortran	Econophysics	(i) 1,000,000 (ii) 150,000	×	×	×	Zaklan et al. (2009) Hokamp and Pickhardt (2010)	Experiments	• Magnetic fields • Temperature
Szabó et al. (2009, 2010) Gulyás et al. (2015) Chapter 6	RePast	Economics	(i) 200^a; 40^b (ii) 500^a; 50^b	×	×	–	–	–	• (Closed) Economy • Job market
Vale (2015)	–	Economics	(i) 10 (ii) 100	–	–	×	–	–	• Social networks • Varying evasion probability
Warner et al. (2015) Rosen et al. (2015) Hemberg et al. (2016) Chapter 10	–	Economics	–	–	–	–	–	–	• Genetic algorithms • Tax evasion schemes

In Szabó et al. (2009, 2010) ^{a}denotes the number of workers and ^{b}denotes the number of employers.

Additional columns describe the software utilized, the domain classifications as an economics or econophysics setting[4] and the magnitude of population considered. The software utilized ranges from open source (for instance, NetLogo, RePast HPC) to restricted and commercial software (e.g., Mathematica, MAT-LAB). With respect to the classification in a domain, the majority of frameworks belongs to economics (28 out of 32), but econophysics seems to gain more and more attractiveness since the seminal paper by Zaklan *et al.* (2009). S. Hokamp and M. Pickhardt contribute to both, the economics and econophysics branch (Hokamp and Pickhardt, 2010; Seibold and Pickhardt, 2013; Hokamp, 2014; Hokamp and Seibold, 2014b; Pickhardt and Seibold, 2014; Bazart *et al.*, 2016). The population size ranges from 10 taxpayers (see Vale, 2015) to 12,000,000 agents (see Cline *et al.*, 2014), while econophysics ABMs deal with up to 1,000,000 taxpayers (e.g., see Zaklan *et al.*, 2009). Notably, Cline *et al.* (2014) were the first to attempt a massive-scale agent-based tax evasion model, without roots in econophysics, to investigate a society of 12,000,000 taxpayers.

Furthermore, Tables 1.2–1.4 allow for jointly discussing and visualizing public goods provision, governmental tasks, back auditing, replication and docking issues, calibration and two highlights for each model. We provide a detailed overview of these issues in the next subsections.

1.6.1   Public Goods, Governmental Tasks and Back Auditing

Mittone and Patelli (2000) proposed the seminal ABM taking tax evasion into account from a micro-perspective. The authors analyze the interplay of tax compliance and public goods provision (Cowell and Gordon, 1988; Myles and Naylor, 1996) through a factor that reflects an agent's group conformity and show that strategic monitoring does not work to repress tax evasion behavior. Antunes *et al.* (2006, 2007a,b) and Balsa *et al.* (2006) present a family of eight agent-based tax compliance settings; Antunes *et al.* (2006) discuss the level of trust in a government (Wintrobe and Gërxhani, 2004) and, in their Ec2 model, they relate such an ethical attitude to a personal surplus received from a central authority taking into account back audits. The authors find that taxpayers "keep a good ethical attitude longer" because of lapse of time effects (Antunes *et al.*, 2006, p. 158). Szabó *et al.* (2009, 2010) and Gulyás *et al.* (2015) draw on Szabó *et al.* (2008) describing a one-product economy along with the shadow economy, tax evasion, and governmental services. The authors consider employees, employers, a tax authority, and a government; the taxpayers base their compliance decision on their experience of

---

[4] Hokamp and Pickhardt (2010) introduced the classification by an individual interaction process to the economic and the econophysics domain of agent-based tax evasion models. If the interaction among taxpayers has roots in physics, we have an econophysics ABM (Hokamp and Pickhardt, 2010; Hokamp and Seibold, 2014a).

satisfaction with the level of governmental services provided. Gulyás *et al.* (2015) analyze the effects of unemployment and figure out that an improvement of governmental services leads to a lower extent of tax evasion.

Pellizzari and Rizzi (2014) investigate a society of citizens and a government, where public expenditures finance public goods, so that the taxpayers take into account the behavior of their neighbors for their own compliance decision. Considering social beliefs the authors discover that individual preferences significantly drive tax evasion behavior. Nicolaides (2014) assumes strategic interaction among taxpayers (Blume, 1993) when contributing to public goods via taxes. The author finds that social norms on tax compliance serve as a substitute for deterrence via an effective audit ability. Méder *et al.* (2012) show in their utopian world without any pecuniary fine (i.e., the implicit audit probability is zero) that the presence (or absence) of prisoner's dilemma situations in the context of public goods provision turns out to be the crucial factor for long-run tax evasion dynamics.[5] The authors find that "higher tax rates induce more evasion" (Méder *et al.*, 2012, p. 187).

The econophysics branch of agent-based tax noncompliance models was invented by Lima and Zaklan (2008) and Zaklan *et al.* (2008, 2009), who assume two states for tax declaration, compliant and full evasion, corresponding to the spins in the Ising model of ferromagnetism (Ising, 1925)[6] and not allowing to consider tax rates. Lima and Zaklan (2008) build upon Wintrobe and Gërxhani (2004) and interpret an external magnetic field as a taxpayer's confidence in governmental institutions. The authors find that enforcement always works to trigger tax compliance; this finding is robust for Barabási-Albert and Voronoi-Delaunay networks.

Hokamp and Pickhardt (2010) demonstrate that back auditing and the evolution of social norms strongly influence tax evasion. Hokamp (2014) shows the counter-intuitive effect that an improvement in public goods provision leads to a higher extent of tax evasion (Alm *et al.*, 1992a; Alm, 2010). Seibold and Pickhardt (2013), Hokamp and Seibold (2014b), and Pickhardt and Seibold (2014) perform an econophysics agent-based approach to tax evasion which rests on the Ising model (cf. Ising, 1925; Zaklan *et al.*, 2009; Hokamp and Pickhardt, 2010). Seibold and Pickhardt (2013) conclude that an increase of the tax relevant periods subject to back auditing, ceteris paribus, lowers the extent of tax evasion. Hokamp and Seibold (2014b) implement public goods via a feedback depending on the rate of fully compliant taxpayers and they find that providing more public goods

---

[5] Garay *et al.* (2012) consider the analytical results by Méder *et al.* (2012) concerning global stability in a nontrivial steady state and provide the related mathematical arguments.

[6] The Ising model is a mathematical model of atomic spins that take one of two states corresponding to the magnetization status (Ising, 1925). In the econophysics tax evasion settings these two states are identified with tax compliance and noncompliance. Furthermore, econophysics models pick up the punishment idea by Davis *et al.* (2003); that is, the notion of penalizing through pre-announced time periods during which the taxpayers are restricted to be fully tax compliant.

decreases tax noncompliance. Picking up the econophysics notion to model the two extreme states of tax compliance, Meacci *et al.* (2012) investigate cross-border effects and the allocation of a budget to combat tax evasion making use of a cellular automata and ordinary differential equations. The authors find that an extra amount spent by the society induces "relevant changes in system behavior both in time and in space" (Meacci *et al.*, 2012, p. 608).

Bertotti and Modanese (2014a,b, 2016a,b) consider social inequality (measured by the Gini index), income redistribution, and tax compliance and they take into account governmental tasks. The authors simulate the direct and indirect interaction of taxpayers by a kinetic model with nonlinear ordinary differential equations. Bertotti and Modanese (2014a) conclude that (i) a fair fiscal policy and (ii) a population with tax compliant behavior for the most part is needed to overcome social inequalities. Bertotti and Modanese (2014b) find that the rich taxpayers benefit most from tax noncompliance, which yields the Gini index to be larger in the presence of tax evasion. Bertotti and Modanese (2016a) show that audits increase tax revenues, which are not affected by an agent's audit experience. Bertotti and Modanese (2016b) confirm the Gini index increases when the level of tax evasion rises (Bertotti and Modanese, 2014b), but it is independent of the spread of tax evasion performed among the income classes. Vale (2015) employs difference equations to investigate social networks within an ABM of heterogeneous taxpayers. The author utilizes back auditing and finds that a significant proportion of agents turn to be either fully compliant or evading. Kim (2003) makes use of decision heuristics combined with numerical simulations to consider tax tariffs and revenue schemes at the macro-level. The author finds that an optimal audit strategy is a concentration of efforts on auditing the lower and middle class taxpayers, if the tax burden is higher for the poor than for the rich. Manhire (2015) investigates tax noncompliance via modeling the role of tax examiners. The author finds that "the audit probability influences individual compliance decisions, it has negligible effects on system-level compliance patterns" (Manhire, 2015, p. 623).

Hashimzade *et al.* (2014, 2015, 2016) and Hashimzade and Myles (2017) consider the interplay of occupational choice and tax evasion in an ABM. Hashimzade *et al.* (2014) demonstrate that via occupational choice a society of heterogeneous taxpayers segregate in homogeneous groups with respect to their risk attitudes.[7] Hashimzade *et al.* (2015) develop two agent-based tax evasion models concerning both, (i) occupational choice associated to risk-taking, and (ii) social networks related to beliefs. The authors claim that the most effective strategy (in the sense of first-order stochastic dominance) is auditing a fixed number of taxpayers from each field of work. Hashimzade *et al.* (2016) show that random auditing generates less tax revenues than predictive analytics based on occupational choice.

---

[7] The segregation behavior due to occupational choice by Hashimzade *et al.* (2014) is in line with racial segregation (Schelling, 1969, 1971).

Hashimzade and Myles (2017) confirm that predictive analytics thrive in raising tax compliance.

1.6.2 Replication, Docking, and Calibration Studies

Andrei *et al.* (2014) and Crokidakis (2014) provide independent replication and docking studies of ABMs on tax compliance, in the economics and econophysics domain, respectively. Andrei *et al.* (2014) employ Korobow *et al.* (2007) and Hokamp and Pickhardt (2010) to show that Erdös-Rényi- and Power-Law-distributed networks reflect the impacts on tax compliance particularly strongly. Crokidakis (2014) draws on Zaklan *et al.* (2009) and makes use of a three-state kinetic opinion exchange model to show tax enforcement works to successfully fight against tax evasion, when a critical threshold is transgressed for the coupling of agents. Furthermore, in this section we draw on Bazart *et al.* (2016), who provide a literature overview on calibration studies in the field of tax evasion and fill in the gap of calibrating an econophysics agent-based tax evasion model.

Bloomquist (2011a) calibrated the very first study of an agent-based tax evasion model (Bazart *et al.*, 2016); his calibration combines experimental data on tax compliance and empirical data by the IRS National Research Program. The author finds that "based on the calibration and simulation results, it appears that both subjects in lab experiments and actual small business taxpayers are risk averse (i.e., overweight the probability of an audit) and taxpayers in the real world do not base their reporting compliance decisions on their neighbors' behavior" (Bloomquist, 2011a, p. 46). Arsian and İcan (2013a,b) draw on Bloomquist (2011a) to analyze tax evasion via an ABM for Turkey. The authors make use of data by the Turkish Revenue Administration and find that "both von Neumann and Moore neighborhoods are reducing compliance behavior of taxpayers considerably" (Arsian and İcan, 2013b, p. 337). Bloomquist and Koehler (2015) calibrate Bloomquist (2011b) with data by the IRS National Research Program and they are "able to model the complexities of real-world tax systems, such as differences in reporting compliance at the line item level and taxpayers' heterogeneous response behaviors" (Bloomquist and Koehler, 2015). Cline *et al.* (2014) draw on Korobow *et al.* (2007), Hokamp and Pickhardt (2010), Bloomquist (2011b, 2012), and Andrei *et al.* (2014) and they employ the US Synthesized Population dataset to present a massive-scale ABM of California. The authors argue that "running models at this (i.e., national) scale will provide the ability to produce policy insights without the fear of hidden scale effects or other issues of drawing national-scale conclusions from city-scale analyzes" (Cline *et al.*, 2014, p. 6).

Nordblom and Žamac (2012) employ a survey of buying black market services in Sweden to reaffirm that tax evasion by the elderly is substantially less as compared to younger taxpayers. Garrido and Mittone (2013) calibrate their ABM with experimental data on tax compliance from Chile and Italy. Given income inequality, the authors point out that the tax authorities might optimize tax revenues by auditing those taxpayers who more likely behave according to the bomb crater effect[8] (Krauskopf and Prinz, 2011).

Magessi and Antunes (2013a,b, 2015) apply to Portugal an agent-based tax evasion model, which rests on Antunes et al. (2006, 2007a,b) and Balsa et al. (2006).Magessi and Antunes (2015) make use of empirical data from a survey to consider the interplay of tax evasion and risk perception; in particular, the authors find that the subjective perception of the audit probability strongly influences the extent of tax evasion. Miguel et al. (2012), Llacer et al. (2013), and Noguera et al. (2014) present an agent-based tax evasion framework for Spain, which allows the investigation of social networks and behavioral aspects. Miguel et al. (2012) describe the basics of the agent-based framework to simulate tax noncompliance. Llacer et al. (2013) find that a consideration of a rational society (consisting of only rational taxpayers) overestimates tax evasion while social interaction allows for generating more plausible extents of tax compliance. Noguera et al. (2014) calibrate the ABM, called SIMULFIS, with empirical data from Spain. The group of authors conducts computational experiments to show that social norms do not always optimize tax compliance.

To fill the gap of calibrating an econophysics ABM of tax compliance[9] Bazart et al. (2016) make use of experimental data of a tax declaration game by Alm, et al. (2009) and Bazart and Bonein (2014). The authors find rather mixed behavior, for example, the taxpayers mix randomness and imitation, than the pure taxpayer behavioral archetypes assumed by Hokamp and Pickhardt (2010). Pickhardt and Seibold (2014) fill the gap of linking econophysics and economics agent-based tax evasion models by a successful replication of the two basic frameworks, Zaklan et al. (2009) and Hokamp and Pickhardt (2010). Lima (2010) provides a replication study of the econophysics agent-based tax evasion model by Zaklan et al. (2009) and adds the majority-vote-model and an Apollonian network (Lima, 2010, 2012a,b) to confirm robustness that enforcement works to trigger tax compliance.

1.6.3   Concluding Remarks on Agent-Based Tax Evasion Models

Agent-based tax evasion models show a broad flexibility to handle the complexity of tax compliance behavior. This is documented by the "highlights" column in Tables 1.2–1.4 presenting two main aspects per agent-based tax evasion

---

[8] Soldiers recognized it is very unlikely that a bomb hits the same place twice in a short time window, and, therefore, a strategy to survive is to hide in a bomb crater. Transferred to tax evasion, taxpayers behave according to the bomb crater effect, if they again evade after being detected as a tax cheater, they assume it is very unlikely to be audited again.

[9] Hokamp and Seibold (2014a) present an econophysics ABM of the shadow economy. The authors utilize experimental data on tax compliance by Bazart and Pickhardt (2011) to figure out that, as compared with Germany, France seems to have a larger fraction of subjects rationally engaged in the shadow economy.

methodology. The highlights cover a variety of topics; for example, social neighborhoods, third-party reporting, political cycles and majority voting. Davis *et al.* (2003) investigate social norms and obtain nearly full tax compliance for comparatively low objective audit probabilities. Bloomquist (2004a,b, 2008) claims that social networks have a significant influence on the extent of tax evasion. Carley and Maxwell (2006) consider a synthetic city to demonstrate that a short-term publicity campaign has the potential to increase the overall participation rates in tax declaration schemes. Korobow *et al.* (2007) conclude that taxpayers with limited knowledge of their neighbor's payoffs show higher tax compliance levels. Warner *et al.* (2015) make use of genetic algorithms from biology to help policy-makers anticipate tax evasion schemes.

Further, the literature review of agent-based tax evasion models reveals a lack of research on public goods and back auditing as well as a lack of studies presenting replications and calibrations. Indeed, some ABMs consider public goods provision but the crucial question remains unclear: how does an improvement of public goods provision change the extent of tax evasion? So far, agent-based tax compliance models have not presented a single clear-cut hypothesis with respect to public goods provision; the findings are even contradictory (Szabó *et al.*, 2009, 2010; Hokamp, 2014). In contrast, the effects of back auditing seem to be clear: a larger lapse of time period leads to a lower extent of tax evasion. Many ABMs neglect back auditing because of its large effect which then may outshine the other effects under consideration. However, back auditing may help solve the tax evasion puzzle "Why do people pay taxes?" (Alm *et al.*, 1992b). For these reasons, ABMs should incorporate back auditing or at least be able to successfully deal with lapse of time effects.

Furthermore, calibration and replication is necessary for ABMs to gain credibility and to avoid the perception of being a "black box"; in particular, the communication of results and the design of experiments is crucial (Lorscheid *et al.*, 2012). While there are now limited calibrations of agent-based tax evasion models, there are much fewer replication attempts. Notably, the very first independent replication of an agent-based tax evasion setting was performed by Andrei *et al.* (2014). Finally, a common protocol for describing agent-based tax evasion models may help provide data needed for independent calibration and replication.

The next section provides an outlook on the structure and presentation of the book and surveys the upcoming chapters that deal with the various aspects of complexity in tax evasion research.

1.7   Outlook: The Structure and Presentation of the Book

Our book is structured as follows. Part I includes this introductory chapter that provides a conceptual background, a review of agent-based modeling, and the literature on tax evasion. It also discusses "dark" economic behavior in general and the methodological link between experiments and agent-based modeling. Part II presents various applications of agent-based modeling to tax evasion, constructed for or calibrated to various countries (the United Kingdom, the United States, Spain, Hungary, and Germany). This is completed by two chapters extending the horizon of computational models of tax evasion. One applies genetic algorithms to co-evolve tax audit strategies with clandestine behavior, in order to "breed" possible novel forms of tax evasion. The other brings an "econophysics" approach to the problem by developing a variant of a classic model of ferromagnetism adapted to tax evasion.

1.7.1 Part I Introduction

The chapter by Prinz (Chapter 2) discusses how the scientific study of and the discourse about clandestine activities should be examined. Since research on such clandestine activities is of great importance for the implementation of new effective policies, not only theoretical speculations, but also experiments and empirical studies are necessary to understand these activities of hidden nature. Theories are abstract and they aim to identify few generalizable behavioral drivers, while empirical or experimental applications try to address specific real-world issues. Thus, according to the author, better scientific understanding and thus better policies need theories counterbalanced by empirical and experimental studies.

The author enumerates a number of different approaches to study clandestine activities and discusses their advantages and disadvantages (e.g., computer simulations calibrated with human-based experiments and empirical data, network analysis, and social media analysis in general). The chapter argues for a hybrid approach, a combination of methods according to the specific question of interest.

The next chapter by Mittone and Saredi (Chapter 3) complies with Prinz's proposal. It discusses model calibration efforts that are aimed to validate ABMs via laboratory experiments. After a summary of the literature on tax evasion modeling, the chapter highlights the recent advances made toward integrating human-subject laboratory experiments with computational agent-based simulations (or Agent-based Computational Economics, ACE). The discussion is put in context by an extensive review of the author's previous work in the same domain, separating three approaches: the macroeconomic, the microeconomic, and the approach of micro-level dynamics for macro-level interactions among behavioral types.

1.7.2 Part II Agent-Based Tax Evasion Models

The chapter by Hashimzade and Myles (Chapter 4) is the first among the ones in Part II reporting on specific ABMs of tax evasion. It provides a summary of the authors' work on ABMs of tax noncompliance. The model presented is based on a basic framework of occupational choice: employment versus self-employment, where the latter is understood as a strategy to optimize taxes. Agents make their choices based on their beliefs formed during social interactions, thus forming endogenous attitudes and beliefs about tax compliance. Computational results show that the strategy of audit target selection is an essential ingredient in order to maximize the effectiveness of audit resources. Moreover, it is demonstrated that predictive analytics dominates random audits in terms of raising tax revenue. This is an important observation, since random audit strategy is a frequently used baseline case in agent-based tax evasion models.

In the following chapter, Llacer *et al.* (Chapter 5) present SIMULFIS, an agent-based simulator, where agent behavior is driven by a set of decision rules ("filters") with the aim of generating realistic behavioral outcomes. Possible "filters" include, for example, rational evaluations, beliefs about the fairness of state policies, and social influence by peers. The focus is on personal choices regarding income declarations, where the incomes themselves and the opportunities for evasion are exogenous. The authors calibrate SIMULFIS to reproduce real-world traits of Spain with a view to investigate scenarios to fight against tax evasion. Their computational results show that audits are more effective than fines to prevent tax noncompliance. On the other hand, the contribution of the social network on tax compliance is less clear, albeit its influence is clearly demonstrated. This underlines the importance of SIMULFIS' capability to analyze the effects of social influence at the micro-level.

The chapter by Gulyás *et al.* (Chapter 6) provides a comprehensive description of TAXSIM, a family of ABMs of tax noncompliance. In TAXSIM the level of noncompliance is a calculated decision of employees and employers, made during contract negotiation, in order to increase net income and to reduce costs, respectively. The agents base their decisions on their estimations of the expected costs of noncompliance and on their satisfaction with government services. The estimations are driven by the agents' individual experiences and by information received through the agents' social networks. The authors discuss computational results to illustrate the capabilities of the model and it provides a brief analysis of the model's behavior space. Among other findings, the importance of adaptive audit strategies is confirmed.

This is followed by the chapter by Bloomquist (Chapter 7) that introduces the IRCM, an ABM to investigate tax (non)reporting behavior taking into account various sources of income, social networks, taxpayer learning, and enforcement methods. Following the introduction, the author presents results considering a community of 85,000 US taxpayers, calibrated using real-world tax reporting behavior based on random audits. The study focuses on the change in taxpayer compliance due to a partial shift in the workforce composition from full-time employees to contingent workers. This application of agent-based tax evasion modeling to the "Gig" economy (i.e., a novel trend, where temporary positions are common and organizations contract with independent workers for short-term engagements) is a unique and interesting contribution to the literature.

The next chapter by Koehler *et al.* (Chapter 8) applies another scale to the computational modeling of tax evasion. It replicates existing and well-known ABMs of tax evasion to test the response of the model to different kinds of networks and scales (Hokamp and Pickhardt, 2010; Andrei *et al.*, 2014). The effect of changing the scale (i.e., the number of agents) is analyzed in connection with the various networks tested. In addition, the authors provide a proof-of-concept demonstration that massive (i.e., realistic) scales can be incorporated and handled within economic ABMs. Using their custom made simulation framework (based on python and capable of multiprocessor runs) to study models with one million agents or more, the authors conclude that the qualitative results are scale-independent for the model under study.

The chapter by Hokamp and Cuervo Díaz (Chapter 9) discusses the agent-based tax evasion framework by Hokamp and Pickhardt (2010) and Hokamp (2013, 2014) providing a comprehensive summary and adding some novel insights. The study considers taxpayers to be classified into four behavioral archetypes (neo-classical A-types, social interacting B-types, ethical C-types and erratic D-types), a tax authority, and a government. As with most ABMs, the focus is on the effects of micro-level composition and interaction on the various evasion and compliance behaviors at the macro-level. The authors calibrate their model to experimental data from a tax declaration game and consider several different situations construed by the presence of the following six model assumptions: lapse of time, subjective audit probability, age heterogeneity, social norm updating, public goods provision, and checking for Pareto-optimality. Computational results show that lapse of time has a significant effect on the extent of tax evasion; a feature frequently neglected in agent-based tax evasion models.

The last two chapters widen the horizon of agent-based modeling of tax evasion and compliance. The chapter by Wijesinghe *et al.* (Chapter 10) deals with the complex transaction networks of tax shelters. These constructs combine multiple business entities, for example, partnerships, trusts, and are designed to reduce and obscure tax liabilities. The complex networks of intricate transactions are manufactured in such a way that their sequences inflate the basis of particular assets under their control. The authors propose an approach to co-evolve the audit actions of the tax authorities with the behaviors of tax evaders, in order to anticipate the likely forms of emerging evasion schemes and thus, to give enforcement agencies a tactical advantage. Computational results reported in the chapter indicate that the proposed combination of agent-based modeling with a heuristic search based on a genetic algorithm is indeed capable of identifying potentially new tax evasion schemes.

The last chapter by Seibold (Chapter 11) adds the flavor of "econophysics" to our collection by discussing the classic Ising model as a means to describe tax evasion dynamics (Ising, 1925). This is done by extending the tax evasion Ising model developed by Zaklan *et al.* (2009). The main contribution is to consider the four behavioral archetypes proposed by Hokamp and Pickhardt

(2010) (discussed earlier) as opposed to the two agent types of the original model (compliant, noncompliant). In addition, the author also proposes an interpretation of the temperature variable in the Ising model as the standard deviation of the "unobserved utility" that corresponds to the "spread in nonmeasurable taste or attitude."

References

Allingham, M.G. and Sandmo, A. (1972) Income tax evasion: a theoretical analysis. *Journal of Public Economics*, **1**, 323–338.

Alm, J. (2010) Testing behavioral public economic theories in the laboratory. *National Tax Journal*, **63** (4), 635–658.

Alm, J. (2012) Measuring, explaining, and controlling tax evasion: lessons from theory, experiments, and field studies. *International Tax and Public Finance*, **19** (1), 54–77.

Alm, J., Jackson, B., and McKee, M. (1992a) Institutional uncertainty and taxpayer compliance. *The American Economic Review*, **82** (4), 1018–1026.

Alm, J., McClelland, G.H., and Schulze, W.D. (1992b) Why do people pay taxes? *Journal of Public Economics*, **48**, 21–38.

Alm, J., Jackson, B.R., and McKee, M. (2009) Getting the word out: enforcement information dissemination and compliance behavior. *Journal of Public Economics*, **93**, 392–402.

Andreoni, J., Erard, B., and Feinstein, J. (1998) Tax compliance. *Journal of Economic Literature*, **36** (2), 818–860.

Andrei, A.I., Comer, K., and Koehler, M. (2014) An agent-based model of network effects on tax compliance. *Journal of Economic Psychology*, **40**, 119–133.

Antunes, L., Balsa, J., and Coelho, H. (2007a) Tax compliance through MABS: the case of indirect taxes, in *Progress in Artificial Intelligence, EPIA 2007*, LNAI, vol. 4874, Springer-Verlag, Berlin, Heidelberg, pp. 605–617.

Antunes, L., Balsa, J., Respício, A., and Coelho, H. (2007b) Tactical exploration of tax compliance decisions in multi-agent based simulation, in *Multi-Agent-Based Simulation VII, International Workshop MABS 2006*, LNAI, vol. 4442, Springer-Verlag, Berlin, Heidelberg, pp. 80–95.

Antunes, L., Balsa, J., Urbano, P., Moniz, L., and Roseta-Palma, C. (2006) Tax compliance in a simulated heterogeneous multi-agent society, in *Multi-AGENT-BASED SIMULATION VI, International Workshop MABS 2005*, LNAI, vol. 3891, Springer-Verlag, Berlin, Heidelberg, pp. 147–161.

Arsian, O. and İcan, Ö. (2013a) An agent-based analysis of tax compliance for Turkey. *Andolu University Journal of Social Sciences*, **13** (2), 143–152.

Arsian, O. and İcan, Ö. (2013b) The effects of neighborhood on tax compliance rates: evidence from an agent-based model. *C.Ü. Sosyal Bilimler Enstitüsü Dergisi*, **22** (1), 337–350.

Axtell, R. (2000a) Why agents? On the varied motivations for agent computing in the social sciences. Brookings Institution Center on Social and Economic Dynamics working paper.

Axtell, R. (2000b) *Effects of Interaction Topology and Activation Regime in Several Multi-Agent Systems*, Springer-Verlag, Berlin, Heidelberg.

Axtell, R. (2005) Three distinct kinds of empirically-relevant agent-based models. Brookings Institution Center on Social and Economic Dynamics working paper.

Axtell, R., Axelrod, R., Epstein, J.M., and Cohen, M.D. (1996) Aligning simulation models: a case study and results. *Computational & Mathematical Organization Theory*, **1** (2), 123–141.

Balsa, J., Antunes, L., Respício, A., and Coelho, H. (2006) Autonomous Inspectors in Tax Compliance Simulation. *Proceedings of the 18th European Meeting on Cybernetics and Systems Research, April 2006, Vienna, Austria*.

Bazart, C. and Bonein, A. (2014) Reciprocal relationships in tax compliance decisions. *Journal of Economic Psychology*, **40**, 83–102.

Bazart, C., Bonein, A., Hokamp, S., and Seibold, G. (2016) Behavioural economics and tax evasion – calibrating an agent-based econophysics model with experimental tax compliance data. *Journal of Tax Administration*, **2** (1), 126–144.

Bazart, C. and Pickhardt, M. (2011) Fighting income tax evasion with positive rewards. *Public Finance Review*, **39** (1), 124–149.

Becker, G.S. (1968) Crime and punishment: an economic approach. *The Journal of Political Economy*, **76** (2), 169–217.

Becker, G.S. (1993) Nobel lecture: the economic way of looking at behavior. *The Journal of Political Economy*, **101** (3), 385–409.

Bertotti, M.L. and Modanese, G. (2014a) Micro to macro models for income distribution in the absence and in presence of tax evasion. *Applied Mathematics and Computation*, **224**, 836–846.

Bertotti, M.L. and Modanese, G. (2014b) Mathematical models for socio-economic problems, in *Mathematical Models and Methods for Planet Earth*, Springer, Switzerland, pp. 124–134.

Bertotti, M.L. and Modanese, G. (2016a) Microscope Models for the Study of Taxpayer Audit Effects, arXiv 1602.08467v1 [q-fin.GN] 18 February 2016.

Bertotti, M.L. and Modanese, G. (2016b) Mathematical models describing the effects of different tax evasion behaviors. *Journal of Economic Interaction and Coordination*, 1–13.

Bloomquist, K.M. (2004a) Modeling Taxpayers' Response to Compliance Improvement Alternatives. *Paper presented at the Annual Conference of the North American Association for Computational Social and Organizational Sciences*, Pittsburgh, PA.

Bloomquist, K.M. (2004b) Multi-agent Based Simulation of the Deterrent Effects of Taxpayer Audits. *Paper presented at the 97th Annual Conference of the National Tax Association*, Minneapolis, MN.

Bloomquist, K.M. (2006) A comparison of agent-based models of income tax evasion. *Social Science Computer Review*, **24** (4), 411–425.

Bloomquist, K.M. (2008) Tax compliance simulation: a multi-agent based approach, in *Social Simulation: Technologies, Advances and New Discoveries*, IGI Global, pp. 13–25.

Bloomquist, K.M. (2011a) Tax compliance as an evolutionary coordination game: an agent-based approach. *Public Finance Review*, **39** (1), 25–49.

Bloomquist, K.M. (2011b) Multi-Agent Simulation of Taxpayer Reporting Compliance. *Invited Paper presented at the International Conference on Taxation Analysis and Research, December 2011*, London, UK.

Bloomquist, K.M. (2012) Agent-based simulation of tax reporting compliance. PhD thesis. George Mason University.

Bloomquist, K.M. (2013) *Incorporating Indirect Effects in Audit Case Selection: An Agent-Based Approach*, The IRS Research Bulletin, Publication 1500, pp. 103–116.

Bloomquist, K.M. and Koehler, M. (2015) A large-scale agent-based model of taxpayer reporting compliance. *Journal of Artificial Societies and Social Simulation*, **18** (2). doi: 10.18564/jasss.2621.

Blume, L.E. (1993) Statistical mechanics of strategic interaction. *Games and Economic Behaviour*, **5**, 387–424.

Buehn, A. and Schneider, F. (2012) Shadow economies around the world: novel insights, accepted knowledge, and new estimates. *International Tax and Public Finance*, **19** (1), 139–171.

Carley, K.M. and Maxwell, D.T. (2006) Understanding Taxpayer Behavior and Assessing Potential IRS Interventions Using Multiagent Dynamic-network Simulation. *Recent Research on Tax Administration and Compliance: Selected Papers Given at the 2006 IRS Research Conference, Statistics of Income Division, Internal Revenue Service, Department of the Treasury*, Washington, DC, pp. 93–106.

Cline, J., Bloomquist, K.M., Gentile, J.E., Koehler, M., and Marques, U. (2014) From Thought to Action: Creating Tax Compliance Models at National Scales. *Paper presented at the 11th International Conference on Tax Administration, April 2014*, Sydney, Australia.

Cowell, F.A. (1992) Tax evasion and inequity. *Journal of Economic Psychology*, **13**, 521–543.

Cowell, F. and Gordon, J.P.F. (1988) Unwillingness to pay - tax evasion and public goods provision. *Journal of Public Economics*, **36**, 305–321.

Crokidakis, N. (2014) A three-state kinetic agent-based model to analyze tax evasion dynamics. *Physica A*, **414**, 321–328.

Davis, J., Hecht, G., and Perkins, J.D. (2003) Social behaviors, enforcement and tax compliance dynamics. *The Accounting Review*, **78**, 39–69.

Dibble, C. (2006) Computational laboratories for spatial agent-based models. *Handbook of Computational Economics*, **2**, 1511–1548.

Epstein, J.M. (2006) *Generative Social Science: Studies in Agent-Based Computational Modeling*, Princeton University Press.

Fortin, B., Lacroix, G., and Villeval, M.C. (2007) Tax evasion and social interactions. *Journal of Public Economics*, **91**, 2089–2112.

Garay, B.M., Simonovits, A., and Tóth, I.J. (2012) Local interaction in tax evasion. *Economics Letters*, **115**, 412–415.

Garrido, N. and Mittone, L. (2013) An agent based model for studying optimal tax collection policy using experimental data: the cases of Chile and Italy. *Journal of Socio-Economics*, **42**, 24–30.

Garson, G.D. (2009) Computerized simulation in the social sciences - a survey and evaluation. *Simulation and Gaming*, **20** (2), 267–279.

Gilbert, N. and Troitzsch, K.G. (2005) *Simulation for the Social Scientist*, Open University Press.

Government Accountability Office (2012) TAX GAP. Sources of Non-Compliance and Strategies to Reduce It. GAO-12-651T, http://www.gao.gov/assets/600/590215.pdf (accessed 6 May 2016).

Grimm, V. *et al.* (2006) A standard protocol for describing individual-based and agent-based models. *Ecological Modelling*, **198** (1), 115–126.

Grimm, V. *et al.* (2010) The ODD protocol: a review and first update. *Ecological Modelling*, **221** (23), 2760–2768.

Grimm, V. and Schmolke, A. (2011) How to Read and Write TRACE Documentations, http://cream-itn.eu/creamwp/wp-content/uploads/Trace-Guidance-11-03-04.pdf (accessed 18 December 2016).

Gulyás, L., Máhr, T., and Tóth, I.J. (2015) Factors to Curb Tax Evasion: Evidence from the TAXSIM Agent-based Simulation Model. *KTI/IE Discussion Papers, MP-DP – 2015/21, April 2015*, Budapest, Hungary.

Harte, J. (2011) *Maximum Entropy and Ecology: A Theory of Abundance, Distribution, and Energetics*, Oxford University Press.

Hashimzade, N. and Myles, G.D. (2017) Risk-based audits in a behavioural model. *Public Finance Review*, **45** (1), 130–145.

Hashimzade, N., Myles, G.D., Page, T., and Rablen, M.D. (2014) Social networks and occupational choice: the endogenous formation of attitudes and beliefs about tax compliance. *Journal of Economic Psychology*, **40**, 134–146.

Hashimzade, N., Myles, G.D., Page, T., and Rablen, M.D. (2015) The use of agent-based modelling to investigate tax compliance. *Economics of Governance*, **16**, 143–164.

Hashimzade, N., Myles, G.D., and Rablen, M.D. (2016) Predictive analytics and the targeting of audits. *Journal of Economic Behavior & Organization*, **124**, 130–145.

Heath, B., Hill, R., and Ciarallo, F. (2009) A survey of agent-based modelling practices (January 1998 to July 2008). *Journal of Artificial Societies and Social Simulation*, **12**, (4), 9.

Hemberg, E., Rosen, J., Warner, G., Wijesinghe, S., and O'Reilly, U.-M. (2016) Detecting tax evasion: co-evolutionary approach. *Artificial Intelligence and Law*, **24** (2), 149–182.

Hokamp, S. (2013) Income tax evasion and public goods provision – theoretical aspects and agent-based simulations. PhD thesis. Brandenburg University of Technology Cottbus.

Hokamp, S. (2014) Dynamics of tax evasion with back auditing, social norm updating, and public goods provision – an agent-based simulation. *Journal of Economic Psychology*, **40**, 187–199.

Hokamp, S. and Pickhardt, M. (2010) Income tax evasion in a society of heterogeneous agents – evidence from an agent-based model. *International Economic Journal*, **24** (4), 541–553.

Hokamp, S. and Seibold, G. (2014a) How much rationality tolerates the shadow economy? – An agent-based econophysics approach, in *Advances in Social Simulation, ESSA 2013*, Advances in Intelligent Systems and Computing, vol. 229, Springer-Verlag, Berlin, Heidelberg, pp. 119–128.

Hokamp, S. and Seibold, G. (2014b) Tax compliance and public goods provision – an agent-based econophysics approach. *Central European Journal of Economic Modelling and Econometrics*, **6** (4), 217–236.

Internal Revenue Service (2007) Reducing the Federal Tax Gap: A Report on Improving Voluntary Compliance, http://www.irs.gov./pub/irs-news/tax_gap_report_final_080207_linked.pdf (accessed 12 April 2012).

Internal Revenue Service (2016a) IRS Tax Gap Estimates for Tax Years 2008–2010, https://www.irs .gov/PUP/newsroom/tax gap estimates for 2008 through 2010.pdf (accessed 23 November 2016).

Internal Revenue Service (2016b) IRS Internal Revenue Manual, Part 25.1.1, https://www.irs.gov/irm/ part25/irm_25-001-001.html#d0e119 (accessed 6 May 2016).

Ising, E. (1925) Beitrag zur Theorie des Ferromagnetismus. *Zeitschrift für Physik*, **31** (1), 253–258.

James, S. and Edwards, A. (2010) An Annotated Bibliography of Tax Compliance and Tax Compliance Costs, http://mpra.ub.uni-muenchen.de/26106/ (accessed 27 April 2012).

Kim, Y. (2003) Income distribution and equilibrium multiplicity in a stigma-based model of tax evasion. *Journal of Public Economics*, **87** (7–8), 1591–1616.

Kirchler, E. (2007) *The Economic Psychology of Tax Behavior*, Cambridge University Press, Cambridge.

Kleven, H.J., Knudsen, M.B., Kreiner, C.T., Perdersen, S., and Saez, E. (2011) Unwilling or unable to cheat? Evidence from a tax audit experiment in Denmark. *Econometrica*, **79** (3), 651–692.

Korobow, A., Johnson, C., and Axtell, R. (2007) An agent-based model of tax compliance with social networks. *National Tax Journal*, **60** (3), 589–610.

Krauskopf, T. and Prinz, A. (2011) Methods to reanalyze tax compliance experiments: Monte-Carlo simulations and decision time analysis. *Public Finance Review*, **39** (1), 168–188.

Lima, F.W.S. (2010) Analysing and controlling the tax evasion dynamics via majority-vote model. *Journal of Physics: Conference Series*, **246**, 1–12.

Lima, F.W.S. (2012a) Tax evasion dynamics and Zaklan model on opinion-dependent network. *International Journal of Modern Physics C: Computational Physics and Physical Computation*, **23** (6). doi: 10.1142/s0129183112500477.

Lima, F.W.S. (2012b) Tax evasion and nonequilibrium model on apollonian networks. *International Journal of Modern Physics C: Computational Physics and Physical Computation*, **23** (11). doi: 10.1142/s0129183112500799.

Lima, F.W.S. and Zaklan, G. (2008) A multi-agent-based approach to tax morale. *International Journal of Modern Physics C: Computational Physics and Physical Computation*, **19** (12), 1797–1808.

Llacer, T., Miguel, F.J., Noguera, J.A., and Tapia, E. (2013) An agent-based model of tax compliance: an application to the Spanish case. *Advances in Complex Systems*, **16**. doi: 10.1142/s0219525913500070.

Lorscheid, I., Heine, B.-O., and Meyer, M. (2012) Opening the "black box" of simulations: increased transparency and effective communication through the systematic design of experiments. *Computational and Mathematical Organization Theory*, **18**, 22–62.

Magessi, N.T. and Antunes, L. (2013a) Modelling agent's risk perception, in *Distributed Computing and Artificial Intelligence, DCAI 2013*, Advances in Intelligent Systems and Computing, vol. 217, Springer-Verlag, Cham, Heidelberg, New York, Dordrecht, London, pp. 275–282.

Magessi, N.T. and Antunes, L. (2013b) Agent's Fear Monitors the Spread of Greed in a Social Network. *Paper presented at the 11th European Workshop on Multi-Agent Systems, December 2013*, Toulouse, France.

Magessi, N.T. and Antunes, L. (2015) Risk perception and risk attitude on a tax evasion context. *Central European Journal of Economic Modelling and Econometrics*, **7** (3), 127–149.

Manhire, J.T. (2015) There is no spoon: reconsidering the tax compliance puzzle. *Florida Tax Review*, **17** (8), 623–668.

Meacci, L., Nuño, J.C., and Primicerio, M. (2012) Fighting tax evasion: a cellular automata approach. *Advances in Mathematical Sciences and Applications*, **22** (2), 597–610.

Meadows, D.H., Meadows, D.L., Randers, J., and Behrens, W.W. III (1972) *The Limits of Economic Growth*, Universe Books.

Méder, Z.Z., Simonovits, A., and Vincze, J. (2012) Tax morale and tax evasion: social preferences and bounded rationality. *Economic Analysis and Policy*, **42** (2), 171–188.

Miguel, F.J., Noguera, J.A., Llacer, T., and Tapia, E. (2012) Exploring Tax Compliance: An Agent-Based Simulation. *Paper presented at the 26th European Conference on Modelling and Simulation, May/June 2012*, Koblenz, Germany.

Mittone, L. and Patelli, P. (2000) Imitative behavior in tax evasion, in *Economic Simulations in Swarm: Agent-based Modelling and Object Oriented Programming*, Kluwer Academic, Dordrecht, Boston, MA, London, pp. 133–158.

Müller, B. *et al.* (2013) Describing human decisions in agent-based models-ODD+ D, an extension of the ODD protocol. *Environmental Modelling & Software*, **48**, 37–48.

Myles, G.D. and Naylor, R.A. (1996) A model of tax evasion with group conformity and social custom. *European Journal of Political Economy*, **12**, 49–66.

Nicolaides, P. (2014) Tax compliance social norms and institutional quality: an evolutionary theory of public good provision. Taxation Papers, working paper 46 – 2014, Luxembourg.

Noguera, J.A., Llacer, T., Miguel, F.J., and Tapia, E. (2014) Tax compliance, rational choice, and social influence: an agent-based model. *Revue Française de Sociologie*, **55** (4), 765–804.

Nordblom, K. and Žamac, J. (2012) Endogenous norm formation over the life cycle – the case of tax morale. *Economic Analysis and Policy*, **42** (2), 153–170.

Oates, L. (2015) Review of recent literature. *Journal of Tax Administration*, **1** (1), 141–150.

Pellizzari, P. and Rizzi, D. (2014) Citizenship and power in an agent-based model of tax compliance with public expenditure. *Journal of Economic Psychology*, **40**, 35–48.

Pickhardt, M. and Prinz, A. (2014) Behavioral dynamics of tax evasion – a survey. *Journal of Economic Psychology*, **40**, 1–19.

Pickhardt, M. and Seibold, G. (2014) Income tax evasion dynamics: evidence from an agent-based econophysics model. *Journal of Economic Psychology*, **40**, 147–160.

Rosen, J., Hemberg, E., Warner, G., Wijesinghe, S., and O'Reilly, U.-M. (2015) Computer Aided Tax Evasion Policy Analysis: Directed Search using Autonomous Agents. *Proceedings of the 14th International Conference on Autonomous Agents and Multi-agent Systems (AAMAS 2015)* (eds R.H. Bordini, E. Elkind, G. Weiss, and P. Yolum), May 4–8, 2015, Istanbul, Turkey.

Sandmo, A. (2012) An evasive topic: theorizing about the hidden economy. *International Tax and Public Finance*, **19** (1), 5–24.

Schelling, T.C. (1969) Models of segregation. *American Economic Review*, **59** (2), 488–493.

Schelling, T.C. (1971) Dynamic models of segregation. *Journal of Mathematical Sociology*, **1**, 143–186.

Schinckus, C. (2013) Between complexity of modelling and modelling of complexity: an essay on econophysics. *Physica A*, **392** (13), 3654–3665.

Seibold, G. and Pickhardt, M. (2013) Lapse of time effects on tax evasion in an agent-based econophysics model, 2013. *Physica A*, **392** (9), 2079–2087.

Slemrod, J. (2007) Cheating ourselves: the economics of tax evasion. *Journal of Economic Perspectives*, **21** (1), 25–48.

Slemrod, J. and Weber, C. (2012) Evidence of the invisible: toward a credibility revolution in the empirical analysis of tax evasion and the informal economy. *International Tax and Public Finance*, **19** (1), 25–33.

Slemrod, J. and Yitzhaki, S. (2002) Tax avoidance, evasion and administration, in *Handbook of Public Economics*, North-Holland, Amsterdam, pp. 1425–1470.

Srinivasan, T.N. (1973) Tax evasion: a model. *Journal of Public Economics*, **2**, 339–346.

Szabó, A., Gulyás, L., and Tóth, I.J. (2008) TAXSIM Agent Based Tax Evasion Simulator. *Paper presented at the 5th Conference of the European Social Simulation Association*, University of Brescia, Italy.

Szabó, A., Gulyás, L., and Tóth, I.J. (2009) Sensitivity analysis of a tax evasion model applying automated design of experiments, in *Progress in Artificial Intelligence, 14th Portuguese Conference on Artificial Intelligence, EPIA 2009, Aveiro, Portugal*, LNAI, vol. 5816, Springer-Verlag, Berlin, Heidelberg, pp. 572–583.

Szabó, A., Gulyás, L., and Tóth, I.J. (2010) Simulating tax evasion with utilitarian agents and social feedback. *International Journal of Agent Technologies and Systems*, **2** (1), 16–30.

Tesfatsion, L. (2006) Agent-based computational economics: a constructive approach to economic theory. *Handbook of Computational Economics*, **2**, 831–880.

Torgler, B. (2002) Speaking to theorists and searching for facts: tax morale and tax compliance in experiments. *Journal of Economic Surveys*, **16**, 657–683.

Vale, R. (2015) A model for tax evasion with some realistic properties. Working Paper available at Social Science Research Network.

Warner, G., Wijesinghe, S., Marques, U., Badar, O., Rosen, J., Hemberg, E., and O'Reilly, U.-M. (2015) Modeling tax evasion with genetic algorithms. *Economics of Governance*, **16**, 165–178.

Wintrobe, R. and Gërxhani, K. (2004) Tax Evasion and Trust: A Comparative Analysis. *Proceedings of the Annual Meeting of the European Public Choice Society, April 2004*, Berlin, Germany.

Wolf, S. *et al.* (2013) Describing economic agent-based models-Dahlem ABM documentation guidelines. *Complexity Economics*, **2** (1), 63–74.

Yitzhaki, S. (1974) A note on income tax evasion: a theoretical analysis. *Journal of Public Economics*, **3**, 201–202.

Zaklan, G., Lima, F.W.S., and Westerhoff, F. (2008) Controlling tax evasion fluctuations. *Physica A: Statistical Mechanics and its Applications*, **387** (23), 5857–5861.

Zaklan, G., Westerhoff, F., and Stauffer, D. (2009) Analysing tax evasion dynamics via the ising model. *Journal of Economic Interaction and Coordination*, **4**, 1–14.

2

How Should One Study Clandestine Activities: Crimes, Tax Fraud, and Other "Dark" Economic Behavior?

Aloys L. Prinz

2.1 Introduction

In this chapter, research methods are studied with respect to their adequacy and appropriateness for the detection, explanation, prediction, and combating of clandestine and dark economic activities. The main result is that there are two pitfalls of such studies, namely "the fallacy of misplaced concreteness" (A.N. Whitehead) and "the fallacy of disregarded abstractness" (M. Schramm). These potential fallacies point to the issue of the adequacy and the appropriateness of research methods and research goals. The more abstract the research objective, the more general its propositions; the more specific the study intentions, the more specific the various considerations. As a consequence, there can be no single method of research for all objectives. At this point, the availability of research methods becomes a major issue. For instance, in the more distant past, the tools consisted of theoretical models and a certain number of empirical methods. Neither laboratory and field experiments nor computer simulation models were on hand. Newly available methods now allow for an expansion of research in two directions toward a

Agent-Based Modeling of Tax Evasion: Theoretical Aspects and Computational Simulations,
First Edition. Edited by Sascha Hokamp, László Gulyás, Matthew Koehler, and Sanith Wijesinghe.
© 2018 John Wiley & Sons Ltd. Published 2018 by John Wiley & Sons Ltd.

deeper understanding of economic and social phenomena, and to different levels of analysis.

Take tax evasion as an example. Theoretically, there are a number of general theories (expected utility theory, psychological theories, etc.) that are useful in formulating hypotheses. With experimental methods and computer simulation models, these theories can now be tested in detail. For instance, one can consider what role social norms may play and how the interaction of people changes their tax-paying behavior. However, if the research objective is to find effective measures to combat, say, VAT fraud, these theories are of limited use. In the latter case, the availability of electronic cash registers, for instance, might be superior to knowing the degree of risk aversion that plays an important role in the relevant economic theory. Moreover, the more realistic and policy relevant a research question becomes, the larger will be the number of intervening variables that are not included in the theory. In the VAT fraud example, the size of the firm, its owners, as also the nature of its business are important variables to consider. Hence, the evaluation of research methods may only provide conditional recommendations as to their applicability and effectiveness, by pondering their advantages and shortcomings, given the research objective.

Another focus of the paper is the complexity of clandestine economic and social interactions. For instance, as is well known, tax evasion, money laundering, and tax havens are interrelated (Tavares, 2013). Consider a firm evading the corporate income tax in its home country via business operations in a tax-haven country. To avoid the immediate detection of this tax fraud, various hiding operations are necessary. In order to spend the money earned in this way on official transactions, laundering activities are required. A large number of transactions and persons are involved in such tax evasion. In addition, the social network structure of communities of people and societies renders clandestine activities particularly complex. As a consequence, research methods must be suitable, appropriate, and applicable concerning the complexity of the different layers of analysis.

The rest of this chapter is organized as follows. The next section deals with why it may be useful to study clandestine (economic) behavior. In Section 2.3, the available tools for such studies are presented and briefly discussed. In Section 2.4, the focus is on the complexity of interactions in clandestine activities. In this context, network effects and the dynamics of interactions are considered. Section 2.5 takes account of the layers of analysis, from the individual, to groups, to the macro level. The adequate application of the available research tools is discussed in Section 2.6. Section 2.7 concludes.

2.2   Why Study Clandestine Behavior At All?

Clandestine behavior results mainly from three very different motivations. Firstly, it provides an area of privacy; secondly, to hide an activity because it is expected to be socially (i.e., by significant others) unacceptable; and thirdly, to conceal illegal behavior. Of course, to prevent and to combat crimes, it is necessary to discover their underlying mechanisms. Nevertheless, clandestine activities, whether legal or illegal, are interesting from a social science perspective. As demonstrated by John List (2006) and Levitt and List (2007), people behave quite differently when they know that they are observed, in comparison to behavior in similar circumstances when they assume they are not observed. This is a general problem for research in social sciences; if people behave differently when observed, how can one study authentic behavior? One possibility is field experiments in which participants do not know that they participate in an experiment, but this is only feasible if participants did not have to give their informed consent (List, 2008). Another possibility is to use social media data, that is, data provided unintentionally by a very large number of users of these media and collected there (Rudder, 2014). This data demonstrates clearly that unobserved behavior varies greatly from that which is observed and pretended (see Rudder, 2014, for instructive examples). Hence, if authentic behavior is to be studied, clandestine behavior may be the key. In effect, social science cannot ignore such behavior if it wants to study genuine human activity.

However, social science is not entirely an end in itself. Although pure science also plays a role, social science results are necessary to provide a basis for better policies. A trial-and-error approach to social problems might be very expensive or even completely ineffective and futile. Knowledge about the mechanisms of crimes, drug trafficking, tax evasion, money laundering, corruption, and other activities may help develop policies and approaches to counter such activities effectively and efficiently. The social costs of illegal clandestine activities are certainly high, but the costs and damages of misconceived policies could be even higher.

In addition to the dependency of human behavior on simply feeling or actually being observed, a crucial question in social science is that of causality. Behavior occurs in a social environment in which action and reaction depend on and cause each other. Consequently, almost all human behavior is determined by the relevant social (sub)system. Therefore, behavior will genuinely suffer from circular causality (see Thomas, 2006). Wherever the circle of behavior is cut, the "causal" starting point is fixed. This means that the concept of causality loses some of its power. However, it may be that not all elements in the social system are of equal importance. Hence, the essence of research is to find those elements that are crucial for the behavior of the social subsystem under study. This requires a very careful investigation of the behavior in question. Without such studies, the chances of finding the drivers of clandestine activities such as crimes are very small. To go beyond symptomatic policies (that are not well-suited to solving social problems), research in the area of clandestine behavior is required. Since there are good reasons to study clandestine behavior and activities, the next question is how to do it.

2.3 Tools for Studying Clandestine Activities

In order to study human behavior in general, the usual tools are theory, empirical data analysis, questionnaires, laboratory and field experiments, and computer simulations. Recently, the huge amounts of data produced in social media and telecommunication networks have also be applied to study human behavior and clandestine activities.

The intended research method depends first on the (scientific and academic) goal of the analysis (Figure 2.1). Generally, three goals are relevant: description, explanation, and prediction (see Bunge, 1967b). This is particularly true for clandestine activities such as crimes and tax evasion. A description is required to separate the particular activity under scrutiny from all others; moreover, it also should provide a method for measuring the dimension of the respective activity. This is the first serious problem with clandestine activities. The next step of explanation consists of reduction and abstraction; an isolated phenomenon is embedded in an abstract and reduced frame in which (causal) relationships can be modeled. The last step of prediction requires a lower level of abstraction in order to anticipate the (potential) results of activities in a real-world scenario.

These scientific objectives require, first of all, theories and models. Theories are highly abstract mental constructs that are based on a few basic axioms that are employed to derive the logic consequences of these axioms. However, theories as such are not testable; they are metaphysical objects (Duhem, 1908/1978, Chapter 1). To become both testable and applicable, specifications are necessary. In this way, a theory is divided into many models, in which "a model is an idealized

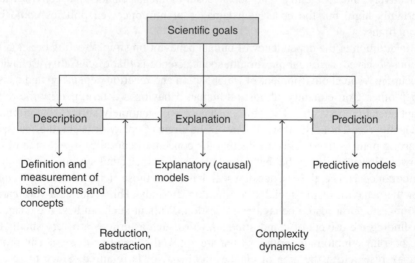

Figure 2.1 Scientific goals of analysis. Source: Own depiction. See for description, explanation, and prediction Bunge (1967b, Part III, Chapters 9 and 10)

representation of a class of real objects" (Bunge, 1967a, p. 386). Moreover, a single model can never represent the entire theory (Leijonhufvud, 1997, p. 193). As depicted in Figure 2.1, at least three different kinds of models can be identified: descriptive, explanatory, and predictive.

To start with the first kind of model, even descriptions and measurement are not possible without a descriptive model as a basis. Although the notions of "crime," "tax evasion," and so on are legally defined, the specification as well as the quantitative measurement of the dimensions of these concepts are not so straightforward because the respective actions are clandestine. Although hidden actions are by nature not directly observable, their consequences become evident sooner or later. A descriptive model has to combine observable outcomes with unobservable activities. Hence, measurement without such a model seems either impossible or merely ad hoc.

To understand a real-world phenomenon in general, an explanatory theory and many explanatory models are not only possible but also necessary. Theories are of necessity highly abstract. Therefore, theories cannot really be considered as "realistic," a failure that Alfred Whitehead called "the fallacy of misplaced concreteness" (Whitehead, 1925/1967, p. 51). Moreover, theories may imply absurd conclusions, but, "I doubt there is any set of assumptions that does not produce absurd conclusions when applied to circumstances far removed from the context in which they were conceived" (Rubinstein, 2006, p. 871). Therefore, whether one likes it or not, theories and clearly specified models are required to explain and understand economic (and noneconomic) phenomena, irrespective of whether they are based on legal or illegal behavior.

Predictive models presuppose even more empirical information that is not necessarily incorporated in explanatory models. As already recognized and analyzed by Duhem (1908/1978) and Quine (1951), testable and applicable models need a number of additional assumptions and specifications that are not strictly part of the theory. Therefore, theories cannot be regarded as true or false, independent of facts. Hence, the commonly supposed dichotomy between analytic and synthetic truth does not exist Quine (1951, p. 20). Put differently, no explanatory or predictive model is useful without the incorporation of empirical facts.

This leads to empirical studies being used as research tools in the area of clandestine activities; with respect to the shadow economy and tax compliance, see Slemrod and Weber (2012) and Feld and Larsen (2012). However, these activities, and sometimes even their results and outcomes, remain undetected by the public and the authorities. Although detected cases of fraud and crimes can be studied empirically, the number of undetected cases is inevitably high. This restricts empirical research approaches. Moreover, questionnaire research is also very limited, since the willingness to admit to socially outlawed behavior such as crimes and fraud, even if not sanctioned, might be rather low. Experimental studies, in the laboratory, as well as in the field, are both possible and useful. As demonstrated by List (2006) and Levitt and List (2007), there is a great difference in the behavior of people when they are observed and when they think that they are not observed. Field experiments are expensive and pose ethical questions, because the condition of "informed consent" would destroy participant perceptions of being unobserved (List, 2008). However, as, for instance, experimental laboratory studies with tax compliance show, such experiments provide some insights into the tax compliance behavior of participants; see Alm and Jacobson (2007), Alm (2012), and Pickhardt and Prinz (2014). Nevertheless, there are clear limitations to the generalizability of these results (Levitt and List, 2007). In this respect, field experiments are of greater value (Kleven *et al.*, 2011), but are not always possible, as, for example, with serious crimes.

Social media and telecommunication data, as indicated above, have become new sources for research. Data sets are also available for crimes, tax fraud, money laundering, and so on. Each activity leaves traces in electronic media and since these traces are the result of unobserved, authentic behavior, they are unbiased. Nowadays, authorities use this data, sometimes by intruding into the private sphere of innocent people, which indicates the ethical limits for this kind of research. Last, but not least, computational studies are both feasible and useful. The advantage of simulations is that the connectedness and social embeddedness of clandestine behavior can be accounted for (see Korobow *et al.*, 2007), which does not violate individual privacy. Moreover, not only the social environment and networks of persons can be incorporated but also the dynamics of behavior and of networks can be studied; for an overview of such studies regarding tax compliance, see Pickhardt and Prinz (2014). Such studies may provide the basis for predictive analyses of clandestine activities.

To sum up, all research tools available can, in principle, be applied to clandestine activities. However, the limited availability of empirical data cannot be overcome completely with the methods at hand, although social media and telecommunication data sources are useful. Nonetheless, the lack of reliable empirical data is one of the major remaining issues of hidden activities.

2.4   Networks and the Complexity of Clandestine Interactions

Although many crimes are committed individually, without interaction with others, in economic crimes and fraud a number of people are usually involved. Moreover, crimes and fraud take place in a social environment that might encourage *or* discourage criminal and fraudulent activities. This implies that clandestine activities exhibit a network structure. The motivation for all kind of activities (not only illegal ones) depends also on the personal social environment, as do the potential activities, as well as their aggregate outcomes.

This network structure makes it difficult to isolate individual actions from the network structure. The theoretical reason is that networks may create nonlinear feedback loops (Heylighen, 2001) between persons and actions, which are not under individual control. According to the sociologist Niklas Luhmann, social systems create and reproduce themselves (Luhmann, 1984, 1996, Chapter 1), and are barely controllable from outside. Although social subsystems reduce complexity, they nevertheless create their own internal complexity (Luhmann, 1984, 1996, Chapter 5). This means that once a social subsystem has created itself, its further development is indeterminate and unpredictable. Self-organization due to voluntary cooperation – known as "catallaxy" in the terminology of Hayek (1976) – is the key notion for such systems and subsystems. Not only is the official economy a "self-organizing system" (Krugman, 1996), but also the shadow economy, including drug trafficking, human trafficking, and so on.

A case in point is tax evasion. In a Networked Agent-Based Compliance Model, Korobow *et al.* (2007) demonstrated the network effects of neighbors at the aggregate level of tax compliance. Given a certain tax law enforcement strategy, the aggregate level of tax compliance was higher when agents knew little of their neighbor's pay-off, and vice versa.

An important characteristic of self-organizing systems is their level of adaptivity (Heylighen, 2001), that is, their ability to adapt themselves to (adverse) changes in the environment; for more on the origin of the adaptation principle in evolutionary biology and psychology, see Buss *et al.* (1998). The level of adaptivity is a central issue in combating illegal activities. For instance, highly adaptable crime networks (systems) are nearly impossible to eradicate or even to restrict. Moreover, the adaptivity of self-organizing systems is also crucial for the study of such systems. What research tools are required and which are available? What can(not) be said of these systems?

It seems that the network structure of clandestine activities – the "dark internet" – is a case in point Bartlett (2014); its self-organizing character, adaptivity, and complexity make it very difficult to analyze and even more difficult to counter.

A more detailed economic analysis of clandestine activities may shed some light on the question of how to detect it. The above-mentioned network structure and self-organizing property of clandestine activities is nevertheless governed by economic incentives; moreover, it has an economic production structure (Figure 2.2). The "black box" of clandestine (economic) activities to a large extent resembles legal and observable economic activities. This means that an input–throughput–output process is at work. Moreover, even the activity of individual firms in open markets can be described and analyzed economically without looking directly into the "black box" of internal production processes. Firms can be defined as legal structures for organizing cooperative production processes, as famously stated by Fama (Fama, 1980, p. 290): "The firm is just the set of contracts covering the way inputs are joined to create outputs and the way receipts from outputs are shared among inputs." For instance, observing inputs and outputs is sufficient to determine a firm's productivity and efficiency. In the context of clandestine activities, irrespective of whether they are carried out at the individual or the firm level, observing inputs and outputs may also reveal much about the "black box" of clandestine activities.

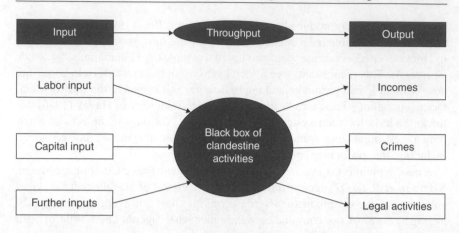

Figure 2.2   Input, throughput, and output of clandestine activities. Source: Own depiction

First of all, at least some output of clandestine activities will always be observable. For instance, crime victims will press charges, drugs are discovered and addictions become evident, the shadow economy leaves its marks in the demand for cash transactions, and so on. Additionally, incomes are observable to a certain degree if they are consumed and money laundering can in principle be detected. Although there are also many strategies to cover the tracks (i.e., by combining illegal activities with legal ones), some part of the outcomes of clandestine activities are observable. If the output is detected, inputs may also be tracked to a certain extent. In this way, observations of outputs and inputs can be employed to measure clandestine activities. If they are measurable, it is possible to "guess" the business model, as well as the network structure of the business. As a case in point, the connection of input and output is applied to measure the size of the shadow economy; the electricity consumption of an economy can be used to estimate the output of unofficial production, given the output of official firms; see Schneider and Enste (2000), on different methods for estimating the size of the shadow economy.

Measuring clandestine activities and re-engineering the network of clandestine activities may not be sufficient to counter it effectively. As indicated above, clandestine activities are self-organizing systems that are (more or less) adaptable to new circumstances as, for instance, police raids. Highly adaptable clandestine activities will reorganize their business model; they may even completely change their network structure and survive the raids. The economic nature of clandestine activities, their illegality, and the pressure of police raid will have an impact. However, whether these activities can be fought effectively is an open question since the development of self-organizing systems is indeterminate. Adaptivity might prove to be the crucial question in this respect. As suggested by the so-called Red Queen Model (Van Valen, 1973, pp. 17 ff.) of evolution (see Roberts and Newman, 1996, for comments, literature and "a model of evolution and extinction"), permanent adaptation to a changing environment is crucial for survival, whatever the species (Carroll, 1872/1993): "Now here, you see, it takes all the running you can do, to keep in the same place"; (Van Valen, 1973, p. 25, endnote 32). This applies to all kinds of legal as well as illegal behavior. In terms of evolutionary game theory, behavior that provides a pay-off at least slightly higher than the average will survive and even expand. To reduce or to eradicate such behavior, its relative pay-off must become (much) lower than the average (Van Valen, 1973).

2.5 Layers of Analysis

The analysis of real-world phenomena requires a number of layers, simply because these phenomena are also layered and have different dimensions. It does not matter much in this respect whether the phenomenon is legal or illegal. The only difference in illegal and clandestine activities is the additional dimension of obscurity and opacity. This should be borne in mind when looking at the levels of activity and analysis depicted in Figure 2.3.

Starting at the individual level, the next layer is a group of individuals who share a common interest and activity. Tax evasion and other (economic) crimes may be committed by individuals (who are themselves usually included in informal and formal networks that are not necessarily criminal), but also by groups that can be characterized as informal firms in the sense of Fama (1980). In the

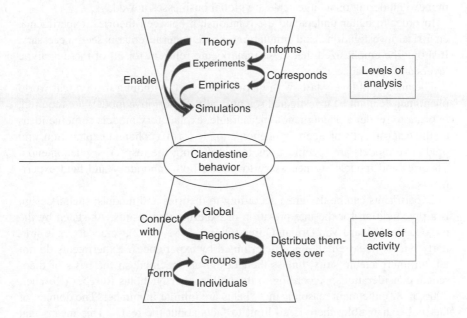

Figure 2.3 Levels of activity and levels of analysis. Source: Own depiction

latter case, crimes are committed by such firms, irrespective of whether they have a formal organizational structure. A case in point is the Mafia with its hierarchical, but informal structure. Another case is a firm that evades a tax; it too has a formal organizational structure. The internal organization is relevant for the criminal activity as well as for the feasibility of police raids and law enforcement. However, it should be clear that the organizational structure may change very quickly under police raids, if it saves the whole business.

Consider again the connections between tax evasion, tax havens, and money laundering. Imagine an individual establishing a dummy firm in a tax haven country with the sole intention of evading the income tax of the home country. Although the intention and the activities are purely individual, they cannot take place without an adequate and conducive environment. First of all, a tax haven country is required that allows for establishing an offshore company *and* that does not transmit data to the home country. Moreover, successful tax evasion creates additional income; in order to spend it safely in the home country, the illegally earned money must be legalized through laundering. This also requires an infrastructure that provides the necessary services.

More generally, individuals, as well as groups, spread themselves in and over regions. The regional distribution depends, for example, on the level of law enforcement, availability of input resources, and proximity of customers (as in the case of drug trafficking, for instance). Last, but not least, the respective activity may be at a global level, as in the above example of tax evasion. Tax evasion and money laundering on a large scale are global businesses nowadays.

In order to better understand the relationship between theories, experiments, empirical investigations, and simulations, some general remarks seem necessary. It should be emphasized that these remarks are relevant for all of these analytic layers and approaches.

The analytical layer starts with theory. As stated, without a theory, real-world phenomena cannot be investigated scientifically. Descriptive models are required, in order to render a phenomenon measurable, explanatory models combine ideas on the determinants of a certain phenomenon to form a coherent explanation, and predictive models are specifications for forecasting the respective phenomenon. Theories and models are necessary to inform both laboratory and field experiments.

Experiments can be designed according to theories and models; such a design is a precondition for the interpretation of experimental results. As stated by the philosopher Willard Van Orman Quine (Quine, 1951, p. 42), science "… is similarly but more extremely underdetermined by experience". Experiments do not tell an interpretable story, unless their design is specified on the basis of theoretical considerations. According to the philosopher Nicholas Rescher (Rescher, 2006, p. 82), the main reason is that "Facts are infinite in number. The domain of fact is inexhaustible: there is no limit to facts about the real.". This means that limiting the meaning of an unlimited number of facts, theories, and models is the prerequisite. Moreover, theories and the models derived from them are also required – for the same reason – to conduct and interpret empirical studies, as well as for running meaningful simulations. Nevertheless, experiments and empirical investigations are indispensable for analysis, because they produce results that enable testing models and even theories, in a kind of feedback-loop. Put differently, without experiments and empirical studies, theories and models would remain in their metaphysical realm; that is, they would not reveal anything about real-world phenomena. In contrast to empirical investigations, experiments allow for a design that makes it feasible to study causal links between certain external changes and human behavior. Although this does not imply that experiments represent the real world in its entirety, in well-designed experiments, certain supposed causal links may be tested.

However, experiments face a scalability limit. Although large-scale laboratory and field experiments can be imagined, their costs are usually unjustifiably high. This is one of the best reasons to use large-scale simulation studies. Their very substantial advantage is that the calculation power of today's computers allow for very large numbers of such simulations at very low cost. Large-scale simulations in fact solve the scalability problem that cannot otherwise be fixed at reasonable and justifiable cost as, for example, by using laboratory or field experiments. Unfortunately, virtual agents in computer simulations differ from human beings, owing to their low degrees of behavioral freedom. However, institutions, norms, and social control of all kinds are among those factors that make human behavior predictable to a certain degree. In addition, "behavioral rules" (Heiner, 1983, p. 561) reduce the flexibility of potential behavior, which then suffers from uncertainty of outcome. After all, human behavior does not seem to be all that unpredictable. It might even be that the network structure of human interactions – in legal and illegal, as well as in observable and clandestine cases – contributes more to the stochasticity of aggregate behavior than underdetermined individual actions. In the latter case, computer simulation can be used in two different ways, firstly, to identify the network contribution to behavioral and outcome fluctuations, and secondly, to forecast aggregate behavior and its outcome.

It should be clear that all layers of analysis, with all available research tools, can contribute quite substantially even to the knowledge on and understanding of clandestine activities. Consider again the example of tax evasion. Theoretical approaches such as the Allingham and Sandmo (1972) and Yitzhaki (1974) model (cf. Chapter 1) try to explain individual tax-evading behavior. The embedding of individual taxpayer behavior in a social environment has also been analyzed theoretically, for instance, by Myles and Naylor (1996), as well as Prinz et al. (2014). Many laboratory experimental studies are available (see Kastlunger et al., 2011), as well as a few field experiments; see Cummings et al. (2009) and Alm (2012). Additionally, a growing number of simulations has been published (see Pickhardt and Prinz, 2014, for a survey), some of them based on econophysics approaches, as, for example, Hokamp and Pickhardt (2010) and Pickhardt and Seibold (2014).

The next question is: How can one apply research tools appropriately to the analysis of clandestine activities? What might reasonably be expected, and what not?

2.6   Research Tools and Clandestine Activities

As indicated above, the available research tools for studying clandestine activities are theories, laboratory and field experiments, empirical investigations, as well as computer simulations. Although all these research tools are indispensable, their potential contribution to the analysis itself is quite different. As a consequence, the question is: what are the specific advantages of the respective method for analyzing clandestine activities?

To start with, some general issues should be discussed. There are several trade-offs in scientific research concerning real-world phenomena. According to Rescher (2006), there is a trade-off between "security and confidence," denoted by s, on the one hand and "definiteness and detail," denoted by d, on the other hand, so that $s \times d \leq constant$ (Rescher, 2006, p. 1). Rescher calls this "Duhem's law of cognitive complementarity," according to his interpretation of Duhem (1908/1978). According to Rescher, "Duhem's law" implies that science can only produce "secure" results at a rather abstract level of reasoning. The more details that are taken into account, the less "secure" the results. Economically, this means that there is a trade-off between the generality of theoretical results and their applicability to real-world problems. In a similar manner, (Roth, 2002) describes the role of the economist as an engineer with respect to designing market mechanisms for solving real problems, in contrast to merely analyzing markets. In order to design market mechanisms, all available tools must play a role. To solve real-world problems, very specific situations and variable constellations have to be accounted for. In contrast, a theoretical analysis attempts to find a few general variables that drive behavior and activities. As Ariel Rubinstein put it: "… economic theory is an abstract investigation of the concepts and considerations involved in real life economic decision-making rather than a tool for predicting or describing real behavior" (Rubinstein, 2001, Abstract). Hence, the role of theory in policy making is to provide an intuitive understanding of the problem and its feasible solution (McAfee and McMillan, 1996, p. 172).

How can one apply research methods adequately to the analysis of (clandestine) activities? There are at least two fallacies in this respect, namely, the "fallacy of misplaced concreteness" (Whitehead, 1925/1967, p. 51) and the "fallacy of disregarded abstractness" (Schramm, 2015, p. 29). Obviously, "science" has other goals than "policy," so that it should not come as a surprise that one gets into trouble if the one is mistaken for the other (Figure 2.4).

As "science" and "abstraction" are very closely related to each other, so too are "policy" (i.e., the intervention in real-world affairs) and "concreteness." As a consequence, general scientific results cannot be expected to be directly applicable.

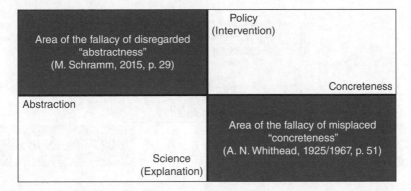

Figure 2.4 Disregarded abstractness versus misplaced concreteness. Source: Own depiction

To try and do so may result in Schramm's "fallacy of disregarded abstractness." By contrast, attempting to make science specific may lead to Whitehead's "fallacy of misplaced concreteness." These fallacies are all the more important, as there is no direct and immediate outcome of "disregarded abstractness" that could demonstrate the fallacy in the case of clandestine activities. Although theories and models are indispensable as analytical research tools, their abstractness renders them poorly suited for a direct application to policy issues of clandestine activities. For instance, countering tax evasion with models could imply an assumption that penalties and audits are perfect substitutes for each other. In practice, they are not. In addition, quantifying detection probabilities and penalties does not seem possible with theoretical models alone.

So far, only unspecific remarks on the applications of research methods have been provided in this section. In the next step, the peculiarities of clandestine activities are considered with respect to research methods. Put differently, the applicability and usefulness of the methods under the specific conditions of *secrecy* and *complexity* are accounted for. Secrecy is a crucial factor, since it only rarely allows for a direct test of hypotheses on "dark" behavior. Moreover, the multilayer network structure of secret activities makes them very complex. Hence, a useful application of research methods to clandestine activities must account for the *complexity of concreteness*, as demonstrated in Figure 2.5. According to this figure, theories (and models) of clandestine activities must provide concepts for their measurement. Moreover, the models must also determine the most important determinants of these activities, as well as concepts concerning the interactions of individuals, groups, and other individuals in these respects. Alternatively expressed, in addition to a static theory of the determinants of (legal and illegal) clandestine activities, theories and models for the dynamics and their network structures are required. For instance, Prinz *et al.* (2014) provide just such a dynamic model for tax compliance with two population

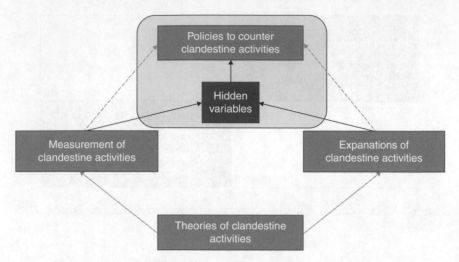

Figure 2.5 Difficulties in countering clandestine activities: the complexity of concreteness. Source: Own depiction

groups. One group is compliance minded and the other evasion minded; that is, the latter group will evade taxes, whenever possible and profitable. The compliance-minded group will not evade taxes, unless the gain from (undetected and unpunished) tax evasion is so high that even compliance-minded people become evasion minded. An interesting result of this dynamic model is that the punishment over and above a certain level may increase rather than decrease tax evasion.

Recently, Perc *et al.* (2013) studied the evolutionary dynamic development of crime with the so-called "inspection game" – see, for instance, Avenhaus *et al.* (2002) and Avenhaus (2004) – that is, a game between criminals, inspectors, and ordinary people, using a spatial variant. Based on simulations, they find a cyclical domination of "criminals," "inspectors," and "ordinary people" in the case of low or moderate inspection costs. They conclude that crime may be evolutionarily recurrent and that it requires "very much counter-intuitive and complex" strategies to contain such crime (Perc *et al.*, 2013). This result is generated by applying evolutionary game theory and computer simulation, which shows how to study the evolutionary dynamics of clandestine (in this case criminal) behavior. Although the direct usability for law enforcement and crime containing is rather limited, the study is nonetheless valuable for the insights it provides.

As explained above, not only theoretical research, but all kinds of experiments as well as computer simulations may prove very helpful, even on the theoretical level; see Rubinstein (2001), on studying the relationship between theory and experiments in economics, and Card *et al.* (2011), on the use of theory in field experiments.

One of the main difficulties for policy-oriented research is, however, the existence of a (probably) high number of hidden variables that interact with the known variables and structures in the real-world of these activities. Hence, without further empirical observations and data, the theoretical analysis is of very limited use in finding effective policies.

Duijn *et al.* (2014) analyze raids on criminal networks (organized cannabis cultivation) based on network and network resilience theory; additionally, they apply data sets from police investigations to test the efficacy of network disruption raids. Although network resilience theory describes potentially useful approaches – see, for example, Sparrow (1991), Klerks (2001), Prinz (2005), and Schwartz and Rouselle (2009) – the study found that police interventions tended to *increase* the efficiency and efficacy of the criminal network. Obviously, such a criminal network demonstrates a very high degree of raid resilience, whereby policy attacks even increase its efficiency. From an evolutionary point of view, the result is not so surprising after all. This study is important though, as it shows that the results of theoretical and simulation studies cannot be converted immediately into successful application. Moreover, it also demonstrates the importance of collecting and analyzing empirical data. This brings us back to the secrecy problem of "dark" (economic) behavior. How can data collection on this clandestine behavior and its results be improved and extended?

Agresten *et al.* (2015) use data sets on telephone contacts among suspects and data on the relationships between individuals involved in criminal offences perpetrated by the Sicilian Mafia. They emphasize that the very high resilience of criminal networks to raids in fact seems to be vulnerable through their phone contacts. Moreover, Tseng *et al.* (2012) describe text-mining methods, in combination with network analysis methods, to detect criminal structures and networks. In this way, social media data may be combined with network analysis to detect criminal networks and activities. Even hidden ties in criminal networks may be detected via network theory, when relevant data sets are available (Isah *et al.*, 2015). Consequently, social media and telecommunication activities provide new and very large data sets that may be used to detect and to measure clandestine activities. Nevertheless, there are two limitations in this respect. Firstly, there are legal and ethical restrictions that protect the privacy of these activities. Secondly, the availability of huge data sets requires concepts and algorithms to decide what to look for. Although such concepts and algorithms are available, the risk of error is high; see Lazer *et al.* (2014), for errors in flu prediction on the basis of Google data.

Another data source is that of solved criminal cases. Tax evasion, money laundering, and so on are detected to a certain extent. These cases automatically produce large data sets through police, court, and administrative activities. As shown by Duijn *et al.* (2014), such data sets can be used successfully for policy-oriented studies. They reveal some secret activities, and this knowledge can be employed to calibrate network and simulation models.

As a research tool, computer simulations offer a wide range of applications for studying clandestine activities. Although the external validity of such simulation cannot be determined without reliable empirical data (see Prinz, 2016, for a detailed analysis of the usability of simulations in tax compliance research), they nevertheless enhance the potential for policy-oriented research. Whether they can be reasonably employed for forecasting the effects of anti-crime policies, for instance, depends on the calibration of the simulation models. However, calibration itself is restricted by the availability of sound empirical data. As pointed out by Dirk Helbing, "… computer simulation can be seen as an experimental technique for hypothesis testing and scenario analysis which can be used complementarily and in combination with experiments in real-life, the lab or the Web" (Helbing, 2012, p. 25). A further step in the development of dynamical and interactive modeling for social and economic processes, whether observable or clandestine, legal or illegal, is "experimental econophysics," as described in detail by Huang (2015). In this approach, laboratory experiments with humans are combined with agent-based simulations in order to study the emergent properties of real-world phenomena (Huang, 2015). Self-organization, as well as complexity, plays a crucial role in this method.

A method that combines the behavioral approach to tax compliance with a social network model and agent-based modeling, called "predictive analytics," has been introduced by Hashimzade *et al.* (2016), in order to investigate tax evasion. Taxpayers form subjective beliefs on tax audit probability in their social interactions with fellow taxpayers and choose their employment (i.e., as an employee or in the form of self-employment) according to their level of risk aversion. The tax authority chooses a nonrandom audit strategy that has an impact on subjectively perceived audit probabilities. In this way, taxpayers are self-selected into groups with different attitudes to tax evasion. This can be used by the tax authority to raise revenue.

Experimental econophysics and predictive analytics are examples of a recent class of hybrid research methods. These methods combine several analytical and computational strategies to develop more specific real-world models and obtain the associated results. Although they are "only" models, their capacity to analyze real-world phenomena in a much more complex social, but still controllable, environment is larger than with single "pure" methods. However, hybrid models too come with a "price tag"; it is no longer possible to say exactly which effect was caused by what. This is less satisfactory from a scientific point of view, but is much more relevant to and suitable for the characteristics of self-organizing systems.

A bundle of further research methods can be found in artificial intelligence. In this respect, machine learning (see Barber, 2012, Chapter 13 ff.) may play a major role. If a huge database is available, computers can be trained to analyze the data thoroughly and automatically. Of course, models are required to train machines in such a manner as to detect the respective structures of clandestine activities. However, it is an open question whether and when machine learning tools reach a level of perfection so that the results are valid and reliable.

Obviously, each research method has its merits and demerits, particularly in the study of clandestine activities. In Table 2.1, an overview of research methods is provided concerning the areas of application with respect to clandestine activities, their advantages, and their shortcomings.

First of all, the methods mentioned in this chapter may all contribute substantially to the study of clandestine activities. The external validity of research (i.e., the meaning of the research results for the respective real-world problem) is not so easy to establish. However, the availability of sound empirical data may improve the situation considerably. Nonetheless, until recently the direct applicability of research results for law enforcement, for instance, seemed rather restricted. The availability of social media and telecommunication data, when combined with other methods, may improve the situation drastically. But even then, it should be borne in mind that the evolutionary stability and robustness of crime and illicit activities may limit the success of enforcement policies.

Having said this, in the second column of Table 2.1, the areas of application for the respective research method are indicated. More or less all kinds of clandestine activities may ultimately be accessible for the various research methods. In the third column, the advantages of the respective research method are briefly described. Theories, models, and laboratory experiments can be put into one class, as they study clandestine behavior in a mainly scientific way, which means that an immediate application of their results may neither be intended nor very successful. The remaining methods, although based on results from the first class of methods, seem to be more appropriate for policy research and application. This is particularly true for hybrid methods. The fourth and last column points to the main shortcomings of the respective method. For the first class of methods, unknown external validity is the most serious problem. The other methods may solve this problem to the extent that valid and reliable data in sufficient amounts is available. Moreover, although social media and telecommunication data may alleviate the data availability problem, they create their own problems, since the quality of the data (validity and reliability) is not as good as expected and the algorithms of data analysis may lead to serious errors. Nevertheless, their advantage is that the data is based on authentic behavior.

Last, but not least, the procedure of scientific studies concerning clandestine activities can be summarized as follows (see Gale and Slemrod, 2001, in the context of the US debate on the estate tax). Theoretical analysis and investigation, in combination with empirical facts, should be used in simulation and hybrid models to obtain at least informed guesses on the mechanics of and policies on clandestine behavior.

Table 2.1 Research methods and their application in the study of clandestine activities (CLAs)

Research method	Areas of application	Advantages	Shortcomings
Theories and models	Determinants and dynamics of CLAs and networks	Focus on a few causal factors and interaction schemes; framework for further analysis	Unknown external validity
Laboratory experiments	Human behavior in CLAs	Measuring the effectiveness of certain incentives on behavior	Unclear external validity; diverse selection biases
Field experiments	Human behavior with real-world conditions	Measuring the effectiveness of certain incentives on behavior	Large-scale effects (population) unclear; long-run effects not observable
Empirics	All areas of CLAs; measurement of inputs, throughput, and output	Measuring real-world effects on a large scale	Data basis small and incomplete
Network analysis and simulations	All areas of CLAs in which social interactions play a substantial role	Modeling of group and network processes with heterogeneous agents	Unclear external validity, if not calibrated with empirical data
Hybrid methods	All areas of CLA research	Synergy of advantages of combined methods	Synergy of shortcomings of combined methods
Social media and telecom data analysis	All areas of CLAs research	Data based on authentic behavior	Privacy concerns; validity and reliability of data and algorithms

2.7   Conclusion

This chapter contains an analysis of how clandestine activities such as crimes, tax fraud, and other "dark" activities can be studied scientifically. The goals of such scientific analysis have been identified as description, explanation, and prediction of the relevant behavior. Moreover, the usefulness of scientific studies for policy interventions is also considered.

It is argued that even the measurement of clandestine activities requires a theory and probably more than one model (in the sense of Mario Bunge's theory of scientific research). That is, without any theory, a scientific study of clandestine behavior – and all other behavior too – is not feasible. However, descriptive models alone are insufficient, as they do not provide rationales for the respective activities; hence, explicative theories and models are required. Such theories, as well as models, must of necessity be parsimonious to be of value. This implies that they are not directly applicable for policy analysis. Moreover, in order to forecast real-world phenomena, predictive models are necessary that encompass all kinds of variables and many specifications that are omitted from explicative theories and models. Explanation requires a high degree of generality, and prediction presupposes a large number of additional variables and facts. As a consequence, there is a generality – specificity gap between explanation and prediction. Policy analysis is all about specificity and, therefore, different and much more specific approaches are essential. The consequence is two potential pitfalls, namely the "fallacy of misplaced concreteness" (Whitehead) for science, and the "fallacy of disregarded abstractness" (Schramm) for policy.

For policy research and evaluations, empirical data are indispensable. This is particularly true for clandestine activities. Recently, social media and telecommunication sources have become progressively more available. This could lead to a major leap in our knowledge about clandestine activities, from drug and human trafficking to tax evasion and other crimes. Although there are ethical concerns involved, communication and trading activities leave traces that can be used even without identifying individuals or firms. This might reduce ethical reservations in using these data empirically. Nevertheless, the validity and reliability of data, as well as of the analytical algorithms employed for their analysis, are far from faultless.

Finally, it should be emphasized that simulations may play a crucial role in theoretical studies and, even more importantly, in policy studies. The reason is that simulations can differentiate between agents as well as between social networks and their interactions. Hence, simulations will be much more complicated than models. Based on theories and models, simulations are theoretical research tools; calibrated with empirical data, they become policy research instruments. The more data on clandestine activities is available, the more attractive and useful simulations will become as policy research tools.

Acknowledgment

I would like to thank two anonymous reviewers for helpful comments on an earlier version of the chapter and Brian Bloch for extensive text editing. However, all remaining errors are mine.

References

Agreste, S., Catanese, S., De Meo, P., Ferrara, E., and Fiumara, G. (2015) Network structure and resilience of Mafia syndicates. *arXiv*: 1509.01608v1 [cs.SI] 4 September 2015.

Allingham, M.G. and Sandmo, A. (1972) Income tax evasion: a theoretical analysis. *Journal of Public Economics*, **1**, 323–338.

Alm, J. (2012) Measuring, explaining, and controlling tax evasion: lessons from theory, experiments, and field studies. *International Tax and Public Finance*, **19** (1), 54–77.

Alm, J. and Jacobson, S. (2007) Using laboratory experiments in public economics. *National Tax Journal*, **60**, 129–152.

Avenhaus, R. (2004) Applications of inspection games. *Mathematical Modelling and Analysis*, **9** (3), 179–192.

Avenhaus, R., Von Stengel, B., and Zamir, S. (2002) Inspection games, in *Handbook of Game Theory with Economic Applications*, vol. **3** (eds R.A. Aumann and S. Hart), North Holland, pp. 1947–1987.

Barber, D. (2012) *Bayesian Reasoning and Machine Learning*, Cambridge University Press.

Bartlett, J. (2014) *The Dark Net. Inside the Digital Underworld*, William Heinemann.

Bunge, M. (1967a) *Scientific Research I: The Search for System*, Springer-Verlag, Berlin, Heidelberg, New York.

Bunge, M. (1967b) *Scientific Research II: The Search for Truth*, Springer-Verlag, Berlin, Heidelberg, New York.

Buss, D.M., Haselton, M.G., Shackelford, T.K., Bleske, A.L., and Wakefield, J.C. (1998) Adaptations, exaptations, and spandrels. *American Psychologist*, **53** (5), 533–548.

Card, D., DellaVigna, S., and Malmendier, U. (2011) The role of theory in field experiments. *Journal of Economic Perspectives*, **25** (3), 39–62.

Carroll, L. (1872/1993) *Through the Looking-Glass: And What Alice Found There*, HarperCollins, New York.

Cummings, R.G., Martinez-Vazquez, J., McKee, M., and Torgler, B. (2009) Tax morale affects tax compliance: evidence from surveys and an artefactual field experiment. *Journal of Economic Behavior & Organization*, **70**, 447–457.

Duhem, P. (1908/1978) *Ziel und Struktur der physikalischen Theorien [La théorie physique, son objet et sa structure]*, Felix Meiner Verlag, Hamburg.

Duijn, P.A.C., Kashirin, V., and Sloot, P.M.A. (2014) The relative ineffectiveness of criminal network disruption. *Nature Scientific Reports*, **4** (4238), 1–15.

Fama, E.F. (1980) Agency problems and the theory of the firm. *Journal of Political Economy*, **88** (2), 288–307.

Feld, L.P. and Larsen, C. (2012) Self-perceptions, government policies and tax compliance in Germany. *International Tax and Public Finance*, **19** (1), 78–103.

Gale, W.G. and Slemrod, J. (2001) Rhetoric and economics in the estate tax debate. *National Tax Journal*, **54** (3), 613–627.

Hashimzade, N., Myles, G.D., and Rablen, M.D. (2016) Predictive analytics and the targeting of audits. *Journal of Economic Behavior & Organization*, **124**, 130–145.

Hayek, F.A. (1976) *Law, Legislation and Liberty: The Mirage of Social Justice*, vol. **2**, University of Chicago Press.

Heiner, R.A. (1983) The origin of predictable behavior. *American Economic Review*, **73** (4), 560–595.

Helbing, D. (2012) *Social Self-Organization. Understanding Complex Systems*, Springer-Verlag, Berlin, Heidelberg.

Heylighen, F. (2001) The science of self-organization and adaptivity. *The Encyclopedia of Life Support Systems*, **5** (3), 253–280.

Hokamp, S. and Pickhardt, M. (2010) Income tax evasion in a society of heterogeneous agents. Evidence from an agent-based model. *International Economic Journal*, **24**, 541–553.

Huang, J.P. (2015) Experimental econophysics: complexity, self-organization, and emergent properties. *Physics Reports*, **564**, 1–55.

Isah, H., Neagu, D., and Trundle, P. (2015) Bipartite network model for inferring hidden ties in crime data. arXiv: 1510.02343v1 [cs.SI].

Kastlunger, B., Muehlbacher, S., Kirchler, E., and Mittone, L. (2011) What goes around comes around? Experimental evidence of the effects of rewards on tax compliance. *Public Finance Review*, **39**, 150–167.

Klerks, P. (2001) The network paradigm applied to criminal organizations: theoretical nitpicking or a relevant doctrine for investigators? Recent developments in the Netherlands. *Connections*, **24** (3), 53–65.

Kleven, H.J., Knudsen, M.B., Kreiner, C.T., Pedersen, S., and Saez, E. (2011) Unwilling or unable to cheat? Evidence from a tax audit experiment in Denmark. *Econometrica*, **79**, 651–692.

Korobow, A., Johnson, C., and Axtell, R. (2007) An agent-based model of tax compliance with social networks. *National Tax Journal*, **60** (3), 589–610.

Krugman, P. (1996) *The Self Organizing Economy*, Wiley-Blackwell, Cambridge, MA and Oxford.

Lazer, D., Kennedy, R., King, G., and Vespignani, A. (2014) The parable of Google Flu: traps in big data analysis. *Science*, **343**, 1203–1205.

Leijonhufvud, A. (1997) Models and theories. *Journal of Economic Methodology*, **4** (2), 193–198.

Levitt, S.D. and List, J.A. (2007) Viewpoint: on the generalizability of lab behaviour to the field. *Canadian Journal of Economics*, **40** (2), 347–370.

List, J.A. (2006) The behavioralist meets the market: measuring social preferences and reputation effects in actual transactions. *Journal of Political Economy*, **114** (1), 1–37.

List, J.A. (2008) Informed consent in social science. *Science*, **322** (5902), 672.

Luhmann, N. (1984) *Soziale Systeme. Grundriss einer allgemeinen Theorie*, Suhrkamp, Frankfurt am Main.

Luhmann, N. (1996) *Social Systems (translated by John Bednarz)*, Stanford University Press.

McAfee, R.P. and McMillan, J. (1996) Analyzing the airwaves auction. *Journal of Economic Perspectives*, **10** (1), 159–175.

Myles, G.D. and Naylor, R.A. (1996) A model of tax evasion with group conformity and social customs. *European Journal of Political Economy*, **12**, 49–66.

Perc, M., Donnay, K., and Helbing, D. (2013) Understanding recurrent crime as system-immanent collective behavior. *PLoS ONE*, **8** (10), e76063.

Pickhardt, M. and Prinz, A. (2014) Behavioral dynamics of tax evasion. *Journal of Economic Psychology*, **40** (1), 1–19.

Pickhardt, M. and Seibold, G. (2014) Income tax evasion dynamics: evidence from an agent-based econophysics model. *Journal of Economic Psychology*, **40** (1), 147–160.

Prinz, A. (2005) The network structure of illegal firms, in *Entscheidungsorientierte Volkswirtschaftslehre* (eds M. Göcke and S. Kooths), Peter Lang, Frankfurt am Main, pp. 465–480.

Prinz, A. (2016) Simulations in tax compliance research: what are they good for? *Review of Behavioral Economics*, **3** (1), 20–46.

Prinz, A., Muehlbacher, S., and Kirchler, E. (2014) The slippery slope framework on tax compliance: an attempt to formalization. *Journal of Economic Psychology*, **40** (1), 20–34.

Quine, W.V. (1951) Main trends in recent philosophy: two dogmas of empiricism. *Philosophical Review*, **60** (1), 20–43.

Rescher, N. (2006) Epistemetrics. Chapter 1: Asking for More Than Truth: Duhem's Law of Cognitive Complementarity, pp. 1–7; Chapter 3: Spencer's Law of Cognitive Development, pp. 15–28; Chapter 7: How Much Can Be Known? pp. 73–94, Cambridge University Press.

Roberts, B.W. and Newman, M.E.J. (1996) A model for evolution and extinction. *Journal of Theoretical Biology*, **180**, 39–54.

Roth, A. (2002) The economist as engineer: game theory, experimentation, and computation as tools for design economics. *Econometrica*, **70** (4), 1341–1378.

Rubinstein, A. (2001) A theorist's view of experiments. *European Economic Review*, **45**, 615–628.

Rubinstein, A. (2006) Dilemmas of an economic theorist. *Econometrica*, **74** (4), 865–883.

Rudder, C. (2014) *Dataclysm: Who We are (When We Think No One's Looking)*, Crown Publishing, New York.

Schneider, F. and Enste, D. (2000) Shadow economies: size, causes, and consequences. *Journal of Economic Literature*, **38** (1), 77–114.

Schramm, M. (2015) How the (Business) World Works. Business Metaphysics & "Creating Shared Value". Annual Meeting of the Working Group on "'Wirtschaftsphilosophie und Ethik'" 2015: "'Creating Shared Value -Concepts, Experience, Criticism'", Zeppelin University (Friedrichshafen).

Schwartz, D.M. and Rouselle, T.D.A. (2009) Using social network analysis to target criminal networks. *Trends in Organised Crime*, **12**, 188–207.

Slemrod, J. and Weber, C. (2012) Evidence of the invisible: toward a credibility revolution in the empirical analysis of tax evasion and the informal economy. *International Tax and Public Finance*, **19** (1), 25–53.

Sparrow, M.K. (1991) The application of network analysis to criminal intelligence: an assessment of the prospects. *Social Networks*, **13**, 251–274.

Tavares, R. (2013) Relationship Between Money Laundering, Tax Evasion and Tax Havens. Special Committee on Organized Crime, Corruption and Money Laundering (CRIM) 2012–2013, European Parliament.

Thomas, R. (2006) Circular causality. *IEE Proceedings - Systems Biology*, **153** (4), 140–153.

Tseng, Y.-H., Ho, Z.-P., Yang, K.-S., and Chen, C.-C. (2012) Mining term networks from text collections for crime investigations. *Expert Systems with Applications*, **39**, 10082–10090.

Van Valen, L. (1973) A new evolutionary law. *Evolutionary Theory*, **1** (1), 1–30.

Whitehead, A.N. (1925/1967) *Science and the Modern World*, The Free Press, New York.

Yitzhaki, S. (1974) A note on income tax evasion: a theoretical analysis. *Journal of Public Economics*, **3**, 201–202.

3

Taxpayer's Behavior: From the Laboratory to Agent-Based Simulations

Luigi Mittone and Viola L. Saredi

3.1 Tax Compliance: Theory and Evidence

As pointed out by Andreoni *et al.* (1998), the problem of tax evasion and noncompliance has been introduced in the economic literature just as an additional "risky asset" to the household's portfolio. The first theoretical representations of individual taxpayer's compliance date back to the 1970s: the most influential, and probably the most criticized, rational choice models have been developed by Allingham and Sandmo (1972) and Srinivasan (1973). As outlined in Chapter 1, these models portray the taxpayer's decision problem as an investment choice involving a sure and a risky lottery, and adopt the formalization of expected utility theory (von Neumann and Morgenstern, 1944). Taxpayers are supposed to choose the extent of income declaration that maximizes their expected utility, defined according to income level, individual risk propensity, audit probability, and monetary punishment, in case of evasion (or noncompliance) detection. These studies focus the effect of such parameters on evasion: for instance, a higher audit probability and/or a punishment proportional to the evaded tax reduces the expected value of evasion, and thus its attractiveness (Yitzhaki, 1974).

However, some issues on the aforementioned models have been raised, and their validity has been questioned, as the portfolio approach fails to address real-world complexity. Taxpayers are assumed to be able to determine the optimal

Agent-Based Modeling of Tax Evasion: Theoretical Aspects and Computational Simulations, First Edition. Edited by Sascha Hokamp, László Gulyás, Matthew Koehler, and Sanith Wijesinghe.

proportion of tax to evade by making burdensome calculations, and having accurate information on the audit strategies adopted by the tax authority. Under these rather unrealistic conditions, such models predict that all taxpayers should evade if the audit probability and the fines most commonly used in reality were adopted. Furthermore, the label of "tax compliance" is generally adopted to refer to a wide variety of behaviors – such as evasion of value added tax,[1] income underreporting, or tax burden reduction – which, though exhibiting remarkably different idiosyncratic characteristics, are treated with no distinction. For a recent review see Muehlbacher and Kirchler (2016).

Graetz and Wilde (1985) suggest that taxpayers' decisions cannot be entirely explained by the level of enforcement, as tax compliance is not only a matter of rates and penalties. Furthermore, as it is not easy to obtain precise information on the actual audit procedure that the tax authority adopt to discover tax evasion, taxpayers may not know the actual risk of being audited, and need to rely on their own estimate of such a risk, in order to make a compliance decision. Such an uncertainty about the probability of getting caught is likely to influence taxpayers' behavior.

This is supported by empirical evidence documented in many countries, coming from different sources such as random audits, surveys, and laboratory and field experiments. Hence, many researchers, behavioral economists included, have tried to find models with a better fit for real taxpayers' behavior, and with a focus on potentially relevant psychosociological factors. This development process has often been built upon an experimental approach: just to list a few examples, the high degree of control and the greater parallelism with the natural world are among the main motivations leading researchers to turn to laboratory experiments, instead of relying only on theoretical analyses. In this sense, one can say that experiments on tax evasion are mainly motivated by economists' dissatisfaction with theoretical models: "Rather than question the experimental method, […] it is perhaps the theory which needs revision" (Baldry, 1987). In fact, experiments can provide a valuable support in studying people's behavior, especially when clandestine activities are involved: the extent of tax evasion as a result of a specific interaction between micro and macro factors cannot be directly measured in an entirely natural and uncontrolled setting. It is not easy to collect evidence on tax compliance, and, even if this was possible, the specific conditions determining tax decisions would not be easily kept under control. In contrast, a detailed investigation of individuals' behavior is allowed by the laboratory approach, which can be considered as the proper system to understand specific real-world phenomena.

---

[1] In contrast, Mittone (2001), considered as a pioneer in the experimental investigation of VAT, recognizes the social nature of this kind of tax and explores it in an artificial market setting. Although VAT and income tax are strictly connected, since self-employed taxpayers evade VAT in order to reduce their tax liability on income, sellers do not decide by themselves but need to collude with their customers. The psychological role, played by the need to find a collusive (illicit) agreement between buyer and seller, is the core issue of VAT evasion.

To this purpose, many extensions based on experimental findings have been proposed in order to integrate theoretical models and make them closer to the intended domain of application: audits are costly for any audited person; the tax authority is distinct from the remainder of the government (Melumad and Mookherjee, 1989); tax collection is delegated (Sanchez and Sobel, 1993); moral and social dynamics, in terms of shame, moral rules, fairness to the tax code and its application (e.g., Erard and Feinstein, 1994; Spicer and Becker, 1980; Benjamini and Maital, 1985; Baldry, 1986; Gordon, 1989; Myles and Naylor, 1996; Torgler, 2002; Eisenhauer, 2006, 2008; Casal and Mittone, 2016), and evaluation of government expenditure and service provision (Cowell and Gordon, 1988) are included; the impact of the decision framework is tested. In this respect, Mittone (2002) finds that the introduction of an environmental structure closer to the one outside the laboratory fosters tax compliance thanks to the creation of social ties among participants. Finally, the investigation of taxpayers' views of the audit probability has shown the inadequacy of expected utility theory for tax evasion (e.g., Friedland, 1982; Spicer and Thomas, 1982; Alm et al., 1992b,c; Hessing et al., 1992; Sheffrin and Triest, 1992; Scholz and Pinney, 1995): people usually exhibit cognitive difficulties in estimating probability relationships and computing expected values (Einhorn and Hogarth, 1985; Casey and Scholz, 1991a,b), and the common uncertainty about the probability of being audited makes taxpayers' decisions more difficult than those made under the full information characterizing the laboratory environment. As traditional economic models rely on the unrealistic assumption that taxpayers have accurate information on auditing strategies, actual taxpayers' behavior cannot be predicted. Therefore, experiments with imprecise information appear to be more realistic.

A further step toward a greater realism in tax research is due to the introduction of bounded rationality (Simon, 1955, 1956). As suggested by Alm (1999), standard theoretical models, grounded on the simplifying assumptions of taxpayers' full rationality and homogeneity, should be revised. These models rely on the adoption of a unique representative agent and disregard the interaction among different types, while human behavior exhibits not only evident anomalies but also a remarkable heterogeneity (Alm et al., 1992c): for instance, some individuals may overweight the occurrence of fiscal audits, or comply because they value what they are financing. Furthermore, laboratory experiments help prove that human beings are not always able to perform complex computations and to choose the utility-maximizing action. They are not making an investment decision in isolation, but are affected by many different "emotional" factors and noneconomic considerations that make their decision process rather complex to model.

However, all these aspects, which have been defined and first investigated at the micro level, as allowed by laboratory experiments and tax theoretical models, may have unexpected consequences and give striking results at an aggregate level. For example, the vast majority of macro-empirical research reports a strong deterrent effect of tax audits on evasion. In contrast, Gemmell and Ratto (2012) empirically explore compliance response to fiscal audit at an individual level, and observe contrasting results, due to many factors, such as the opportunity to underreport and past audit experience. This implies that, in order to obtain relevant policy suggestions, neither of the two dimensions has to be disregarded. In this respect, as reviewed by Alm (2010), a growing number of researchers have adopted behavioral techniques, which rely on both human-based experimental economics and agent-based modeling, in order to address this micro–macro issue, and to gain new insights on taxpayers' behavior, which could not be observed otherwise. Micro-level experimental findings have widely shown that human agents are not rational, as assumed in theoretical models; on the contrary, they are guided by emotions, psychological and moral constraints, which might be mimicked in a computational simulation, as agents are calibrated according to human-based experimental evidence. In such a way, interesting and useful considerations can derive from agent-based modeling: on the one hand, it allows the implementation of a rather realistic system of individuals, with the intent of uncovering and testing specific cognitive aspects of taxpayers' decision process; on the other hand, the societal evolution, as due to an interaction among heterogeneous agents, can be studied from a macro perspective.

Hence, we point out that a synergic adoption of a human- and a computer-based approach gives the opportunity of gaining a deeper understanding of empirical phenomena or behavioral patterns, by scientifically studying a *valid* representation of these in the laboratory. Thanks to the combination of these two approaches, compliance decisions are studied both at an individual and at a collective level: the exploration of the overall behavior of the society requires taking into account social interactions among heterogeneous and boundedly rational human beings. Agent-based simulations might provide valuable support, since they can rely on realistic assumptions – that is, behavioral regularities previously observed in the laboratory – and allow the implementation of complex settings in which both micro and macro factors interact and affect agents' behavior, as it usually happens outside the walls of the experimental laboratory.

3.2   Research on Tax Compliance: A Methodological Analysis

According to the previously presented evolution of research on tax compliance, an apparent challenge between economic theoretical models and the experimental approach seems to emerge. On the one hand, experimenters claim that anomalies observed in the laboratory are an important proof of the failure of theory in describing and predicting taxpayers' behavior in an accurate way. On the other hand, however, theorists reply that their models are intended to address phenomena taking place in the real world and not in the artificial environment of the laboratory. Alm *et al.* (1992b) point out that " […] experimental results can contribute significantly to policy debates, as long as some conditions are met: the payoffs, and the experimental setting must capture the essential properties of the naturally occurring setting that is the object of investigation. Laboratory methods may offer the only opportunity to investigate the behavioral responses to policy changes."

In such a framework, the necessity of external validity of grounding experiments is evident: researchers claim that results are not always generalizable, that is, applicable to the real world, because the environment reproduced in the laboratory is too simplistic and does not take into account many relevant variables. For instance, Webley (1991) argues that " [experimental] results may reflect a person's understanding of economics rather than the behavior that would be displayed in the real situation." The experimental setting might be perceived as too artificial and far from the environment outside the laboratory, if it is not a perfect replica of the real world. The experimental system needs an external validity hypothesis, which maps laboratory elements onto elements of the phenomenon observed in the field. Only if this hypothesis holds, can researchers draw valuable inferences on individuals' decision process from the laboratory and move to the world outside (Guala, 2002).

The present chapter enters such a debate and explicitly focuses on the problem of external validity of tax experiments, adding to the small literature available on the topic. On the contrary, it disregards the internal validity issue, which has already received much attention in the experimental literature. Specifically, the novelty of this methodological review resides in the proposal of a synergic approach, involving both human-based observations and agent-based simulations as a valuable tool aimed at solving the problem of external validity of tax experiments.

In a very recent contribution, Muehlbacher and Kirchler (2016) address this methodological issue, providing an interesting review of both experimental and empirical research on tax compliance. The authors point out that little is known about the external validity of tax experiments and identify a number of criticisms: in addition to a rather general critique on artificiality, participants' self-selection, experimenter effect, social desirability, and social blaming, a more detailed methodological review on tax research is offered. For instance, as already well documented in previous studies, income-reporting decisions in a tax setting systematically differ from those in an abstract setting (Baldry, 1986; Alm et al., 1992c; Mittone, 2006; Choo and Fonseca, 2016); the introduction of a redistribution mechanism strongly affects taxpayers' decisions (Alm et al., 1992a,c; Mittone, 2006); students might not be representative because they have no experience in paying taxes (Webley, 1991). Furthermore, compliance depends on the way in which subjects' income is provided (Boylan and Sprinkle, 2001; Boylan, 2010; Durham et al., 2014). Finally, in reality, there is a temporal distance between compliance decisions and audits, which might have a significant effect on actual compliance decisions, and make experiments disregarding this issue less reliable (Kogler et al., 2016). Based on this, Muehlbacher and Kirchler (2016) suggest that experimental investigations in the laboratory should induce the same psychological mechanisms taxpayers adopt outside, and take into account possible interactions between treatment factors and setting characteristics. If such requirements are met, experimental findings can be applied also outside the laboratory, and thus provide useful insights for policy interventions.

According to Alm *et al.* (1995), "a government compliance strategy based only on detection and punishment may well be a reasonable starting point but not a good ending point. Instead, what is needed is a multifaceted approach (…) Put differently, explaining tax compliance requires recognizing the myriad factors that motivate individual behavior, factors that go much beyond the standard economics-of-crime approach to include theories of behavior suggested by psychologists, sociologists, and other social scientists. Until this effort is made, it seems unlikely that we will come much closer to unravelling the puzzle of tax compliance." Following the same approach, Guala and Mittone (2005) get into this debate on the role and external validity of experiments, suggesting that experiments might help theoretical models to get closer to real-world phenomena, and thus answer specific questions about causal relationships.

From this viewpoint, Guala and Mittone (2005) claim that experiments serve as *epistemic mediators* between theoretical models and empirical economic phenomena. In fact, theory and experiments are not considered as two distinct entities: they both require initial hypotheses and inference; they are two useful and complementary structures to study and subsequently understand economic behavior. Figure 3.1 shows this relationship as presented by Guala and Mittone (2005): They identify a gap between theoretical models and the intended domain of application, and experimental systems occupy the middle ground between the two. Nevertheless, both experiments and targeted economic phenomena belong to the same "real world": in fact, according to the authors, compared to theoretical models, experimental systems are closer to the target, since they actually allow the collection of observations of real people's behavior under specific conditions, although in an environment that has been artificially manipulated by the experimenter.

Figure 3.1 also refers to the way in which the gap – or better, the gap between theory and experiments, and the one between these and the target – can be closed. On the one hand, *internal validity* – in terms of testing different hypotheses

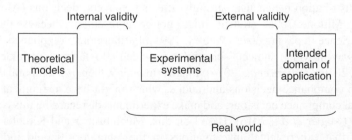

Figure 3.1   Experimental systems as mediators between theoretical models and economic phenomena

in isolation by controlling for confounding variables and ruling out undesired effects – bridges the gap between theoretical models and experiments. On the other hand, *external validity* – in terms of laboratory identification of mechanisms that characterize also the targeted phenomena – is intended to bridge the gap between experimental systems and the specific domain of application. While the former has received much attention in the economic literature, little can be found on the analysis of external validity (Muehlbacher and Kirchler, 2016). The problem of internal validity can be "easily" solved by adopting a number of techniques allowing the identification of causal relationships. However, even a high degree of internal validity does not ensure that the external validity requirement is met. Experiments provide a higher degree of concreteness with respect to theoretical models, by including features that could be reasonable for externally valid inferences. However, they are still artificially isolated from the world outside the walls of the laboratory, in which a wide variety of micro and macro factors – tax audit plans, risk preference, reasoning biases, moral constraints, social norms, social comparison, interaction and imitation, fairness, trust, just to name a few – interact in determining actual taxpayers' behavior. Experiments try to implement these factors, yet under the constraint of balancing between internal and external validity: an excessively complicated experimental setting impairs the identification of clear causal effects, and makes experimental results harder to interpret (Cowell, 1991). Therefore, most laboratory experiments are not able to perfectly replicate the specific targeted phenomena, and feed back into the theoretical literature. Experiments can help at an intermediate stage, as they cannot bridge the gap between the target and the theoretical model: the highly controlled experimental setting is aimed to determine which theory better explains a certain pattern of data, but this explanation might not be valid outside the laboratory (Guala, 1998, 1999, 2003).

Nevertheless, as suggested by Guala and Mittone (2005), experiments might also be intended to discover new real and robust empirical phenomena, not necessarily explained by existing theories to be tested in the laboratory. These phenomena might include generic psychological effects, biases, and heuristics to be applied to specific empirical situations. In such a case, experiments do not need to perfectly reproduce the target but may contribute to the creation of a *library of phenomena* (Guala and Mittone, 2005): they simply discover new facts useful from a policy perspective.

In addition, Guala and Mittone (2005) propose interesting examples of robust biases involved in probabilistic reasoning and the effects of uncertainty that have been identified and extensively studied in the laboratory and in the field, also as strictly related to research on tax compliance. Both Sheffrin and Triest (1992) and Scholz and Pinney (1995) perform an econometric analysis of the influence of taxpayers' perceived probability of detection on compliance decisions. The former analysis finds that individuals who perceive a higher audit probability expect significantly less evasion in the population, and those not trusting others or the government engage in more evasion. Nevertheless, such an analysis solely relies on survey data, and therefore, as suggested by Andreoni *et al.* (1998), the results could be biased by the coherent image individuals tend to convey in surveys. In contrast, Scholz and Pinney (1995) also collect tax-return data, and their analysis is intended to investigate the extent of people's guilt and moral obligation, by testing the duty heuristic hypothesis: if taxpayers have no accurate information on the probability of detection, they can rely on heuristics to derive subjective estimates of the risk and to make their compliance choice. The authors observe a significant positive relationship between subjective probability and duty, which in fact leads to an overestimation of the risk of getting caught, and therefore to a higher degree of compliance. Such an effect is even strengthened by people's tax knowledge and previous contacts with the authority. This evidence is supported also by Hessing *et al.* (1992): although, according to Andreoni *et al.* (1998), their results seem to partially contradict those reported by Scholz and Pinney (1995), it emerges that the duty is fostered by mere contacts between taxpayers and the tax authority, while it is impaired by previous audits and fines. In fact, traditional enforcement activities built on coercive power seem to negatively affect taxpayers' sense of duty (Kirchler *et al.*, 2008); tax agencies prefer to adopt a horizontal monitoring approach, by treating taxpayers as customers to whom they can provide useful services.

Friedland (1982), Spicer and Thomas (1982), and Alm *et al.* (1992b) manipulate the quality and the accuracy of information on fines and probabilities in the laboratory: they observe that a higher degree of informational ambiguity enhances compliance. Nevertheless, as theoretically proved by Snow and Warren (2005), such an effect strictly depends on individual ambiguity aversion.

From a similar viewpoint, Bernasconi (1998) suggests that the portfolio approach needs to be integrated with subjects' probability weighting. The nonlinear weight function proposed according to rank dependent utility models (Quiggin, 1982), and prospect theory (Kahneman and Tversky, 1979) may describe the higher degree of compliance actually observed, compared to the theoretically predicted low level. In this respect, prospect theory provides new approaches to modeling tax evasion decisions (Schepanski and Shearer, 1995; Dhami and Al-Nowaihi, 2007; Ping and Tao, 2007; Trotin, 2010; Piolatto and Rablen, 2014; Piolatto and Trotin, 2016), by taking into account probability weighting and reference dependence (Copeland and Cuccia, 2002; Bernasconi and Zanardi, 2004; Watrin and Ullmann, 2008).

Also Erard and Feinstein (1994) underline the significant impact of probability weighting on taxpayers' decisions: in order to provide useful and reliable behavioral insights, fiscal models have to take into account the difference between actual audit probabilities and estimates. In support of the occurrence of this probability weighting process, Spicer and Hero (1985) build a repeated measurement setting and find that the extent of underreporting diminishes as the number of audits previously undergone increases. This evidence has been explained by the *availability heuristic*: people tend to rely on immediate examples they recall when evaluating a decision problem (Tversky and Kahneman, 1973). An alternative explanation is the *target effect*, according to which people assume that a fiscal investigation is likely to be followed by another one (Hashimzade *et al.*, 2013).

In contrast, Mittone (1997) reports on an experiment investigating the difference between probability subjective estimation and weighting: taxpayers exhibit overestimation when simply asked to judge the probability of being audited, and underweighting when asked to actually make a decision. Specifically, according to their estimated probability, compliance is expected to ensure a higher expected value than evasion does; however, in the compliance decision, evasion is the predominant choice.

More detailed analyses of the dynamics underlying taxpayers' decisions in a repeated measurement framework, mimicking a "taxpayer's life span," are provided by Mittone (2006) and Kastlunger *et al.* (2009). In contrast to Bayesian updating, such that audited taxpayers have higher estimates of audit probability than nonaudited taxpayers, and are more deterred from evasion, these authors observe that the occurrence of an audit seems to make taxpayers more prone to evade. This result is commonly referred to as the *bomb crater effect*: the probability of observing compliance decreases if a taxpayer has just undergone a fiscal audit. According to Guala and Mittone (2005), this phenomenon observed in the laboratory has to be tested under a variety of conditions in order to verify whether it exhibits robustness and external validity. As for the former property, Kastlunger *et al.* (2009) find similar results and report that the decrease in compliance after an audit is very rarely due to loss-repair tendencies: the decrease in compliance seems not to depend on whether the taxpayer is fined in the previous round or found to be compliant. As for the latter, it might not be easy to observe the bomb crater effect outside the laboratory: in many countries, variability in declarations increases the probability of being investigated. Nevertheless, such an effect might emerge under specific conditions. Studies about the impact of audits on subsequent compliance have shown that the decline in compliance after an audit can also be observed in real taxpaying situations (DeBacker *et al.*, 2015), and not only with respect to income tax. Bergman and Nevarez (2006) analyze VAT data from individual tax return information in Argentina and Chile and identify the effect; however, the authors also argue that taxpayers who evade more tend to be less deterred by audits.

The so-called *echo effect* is another laboratory phenomenon Guala and Mittone (2005) deal with. Mittone (2006) studies the effect of different patterns of audits over time, and finds that frequent audits experienced early in "tax life" may lead to higher compliance at later stages. Guala and Mittone (2005) report that this phenomenon is robust to changes in the experimental setting, and suggest that this laboratory evidence is supported by a number of real life examples: for instance, fare evasion on Italian public transport is increased by the experience of infrequent controls. Therefore, it seems reasonable to assume that taxpayers evaluate or weight the audit probability according to their experience: repeated audits may lead to a decrease in evasion even in the long run because of chance misperception. Taxpayers learn that the likelihood of audits is higher than the objective probability when these are rather frequent in the beginning; therefore, they rely on this sample to form their probability evaluations and stick to a high compliance level even when the frequency of investigations diminishes.

In summary, our analysis starts recalling the approach by Guala and Mittone (2005) who identify the mediator role of economic experiments in the study of empirical phenomena. Experiments rely on hypotheses and allow the investigation of specific, framed and concrete settings: individuals' decisions are real, although in an artificial environment. We recognize the undeniable relevance of experimental systems in supporting theoretical models and providing better insights on empirical regularities, which otherwise could not be studied and understood so clearly outside the laboratory. For this reason, experiments call for internal validity, while an a priori external validity is not necessary: as previously pointed out, experimental investigations might contribute to the identification of robust economic and psychological phenomena that can be borrowed and applied to specific cases inside or outside the laboratory. At the same time, however, it is also true that, in order to increase their reliability, experimental findings might need to be further tested before valid inferences are drawn.

In this respect, we provide a novel contribution to the literature on experimental methodology in tax research, by extending the framework presented by Guala and Mittone (2005) and claiming that agent-based simulations offer valuable support. In fact, both theoretical economic models and related experiments are mainly defined in a microeconomic setting and they address empirical issues with a high degree of specificity. In addition to this, experiments cannot control for all cognitive drivers involved in the decision process of tax compliance, but only for those specifically targeted and isolated by the experimental design. In this framework, a computational approach to the study of tax evasion tests not only the robustness of experimental findings but also their external validity. On the one hand, agent-based simulations may provide valuable insights of cognitive nature, which an experimenter would not be able to get by simply observing the behavior of a limited sample of human subjects in the laboratory. Human-based experiments contribute to the library of phenomena; computer-based simulations aim at validating laboratory findings, and help understand complex cognitive processes involving psychological biases and heuristics. On the other hand, simulations allow the combination of micro- and macro-level factors actually interacting outside the laboratory and determining people's compliance.

3.3   From Human-Subject to Computational-Agent Experiments

From the previous analysis, it is evident that a pure theoretical approach may offer an "unrealistic [or better, incomplete] picture of human decision-making," which is neither based on nor confirmed by empirical evidence (Selten, 2001). Therefore, it requires to be mediated by an experimental approach, in order to effectively target the empirical domain of interest. This implies the adoption of heterogeneous and less strict assumptions on individual behavior. Nevertheless, in spite of helping theoretical models target specific phenomena, some experiments might still lack external validity. On the one hand, decisions observed in the laboratory are real and the choice setting is specifically intended to address the issue of interest; on the other hand, human samples usually are rather small, and the setting might turn out to be too simple and prevent valid inferences to be transferred outside the laboratory. According to the *materiality thesis* by Guala (2002), experiments may not display a formal similarity to the complex framework of the target system, although being able to replicate almost the same causal processes taking place in the real world outside the laboratory. Therefore, relying on the assumption that human beings are basically the same inside and outside the laboratory, it is possible to identify a correspondence at a "material" level between the experimental and the target system, but not necessarily at a "formal" and "abstract" level, which might hinder the external validity of experiments.

In this framework, agent-based computational economics (ACE) may significantly contribute to the development of more realistic decision-making models; it helps bridge the gap between economic models and the intended domain of application. Similar to theoretical and experimental approaches, simulations require a formal definition of behavioral types. As a matter of fact, ACE is not intended to disregard theoretical considerations, as relationships describing human behavior need to be known in advance for the calibration of agents. Nevertheless, simulations can rely on behavioral assumptions (and experimental observations) so that different agent types are defined, the standard neoclassical economics idea of a homogeneous representative agent is overcome, and a realistic replication of the world is provided. In this sense, both the theoretical and the experimental analysis are enriched by the introduction and the robustness check of heterogeneous behavioral patterns, which might be designed according to previously collected empirical and experimental evidence.

Nevertheless, in contrast to laboratory experiments, according to the ontological analysis by Guala (2002), simulations rely on a process of abstraction: the external validity requirement might be hardly met at a material level, as the correspondence between the simulating and the target system is of a more "formal" kind. ACE agents are virtual entities endowed with specific attributes, purposes, and behaviors; they interact with a rather complex landscape – it consists of institutions, enforcement rules, social networks, and so on – which, in general, resembles the real world but cannot be replicated in human experiments; they receive an input and, based on this, select an action allowing them to reach their predefined goals, such as wealth, happiness, or honesty.

Nowadays, in the field of economic behavior, the spectrum of possible experimental methodologies is quite broad and ranges from 100% human-subject to

100% computational-agent experiments. These two extremes were first thought to be either in opposition or completely unrelated: until a few years ago, the majority of ACE researchers did not consider human-subject experiments as a valuable and real source of information and results in order to build and calibrate simulation models, as reviewed by Duffy (2006). Similarly, the great majority of experimentalists tend to exclusively rely on human-based tests or explorative investigations, without trying to increase the potential and the extent of their experimental results by means of computational simulations.

However, these two methodologies tend to converge: half way, different techniques, such as a mixture of human and computer agents interacting with each other, human-calibrated computer agents, and computer agents with real-world data streaming, are gaining relevance. Researchers admit that the laboratory with human subjects is a rather artificial context: time is compressed, subjects are asked to make unnatural repeated decisions so that lifetime span can be mimicked, and the landscape is fully controlled and manipulated by the experimenter. Experimental design factors, such as round numerosity, are strictly related to the specific aim of the investigation: even a few rounds are enough to study some simple learning processes, while a higher number of repetitions are necessary if more complex behavioral dynamics are investigated. Nevertheless, an excessive increase in the number of rounds can often harm the results' reliability, as participants get bored. Therefore, from this viewpoint, well-designed experiments allow researchers to carefully study and deeply understand simple dynamics and individual behaviors. In a complementary way, simulations permit to disregard the boredom issue and analyze more complex and dynamic behavioral processes over an extended period of time and among heterogeneous agents: in order to see the emergence and the evolution of behaviors over time and investigate cognitive processes, ACE researchers implement artificial agents that make decisions and react to consequences and signals. As this approach is based on heterogeneous and predominantly boundedly rational agents acting within a dynamic environment, it extends the idea of the representative agent that does not evolve, is fully rational, and is endowed with an unlimited computational power.

In doing this, simulation models may rely on data from human-subject experiments. The agent-based methodology can be used to understand results from human-based studies, since it allows the exploration of the decision process in a more complex economic environment, by replacing humans with agents. The potential of experimental results can be increased by means of these computational tests: it is possible to explore the psychological mechanisms giving rise to phenomena whose robustness and external validity can be checked. Simulations relate the micro-level (i.e., agent-level) behavior to macro-level (i.e., system-level) dynamics, represent multiple scales of analysis in a natural way, and investigate adaptation and learning. Agents are built on experimental evidence, and behave according to actually observed heuristics; in addition, the implementation of an interaction among different agents over time provides insights into macro evolution, which could not be investigated in a simpler human-based experiment. The "formal" similarity ensured by simulations is combined with the "material" one provided by laboratory experiments in a complementary way. The potential of both methodologies is exploited in order to meet the external validity requirement (Guala, 2002). Therefore, from this viewpoint, not only simulations contribute to the external validity hypothesis of experimental systems, but, in turn, experiments increase ACE studies' validity, which is considered as one of the key aspects to judge the performance of a computational model (Taber and Timpone, 1996). In fact, simulation results can be tested in the laboratory in order to better grasp human behavior in computational settings and observe whether and why computer and human behaviors differ. Collected data can feed the software model and contribute to ameliorate agent-based predictions of real-world economic behaviors, and ground them on a material basis, rather than a merely formal one.

Based on this synergic approach, agents' behavioral traits are no longer defined only according to simplifying theoretical assumptions but according to observations actually taken from the real world: behavioral regularities discovered in economics and psychology experiments (e.g., Andreoni *et al.*, 1998; Mittone, 2002, 2006; Kirchler, 2007) can be used to calibrate and/or test simulation models, which, in turn, can help check and explain experimental results. Therefore, both approaches gain in external validity: the high number of degrees of freedom in agent-based models can be managed with human calibration, and human-based experiments, which are not always able to perfectly manipulate subjects' behavior and control their cognitive processes, can find a further confirmation in simulations.

In light of the above, the complementarity of simulations and laboratory experiment also emerges as a support to external validity: the "formal" similarity between simulations and real-world phenomena can be combined with the "material" similarity characterizing the relationship between the experimental and the target systems. On the one hand, the simulations' need for a relevant background knowledge can be met by means of an experiment-based calibration: evidence on human decision processes are collected in the laboratory and used to feed simulated agents, so that they can resemble real decision makers also at a more material level.[2] On the other hand, the mere materiality of laboratory experiments is enriched by simulations' formal correspondence to reality: people's behavior is first observed in a rather simple and artificial laboratory setting; then, it is further investigated and tested in a more realistic environment, in which human-calibrated agents interact. Therefore, it seems possible to conclude that none of the two methodologies has epistemic privilege over the other (Parke, 2014): "material" and "formal" correspondence should be used in a complementary manner, so that each methodology can take advantage from the other while addressing the common external validity issue.

---

[2] According to Winsberg (2009), the main difference between simulations and experiments depends on the prior "knowledge that is invoked to argue for the external validity of the research."

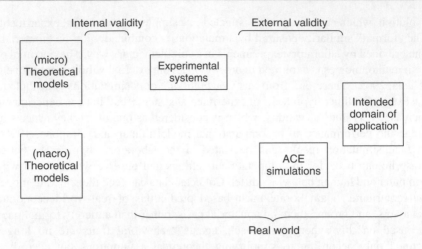

Figure 3.2   ACE simulations and experiments

A graphical analysis of this relationship between experiments and ACE models is provided in Figure 3.2: the framework adopted by Guala and Mittone (2005) is extended in order to include agent-based simulations as a support for experiments in studying empirical economic phenomena and bridging the external validity gap. The figure shows that ACE simulations rely on both macro and micro theoretical models: in fact, they allow the investigation of the evolution over time of network systems involving heterogeneous individuals, and this heterogeneity is built upon micro experimental evidence. In this sense, simulations also belong to the real world as experimental observations are used for agents' calibration, and complex and more complete settings can be implemented to the purpose of making experimental results more likely to be externally valid. Agent-based models rely on preference assumptions but they exhibit a high degree of complexity with respect to human-based experiments, as they mimic societies made of heterogeneous individuals. Not only can human-calibrated agents be endowed with diverse attributes, such as income level, risk propensity, compliance preferences, norm adherence, heuristics, biases, and so on, but various policy parameters and the effect of these on the interaction among agents can also be taken into account.[3] This allows both the investigation of taxpayers' cognitive process and the combination of micro-level evidence and macro dynamics among heterogeneous agents in a unique decision setting resembling the economic environment of interest. The analysis of tax compliance dynamics in a pretty realistic, though complex, system may lead to discover new and efficient policy options (Garrido and Mittone, 2013;

---

[3] As extensively reviewed in Chapter 1, models either belong to the economic domain or fall in the domain of econophysics, if agents' behavior is described as a stochastic process affected by the changing balance between individual's autonomy and influence from neighbors' behavior. This chapter focuses only on the economic domain.

Pickhardt and Seibold, 2014), which could take into account the variety of reactions emerging in a population of heterogeneous taxpayers.

3.4 An Agent-Based Approach to Taxpayers' Behavior

With a closer focus on tax experiments, this section proposes two separate ACE approaches intended to pursue the aforementioned goal of filling the external validity gap: both of them are aimed at tackling the limitations of full rationality and behavioral homogeneity, which impair the external validity of theoretical and experimental claims.

Firstly, agent-based models may analyze the interaction among *types* and study the subsequent emerging macro dynamics; this is mainly based on the implementation of recurring behavioral styles in the population of taxpayers previously identified in laboratory experiments (Mittone and Patelli, 2000; Davis *et al.*, 2003; Antunes *et al.*, 2007; Hokamp and Pickhardt, 2010; Hokamp, 2014). Owing to their scope, these models are usually characterized by a modest degree of granularity: they try to tackle the unrealistic theoretical assumption of a lack of heterogeneity in taxpayers' behavior, yet without always addressing the bounded rationality issue. They are not intended to explore individuals' cognitive dynamics; therefore, behavioral types are specified as rather simple agents. This interest in the identification of groups of taxpayers dates back at least to the 1990s. Building on Cowell (1991), Hessing *et al.* (1992) identify three behavioral types according to willingness to comply, and underline the importance of behavioral heterogeneity to evaluate the extent of efficiency and effectiveness of different policy instruments: some auditing strategies might have the negative impact of crowding out honesty, and thus reducing individual willingness to comply; in contrast, an efficient strategy might fight tax evasion by sustaining honesty and compliance. In this respect, contrary to human-based experiments, agent-based simulations allow the implementation and manipulation of population hetero-geneity in a highly controlled manner, so that this, and its interaction with other variables, can be treated and analyzed as a determinant of the efficacy of policies. In a synergic view, such simulation results can be subsequently tested on human subjects.

Secondly, simulations may also look into *micro behavioral patterns* that go beyond the macro type specification. Therefore, they are characterized by a higher granularity, since agents are more complex in their attributes. In fact, in this case, human behavior is first investigated at an individual level in the laboratory and then reproduced by means of artificial agents (Bloomquist, 2006; Garrido and Mittone, 2008; Méder *et al.*, 2012): simulations help uncover and understand human cognitive processes and psychological drivers, which cannot be fully investigated in a purely human setting. Therefore, this kind of analysis is well suited to the implementation of boundedly rational decision makers, that are intended to mimic human subjects, and choose according to a restricted set of information.

The following sections provide an exemplification of the methodological validity of combining human- and agent-based techniques in the study of tax phenomena. To this purpose, the aforementioned distinction between mainly macro or micro computational analysis is adopted; in addition, some attempts of reconciling such a distinction are analyzed (e.g., Korobow *et al.*, 2007; Garrido and Mittone, 2013; Mittone and Jesi, 2016). The analysis of these simulation examples is aimed at offering guidance in the implementation of the synergic approach, involving both human- and agent-based models, with the intent of filling the external validity gap of economic experiments. Providing an extensive review of research with human-calibrated models is, instead, beyond the scope of the present chapter.

3.4.1   The Macroeconomic Approach

Mittone and Patelli (2000) carry out a dynamic simulation in order to model a fiscal environment in which different types of taxpayers interact and, according to their degree of compliance, a public good is provided.[4] The two authors investigate taxpayers' psychological and moral motives by using human-calibrated simulation models. The idea of studying specific taxpayers' behavioral traits is developed in the seminal work by Mittone (2002): he categorizes behavioral regularities and identifies classes of subjects reacting in a similar way to certain *economic* and *moral factors*.[5]

Mittone (2002) verifies whether subjects' behaviors can be captured and classified in homogeneous categories by performing a cluster analysis.[6] He finds four main clusters:[7] the great majority of subjects do not exhibit a stable behavior, in line with the intuition that previous experience affects taxpayers' decisions. Behavioral clusters are almost identical across experimental conditions.

---

[4] The two authors specify that, given their focus, what they refer to as "agent" or "taxpayer" (taken as synonyms in this context) is a "taxpaying behavior," defined as a form of reaction to the introduction of a tax.

[5] The former are exemplified by income level, tax rate, audit probability, and fees, which enter taxpayers' utility function. Moral factors are investigated by manipulating the decision context in a between-subject manner: the effects produced by either the introduction of a tax yield redistribution, which depends on all taxpayers' compliance decisions, or the lack of any reference to the fiscal environment are tested. This allows the researcher not only to study the role of moral constraints but also to check the robustness of the emerging categories over different settings.

[6] For this kind of analysis, the author adopts the average linkage between groups method (also called UPGMA – unweighted pair-group method using arithmetic averages), and uses standardized variables. According to this method, the distance between two groups is the average of the distances between all pairs of individuals (i.e., by taking one individual for each of the two clusters).

[7] The four styles are (i) the (pure) "absolute stability" of subjects always paying the tax due; (ii) the "relative stability" of subjects always evading, yet at different extents; (iii) the "oscillatory behavior" between full compliance and partial evasion; (iv) the "mixed behavior" of subjects adopting the oscillatory behavior in the first half of the session, and fully evading in the second half. The large majority of subjects exhibited a stable behavior: it was rare to observe a drastic change in their "style."

Results confirm the difficulty of modeling and explaining the actual dynamics of taxpayers' behavior by simply relying on the traditional expected utility approach: refinements based on empirical and experimental observations are necessary, in order to understand the interaction between behavioral heterogeneity and enforcement policies. As pointed out by Mittone (2002), contrary to rational predictions, participants seem not to be comfortable with repeated choices under risk, and alternate opposite choices, probably because the ongoing interaction with the environment leads them to weight probabilities, and not to stick with a predetermined pure strategy. However, he also reports that tax yield redistribution triggers honesty. This change in the composition of the population due to the institutional setting might have serious policy implications: these experimental results seem to suggest that the policy maker should implement fiscal plans designed according to the institutional setting, and subsequently exploit the composition of taxpayers' population to foster honesty imitation, as proposed by Hessing *et al.* (1992). To this purpose, a valid support is offered by ACE simulations, which test the efficiency and the efficacy of different enforcement strategies on a large population consisting of a realistic variety of human-calibrated types. This computational approach adds to the experimental one, since it manipulates the composition of the population, and thus controls for the macro effects deriving from the interaction among different behavioral types under given institutional and fiscal settings.

Building on this, Mittone and Patelli (2000) study the true nature of tax compliance, by focusing not only on the effect of tax authority enforcement but also on social interaction and moral concerns. They focus on the coexistence of three different behavioral styles that require agents' interaction in order to evolve, and adopt a computational approach that might contribute to the validity of experimental findings. The work by Mittone and Patelli (2000) can be considered as a good example of the synergic approach of ACE simulations and human-based experiments aimed at providing greater realism and concreteness for the subsequent application of results to specific policy targets. In fact, Mittone (2002) identifies in the laboratory some behavioral regularities; Mittone and Patelli (2000) not only test the robustness of laboratory findings but also study the macro evolution of a dynamic and heterogeneous population facing different enforcement systems. Behavioral types identified with the experimental micro approach are used to calibrate agents, whose behavior is analyzed from a macroeconomic viewpoint: type imitation and population evolution are the main scopes of this ACE investigation. These types include the *honest taxpayer*, the *imitative taxpayer*, and the *perfect free-rider*.[8] Agents share a decision algorithm, led by utility maximization, while each type has a unique utility function that specifies its behavior.[9] This allows the implementation of heterogeneous agents,

---

[8] This classification resembles the one identified and investigated by Fischbacher *et al.* (2001) and Burlando and Guala (2005): unconditional cooperators, conditional cooperators, and free-riders.

[9] The utility function is built on the work by Myles and Naylor (1996), in which tax compliance is assumed to be affected by social customs and group conformity. The *honest agent* derives additional

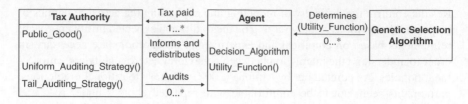

Figure 3.3 System structure diagram representing the system by Mittone and Patelli (2000)

and aims to extend microeconomic models toward the investigation of behavioral evolution and evasion activity in a population of interacting taxpayers, who have different preference structures.

In such a system, at regular intervals, a genetic algorithm can be activated in order to update the composition of the population, without modifying the overall number of agents. The two authors set an initial scenario, and observe how a given population composition evolves over time: taxpayers initially belong to one of the three categories, but then they can decide to switch to another type, according to the degree of success of their style in pursuing the goal of utility maximization.

This optimization strictly depends on tax-payment decision and the risk of being investigated. In fact, in each round, a fixed number of agents are audited according to either a uniform auditing (all agents have the same probability of being investigated) or a low-tail auditing strategy (agents who report the lowest amount of tax have a higher probability of being audited). This diversification is intended to investigate the effect that different auditing policies produce on taxpayers' behavior, also depending on the specific degree of population heterogeneity. Figure 3.3 shows the functioning of the simulated economy: it is evident that agents' decisions are determined by both social interaction and the enforcement activity of the tax authority.

In summary, Mittone and Patelli (2000) study the macro experimental interaction among behavioral types identified by means of a microeconomic approach; they test the efficacy of different audit strategies in fighting tax compliance when population heterogeneity is not the result of abstract assumptions but of real-world observations. Such a controlled exploration of the interaction of different population compositions with the environment would not be easily implementable in a purely human-based setting. This justifies the adoption of an agent-based approach, which allows the observation of macro behavioral dynamics, but needs human calibration for the implementation of realistic taxpayers.

---

utility, proportional to the percentage of honest taxpayers in the population, from behaving in accordance to the social norm of compliance. The *imitative agent*'s utility function depends on the amount of tax he should pay and on the average amount of tax paid by the population. Finally, the *free-rider agent* derives a positive utility from adopting an opportunistic behavior. Moreover, they all get a non-negative utility from the public sector: the model hypothesizes the existence of a single public good, which is produced every round thanks to the tax yield previously collected.

Results show that a uniform auditing strategy is more effective than a low-tail one in fostering compliance; imitating the honest behavior is a winning strategy when low-tail auditing is implemented. Finally, genetic selection favors honesty, as frequently observed also inside and outside the laboratory when moral concerns on contribution are involved in the decision process. When taxpayers are aware that they will actually benefit from their fiscal contribution, they appear to be more prone to comply.

Hence, if combined with theoretical models and laboratory experiments, agent-based simulations can help understand and explain behavioral processes underlying tax payment decisions. The novelty of this study resides in the implementation of a simulation model investigating the relationship between enforcement activity and social interaction among different behavioral styles, which have emerged as regularities in previous human-based experiments. Taxpayers are heterogeneous and their behavior is described by utility functions; furthermore, they can switch to a different type according to the "satisfaction" they are able to derive from the behavior of their own category. Nevertheless, agents' decision-making process still relies on optimization, and depends only on the information they receive from the system. In this sense, they are myopic, since they are not designed to take into account either intertemporal or strategic expectation on the evolution of the environment. For this reason, a further development in this direction might include a distinction between naive and sophisticated agents, where the latter should be modeled in order to mimic agents capable of making efficient predictions about their future behavior and that of their mates.

3.4.2   The Microeconomic Approach

Besides this macro approach, mainly focused on the analysis of the effect of behavioral heterogeneity, a parallel line of research based on the assumption of boundedly rational agents has gained relevance too: the micro dimension of individual history serves as a base for ACE simulations aimed at understanding and explaining decision makers' cognitive process. Under this view, decisions are expected to vary according to individuals' *state*, which is determined by external environment and past experience, and is translated in a "local" set of information the agent may use to decide. For instance, in the fiscal context, evasion might be more likely when an individual has been audited either during the previous round (bomb crater effect) or at the beginning of his fiscal life (echo effect). Human-based experiments show whether different experiences lead to states characterized by diverse levels of willingness to evade, and thus, more in general, whether subjects modify their behavior according to their current condition. Thanks to agent-based simulations, it is possible to systematically analyze and explain human behavior in order to check the robustness and the external validity experimental phenomena on a larger population. For instance, the standard theoretical approach could be replaced by a setting closer to the aspiration adaptation theory by Selten (1998): agents have a limited set of decision dimensions, and they can select even opposite actions, depending on their specific current state, which affects probability evaluation and weighting.

This micro approach is well exemplified by Garrido and Mittone (2008), who use the theory of finite automata (Rubinstein, 1986; Romera, 2000) to interpret Italian and Chilean experimental data on tax compliance. They report that the behavior of the great majority of subjects can be explained by either unconditional honesty or the bomb crater effect, which is part of the library of phenomena (Guala and Mittone, 2005; Mittone, 2006; Kastlunger *et al.*, 2009).

Recalling the original notion of boundedly rational approach, Garrido and Mittone (2008) consider individuals as limited in their computational power: each taxpayer can rely on a restricted set of information, in order to decide whether to comply or fully evade. The two authors assume that the probability of evading depends on the current state of the taxpayer (referred to as "locally determined decision maker"): this state may change according to external events, such as the occurrence of a fiscal audit. Every artificial agent consists in a finite state automaton (Moore, 1956; Sipser, 2006), whose binary stochastic output (compliance vs evasion) does not depend only on the current state (for instance audited in the previous period) but also on the probability of evasion associated to that specific state. Garrido and Mittone (2008) collect human-subject data: experimental results show the bomb crater effect, which however turns out to be less evident at an aggregate level. For this reason, the experiment is followed by an agent-based simulation aimed at identifying the specific micro determinants of taxpayers' behavior, that is, to identify the automaton with the highest success ratio in predicting human subjects' decisions.

According to our view, this application of an agent-based model helps understand how simulations can support the experimental approach in the field of tax research. Human-based experiments provide more or less clear insights on taxpayers' behavior, and an agent-based system enriches our understanding of human behavioral regularities, by testing many cognitive drivers and inner motives that are supposed to be involved in the decision process. This synergic approach might resemble theory testing in the laboratory: as a matter of fact, just as experiments help identify which theory best explains human behaviors observed in the laboratory, simulations allow the identification of the main cognitive drivers explaining human behavior and thus check its validity outside the laboratory.

Specifically, in their work, Garrido and Mittone (2008) propose a set of seven hypotheses,[10] which might explain experimental findings by testing the

---

[10] The decision of evading depends on whether the subject is (H_0) audited in the previous period; (H_1) audited in the previous period, and caught; (H_2) audited during the previous two periods; (H_3) audited in the previous two periods, and caught; (H_4) audited in the previous three periods; (H_5) audited in the previous four periods; (H_6) audited in the previous three periods, and caught.

robustness of the bomb crater against the loss-repair effect.[11] Each hypothesis is translated into an automaton, whose states map the characteristics of the hypothesis itself. This tests different behavioral motives and helps identify the best one in explaining patterns observed in the laboratory. From this perspective, it is again evident how ACE simulations can provide a valuable support also at a micro level: they help increase the potential of experimental evidence and fully understand the psychological and cognitive drivers characterizing the individual decision process.

The hypothesis that gives the most detailed description and prediction of subjects' behavior is the one involving the bomb crater effect; the only other relevant automaton is the one describing unconditionally honest agents, that is, those who fully comply, irrespective of their current state. These results confirm, on the one hand, the robustness of the bomb crater effect as a common behavioral trait, and, on the other hand, the existence of an honest type (Mittone and Patelli, 2000).

In addition, Garrido and Mittone (2008) test three further hypotheses, in order to control for the effect of different audit sequences.[12] In line with the expectations of robustness and external validity of the echo effect (Guala and Mittone, 2005), computational results confirm that human subjects' behavior can be explained by means of a rather simple hypothesis: repeatedly auditing subjects at the beginning of their fiscal life has a positive impact on compliance over a certain time period, because of a wrong probability evaluation people form when relying on sampled experience (Guala and Mittone, 2005; Mittone, 2006; Kastlunger et al., 2009).

Hence, two main behavioral patterns are identified: about 70.3% of the entire experimental pool consists of subjects who never evade and subjects who evade strategically according to the bomb crater effect. Honest subjects exhibit an evading probability close to 0, irrespective of their state; in contrast, in case of strategic evaders, the likelihood of evasion is low only in the "not audited" and in the initial state. The remaining 29.7% does not exhibit a clear behavioral pattern, since, in every state, they evade with a probability close to 0.5. Therefore, the adoption of ACE modeling leads to conclude that even a simple behavioral hypothesis, which might be modeled as a heuristic, can explain a large proportion of subjects' decisions. Nevertheless, such a comprehension of human behavior can be achieved thanks to agent-based investigations, as the mere observation of human subjects might not be sufficient to draw valid conclusions.

---

[11] Both effects imply an immediate decrease in compliance after an audit. However, the former is due to chance misperception and depends only on the occurrence of an audit, and not on actual evasion detection. In contrast, the loss-repair effect emerges only when a taxpayer is found to be an evader.

[12] They assume that taxpayers' evasion might depend (H_7) positively on being audited in the previous period, and negatively on experiencing an audit in the first five periods of the experiment. This hypothesis is extended in H_8, which also considers the effect of being caught during the latest audit. Finally, they test whether (H_9) subject's decision depends on being audited during the previous two periods, and on experiencing an investigation during the first five periods of the experimental session.

3.4.3   Micro-Level Dynamics for Macro-Level Interactions among Behavioral Types

This section deals with ACE models that combine micro behavioral aspects and macro dynamics, with the intent of providing a better understanding of both experimental evidence and economic phenomena taking place outside the laboratory. Therefore, this kind of comprehensive analysis can be of great relevance for policy implications, by relying on the implementation of human-calibrated agents.

The first example is the work by Garrido and Mittone (2013), which analyzes how the efficiency and the efficacy of an enforcement strategy – defined in terms of audit frequency and targeting – can be considered as a function of the population composition. However, contrary to the macro analysis by Mittone and Patelli (2000), behavioral types are defined according to income distribution and specific traits that characterize individual cognitive process (Garrido and Mittone, 2008). Taxpayers are endowed with a decision function, and, in each round, they choose whether to evade. Honest taxpayers tend to comply in any case, irrespective of their current state; strategic evaders behave according to the bomb crater effect. Right after all taxpayers make their decisions, the policy maker applies an optimizing selection rule that targets a subset of agents to audit: on the one hand, collected tax increases revenues; on the other hand, audits are costly and not always successful.[13]

Garrido and Mittone (2013) conclude that the optimal audit scheme must take into account income distribution, the possibility of identifying behavioral patterns with micro foundations, and the specific fiscal history of individuals. Micro-level behavioral regularities emerging in laboratory experiments turn out to be fundamental in designing an auditing strategy: being aware of some cognitive biases can help predict people's behavior; agent-based simulations built on these biases are useful to plan a coherent and efficient fiscal policy. As income inequality increases, the optimal plan targets the richest taxpayers, and frequently repeats two consecutive audits as a strategy against the bomb crater effect. In contrast, as income distribution becomes more uniform, the optimal plan suggests spreading audits throughout the entire population. Every time an agent is investigated, his last four declarations are verified: if the tax authority audits each agent every four periods, also strategic evaders are caught.

Nevertheless, despite the relevant contribution of this study in understanding how the policy maker can address the issue of tax evasion when realistically dealing with a heterogeneous population, results are partially due both to the rationality assumption on the tax authority and to the main characteristics of taxpayers' choice function (no intensive decision is allowed, and no actual learning is implemented). This simplifying decision process might lead to partially misleading behaviors: in fact, in their laboratory experiment, authors allow for intensive decisions, and observe that rich individuals often prefer to

---

[13] The degree of efficiency takes its maximum value 1 only when each audit catches an evader.

evade a small amount of tax with the aim of reducing the probability of being targeted. In contrast, the simulation by Garrido and Mittone (2013) disregards this important aspect, and the optimal plan might target the richest taxpayers so that expected revenues of the tax authority are maximized.

The second example of a computational analysis combining the macro and the micro approach is the one by Mittone and Jesi (2016). By extending the agent-based analysis by Garrido and Mittone (2013) and Mittone and Patelli (2000), they build a complex adaptive system in which a variety of behavioral types coexist. However, contrary to Mittone and Patelli (2000), these types are based on the definition of simple heuristics, and not of utility functions to optimize. In addition, with the intent of overcoming the limitations of the model by Garrido and Mittone (2013), they also allow for intensive decisions, learning, different risk perceptions, and for probability weighting as a common feature of individuals' decision process. Specifically, Mittone and Jesi (2016) investigate the functioning and the evolution of a system where boundedly rational agents cope with a public good that might be consumed and created by the agents themselves. The authors build a self-reproducing economy as a setting for the study of the emergence of a responsible behavior in managing a renewable resource. They study the necessity of an exogenous mechanism of auditing in order to achieve a sustainable setup.

In every period, agents extract their private endowment from the good; then, they contribute by paying their tax due. These actions are carried out according to a limited set of heuristics (basically either imitative behaviors or habits) and to the employment type of the agent (employee vs self-employed):[14] heuristics suggest the amount to take in order to perform a satisfying extraction, but the agent can try to extract more resources. In the beginning, each agent is randomly assigned a type and one of the five available heuristics; then, in order to keep the system dynamic, new agents are injected into the economy, and individuals can switch from one heuristic to another according to the achieved satisfaction level: the higher the level of sadness (i.e., the lower the degree of satisfaction), the higher the probability that an agent opts for switching to another heuristic. This sadness is mainly determined by the level of the extracted endowment, and the proportion of agents actually contributing to the public good. In addition, irrespective of the individual heuristic adopted, agents share the bomb crater effect as a common micro-founded psychological trait: as widely observed in human-based experiments, after an investigation occurs, the audited agent evades, underestimating the probability of a repeated audit. Such a characterization makes agents closer to human beings: they are not supposed to be rational, but rather emotional and biased in their decision process. For this

---

[14] Agents can be categorized into two different cross-sectional sets according to institutional constraint introduced: agents subject to the high constraint (employees) cannot evade more than 10% of the tax due, while those subject to the low constraint (self-employed) are free to evade up to 50% of the tax due.

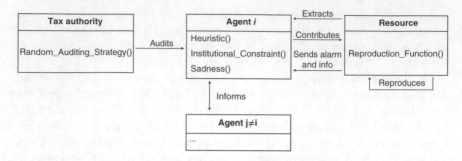

Figure 3.4   System structure diagram representing the system by Mittone and Jesi (2016)

reason, results can be of great interest and relevance for externally valid policy suggestions.

At the end of every period, the tax authority performs random audits and the good reproduces itself so that it cannot extinguish. Before the reproduction takes place, the good triggers a signal if the critical status in terms of quantity is reached. Agents react to this alarm according to their sensitivity level, that is, to their propensity to take risk and their adopted heuristic. See Figure 3.4 for a comprehensive representation of the overall system structure.

Thanks to the implementation of this complex framework with a micro-founded behavioral heterogeneity, Mittone and Jesi (2016) identify the extent of behavioral heterogeneity that is able to trigger a responsible behavior, and thus find interesting results from a normative point of view. In fact, as already suggested by Hessing *et al.* (1992), they claim that selfishness, and thus evasion, can be effectively counterbalanced if other behavioral types are more attractive for taxpayers. In their model, an efficient fiscal policy can tremendously decrease tax evasion not only by means of audit deterrence, but also by sustaining the advantage that a taxpayer can get adopting an honest behavior. From this perspective, an efficient policy should exploit behavioral heterogeneity and induce taxpayers to the imitation of honest agents by making compliance more attractive for both employees and self-employed workers.

Hence, also in this case, the validity of the synergic approach of human-based and agent-based experiments is undeniable: human evidence serves as a basis to build behavioral types, and simulations allow the manipulation of population heterogeneity as a treatment variable, with the intent of leveraging the full potential and overcoming the limits of human-based experiments. This results in a more complete and deeper analysis of tax payment decisions: useful policy suggestions can be derived, as it is possible to implement a rather realistic system in which different fiscal strategies are tested on a dynamic and heterogeneous population of interacting agents.

3.5   Conclusions

Since the appearance of the first theoretical models in the early 1970s, the study of tax compliance has moved a long way toward the development of new models, taking into account psychological regularities and anomalies of decision making. The increasing success of the application of behavioral economics has shown the importance of relying on empirical and experimental data in order to integrate theoretical analyses and overcome the traditional limit of the representative agent. In fact, recent evidence from laboratory experiments and surveys underlines the impact of noneconomic considerations in determining individuals' behavioral heterogeneity in real-world compliance, and the relevance of understanding taxpayers' behavior and the underlying cognitive process, in order to provide useful normative policies able to sustain compliance and deter evasion.

From our perspective, much interest and effort need to be devoted to the combination of experimental techniques and agent-based models, in order to investigate the interaction between taxpayers' cognitive process and the surrounding environment. This would not only contribute to the external validity of experimental findings by testing the robustness of human subjects' behavior in systems with an increased degree of complexity, but it would also allow the integration of a micro-level perspective with macro-level considerations: dynamics at an aggregate level can be studied starting from micro-level observations.

This chapter explains how simulations can increase external validity of tax experiments in two different ways. On the one hand, agent-based models consist in the implementation of a set of human-based behavioral types and extend experimental analyses by manipulating the composition of agents' population. This provides a greater adherence to the environment outside the laboratory and tests the effects of a variety of policies on a heterogeneous population. In fact, agent-based models may define different macro behavioral types interacting with diverse policy solutions adopted by the tax authority, and this heterogeneity can be based on the identification of micro-level behavioral dynamics emerging from psychology, economics laboratory experiments, and empirical studies. On the other hand, external validity of experiments can be increased by identifying the main cognitive drivers that explain phenomena observed in the laboratory. Human-based experiments contribute to the library of phenomena, by simply searching for facts and regularities, while agent-based simulations analyze and test these phenomena, so that they can be applied to specific cases to a normative purpose.

Overall, this kind of innovative approach adds to the ongoing discussion about the inclusion of behavioral realism into theoretical studies in the literature on tax evasion. Therefore, it supports a greater parallelism with the natural world, yet without denying the importance of model development: the synergic combination of theoretical analyzes and human-calibrated simulations may help shed new light on the issue of tax evasion, since it focuses on the specific problem of new policy implementations in a rigorous way and in a realistic environment, before actual application in the field.

References

Allingham, M.G. and Sandmo, A. (1972) Income tax evasion: a theoretical analysis. *Journal of Public Economics*, **1**, 323–338.

Alm, J. (1999) Tax compliance and administration. *Public Administration and Public Policy*, **72**, 741–768.

Alm, J. (2010) Testing behavioral public economics theories in the laboratory. *National Tax Journal*, **63** (4), 635–658.

Alm, J., Jackson, B.R., and McKee, M. (1992a) Estimating the determinants of taxpayer compliance with experimental data. *National Tax Journal*, **45**, 107–114.

Alm, J., Jackson, B.R., and McKee, M. (1992b) Deterrence and beyond: toward a kinder, gentler IRS. *Why People Pay Taxes*, **1**, 311–329.

Alm, J., McClelland, G.H., and Schulze, W. (1992c) Why do people pay taxes? *Journal of Public Economics*, **48** (1), 21–38.

Alm, J., Sanchez, I., and De Juan, A. (1995) Economic and noneconomic factors in tax compliance. *Kyklos*, **48** (1), 1–18.

Andreoni, J., Erard, B., and Feinstein, J. (1998) Tax compliance. *Journal of Economic Literature*, **36** (2), 818–860.

Antunes, L., Balsa, J., Respício, A., and Coelho, H. (2007) *Tactical Exploration of Tax Compliance Decisions in Multi-Agent Based Simulation*, Multi-Agent-Based Simulation vol. **VII**, Springer-Verlag, pp. 80–95.

Baldry, J.C. (1986) Tax evasion is not a gamble: a report on two experiments. *Economics Letters*, **22**, 333–335.

Baldry, J.C. (1987) Income tax evasion and the tax schedule: some experimental results. *Public Finance= Finances Publiques*, **42**, 357–383.

Benjamini, Y. and Maital, S. (1985) Optimal Tax Evasion & Optimal Tax Evasion Policy Behavioral Aspect, in *The Economics of the Shadow Economy*, vol. **4** (1), Springer-Verlag, pp. 245–264.

Bergman, M. and Nevarez, A. (2006) Do audits enhance compliance? An empirical assessment of VAT enforcement. *National Tax Journal*, **59** (4), 817–832.

Bernasconi, M. (1998) Tax evasion and orders of risk aversion. *Journal of Public Economics*, **67** (1), 123–134.

Bernasconi, M. and Zanardi, A. (2004) Tax evasion, tax rates, and reference dependence. *FinanzArchiv: Public Finance Analysis*, **60** (3), 422–445.

Bloomquist, K.M. (2006) A comparison of agent-based models of income tax evasion. *Social Science Computer Review*, **24** (4), 411–425.

Boylan, S.J. (2010) Prior audits and taxpayer compliance: experimental evidence on the effect of earned versus endowed income. *Journal of the American Taxation Association*, **32** (2), 73–88.

Boylan, S.J. and Sprinkle, G.B. (2001) Experimental evidence on the relation between tax rates and compliance: the effect of earned vs. endowed income. *Journal of the American Taxation Association*, **23** (1), 75–90.

Burlando, R.M. and Guala, F. (2005) Heterogeneous agents in public goods experiments. *Experimental Economics*, **8** (1), 35–54.

Casal, S. and Mittone, L. (2016) Social esteem versus social stigma: The role of anonymity in an income reporting game. *Journal of Economic Behavior & Organization*, **124**, 55–66.

Casey, J.T. and Scholz, J.T. (1991a) Beyond deterrence: behavioral decision theory and tax compliance. *Law and Society Review*, **25** (4), 821–843.

Casey, J.T. and Scholz, J.T. (1991b) Boundary effects of vague risk information on taxpayer decisions. *Organizational Behavior and Human Decision Processes*, **50** (2), 360–394.

Choo, L., Fonseca, M.A., and Myles, G.D. (2016) Do students behave like real taxpayers? Experimental evidence on taxpayer compliance from the lab and from the field. *Journal of Economic Behavior & Organization*, **124**, 102–114.

Copeland, P.V. and Cuccia, A.D. (2002) Multiple determinants of framing referents in tax reporting and compliance. *Organizational Behavior and Human Decision Processes*, **88** (1), 499–526.

Cowell, F. (1991) *Tax-Evasion Experiments: An Economist's View*, Cambridge University Press.

Cowell, F.A. and Gordon, J.P.F. (1988) Unwillingness to pay: tax evasion and public good provision. *Journal of Public Economics*, **36** (3), 305–321.

Davis, J.S., Hecht, G., and Perkins, J.D. (2003) Social behaviors, enforcement, and tax compliance dynamics. *The Accounting Review*, **78** (1), 39–69.

DeBacker, J.M., Heim, B.T., Tran, A., and Yuskavage, A. (2015) Legal enforcement and corporate behavior: an analysis of tax aggressiveness after an audit. *Journal of Law and Economics*, **58** (2), 291–324.

Dhami, S. and Al-Nowaihi, A. (2007) Why do people pay taxes? Prospect theory versus expected utility theory. *Journal of Economic Behavior & Organization*, **64** (1), 171–192.

Duffy, J. (2006) Agent-based models and human subject experiments. *Handbook of Computational Economics*, **2**, 949–1011.

Durham, Y., Manly, T.S., and Ritsema, C. (2014) The effects of income source, context, and income level on tax compliance decisions in a dynamic experiment. *Journal of Economic Psychology*, **40**, 220–233.

Einhorn, H.J. and Hogarth, R.M. (1985) Ambiguity and uncertainty in probabilistic inference. *Psychological Review*, **92** (4), 433–461.

Eisenhauer, J.G. (2006) The shadow price of morality. *Eastern Economic Journal*, **32** (3), 437–456.

Eisenhauer, J.G. (2008) Ethical preferences, risk aversion, and taxpayer behavior. *The Journal of Socio-Economics*, **37** (1), 45–63.

Erard, B. and Feinstein, J.S. (1994) Honesty and evasion in the tax compliance game. *The RAND Journal of Economics*, **25** (1), 1–19.

Fischbacher, U., Gächter, S., and Fehr, E. (2001) Are people conditionally cooperative? Evidence from a public goods experiment. *Economics Letters*, **71** (3), 397–404.

Friedland, N. (1982) A note on tax evasion as a function of the quality of information about the magnitude and credibility of threatened fines: Some preliminary research. *Journal of Applied Social Psychology*, **12** (1), 54–59.

Garrido, N. and Mittone, L. (2008) A description of experimental tax evasion behavior using finite automata: the case of Chile and Italy. Cognitive and Experimental Economics Laboratory (CEEL) working papers, 809.

Garrido, N. and Mittone, L. (2013) An agent based model for studying optimal tax collection policy using experimental data: the cases of Chile and Italy. *The Journal of Socio-Economics*, **42**, 24–30.

Gemmell, N. and Ratto, M. (2012) Behavioral responses to taxpayer audits: evidence from random taxpayer inquiries. *National Tax Journal*, **65** (1), 33–58.

Gordon, J.P.P. (1989) Individual morality and reputation costs as deterrents to tax evasion. *European Economic Review*, **33** (4), 797–805.

Graetz, M.J. and Wilde, L.L. (1985) The economics of tax compliance: facts and fantasy. *National Tax Journal*, **38**, 355–363.

Guala, F. (1998) *Experiments as Mediators in the Non-Laboratory Sciences*, Philosophica-Gent, pp. 57–76.

Guala, F. (1999) The problem of external validity (or "parallelism") in experimental economics. *Social Science Information*, **38** (4), 555–573.

Guala, F. (2002) Models, simulations, and experiments, in *Model-based reasoning: Science, Technology, Values* (eds L. Magnani and N.J. Nersessian), Kluwer, New York, pp. 59–74.

Guala, F. (2003) Experimental localism and external validity. *Philosophy of Science*, **70** (5), 1195–1205.

Guala, F. and Mittone, L. (2005) Experiments in economics: external validity and the robustness of phenomena. *Journal of Economic Methodology*, **12** (4), 495–515.

Hashimzade, N., Myles, G.D., and Tran-Nam, B. (2013) Applications of behavioural economics to tax evasion. *Journal of Economic Surveys*, **27** (5), 941–977.

Hessing, D.J., Elffers, H., Robben, H.S.J., and Webley, P. (1992) Does deterrence deter? Measuring the effect of deterrence on tax compliance in field studies and experimental studies, in *Why People Pay Taxes: Tax Compliance and Enforcement* (ed. J. Slemrod), University of Michigan Press, Ann Arbor, MI, pp. 291–305.

Hokamp, S. (2014) Dynamics of tax evasion with back auditing, social norm updating, and public goods provision–an agent-based simulation. *Journal of Economic Psychology*, **40**, 187–199.

Hokamp, S. and Pickhardt, M. (2010) Income tax evasion in a society of heterogeneous agents–evidence from an agent-based model. *International Economic Journal*, **24** (4), 541–553.

Kahneman, D. and Tversky, A. (1979) Prospect theory: an analysis of decision under risk. *Econometrica: Journal of the Econometric Society*, **47** (2), 263–291.

Kastlunger, B., Kirchler, E., Mittone, L., and Pitters, J. (2009) Sequences of audits, tax compliance, and taxpaying strategies. *Journal of Economic Psychology*, **30** (3), 405–418.

Kirchler, E. (2007) *The Economic Psychology of Tax Behaviour*, Cambridge University Press.

Kirchler, E., Hoelzl, E., and Wahl, I. (2008) Enforced versus voluntary tax compliance: the "slippery slope" framework. *Journal of Economic Psychology*, **29** (2), 210–225.

Kogler, C., Mittone, L., and Kirchler, E. (2016) Delayed feedback on tax audits affects compliance and fairness perceptions. *Journal of Economic Behavior & Organization*, **124**, 81–87.

Korobow, A., Johnson, C., and Axtell, R. (2007) An agent–based model of tax compliance with social networks. *National Tax Journal*, **60** (3), 589–610.

Méder, Z.Z., Simonovits, A., and Vincze, J. (2012) Tax morale and tax evasion: social preferences and bounded rationality. *Economic Analysis and Policy*, **42** (2), 257–272.

Melumad, N.D. and Mookherjee, D. (1989) Delegation as commitment: the case of income tax audits. *The RAND Journal of Economics*, **20** (2), 139–163.

Mittone, L. (1997) Subjective versus objective probability: Results from seven experiments on fiscal evasion. CEEL Working Paper 4-97.

Mittone, L. (2001) Vat evasion: an experimental approach. University of Trento, Department of Economics and Management – Discussion Paper 5-01.

Mittone, L. (2002) Individual styles of tax evasion: an experimental study. CEEL Working Paper 2-02, University of Trento.

Mittone, L. (2006) Dynamic behaviour in tax evasion: an experimental approach. *The Journal of Socio-Economics*, **35** (5), 813–835.

Mittone, L. and Jesi, G.P. (2016) *Heuristic Driven Agents in Tax Evasion: An Agent-Based Approach*, Cognitive and Experimental Economics Laboratory, Department of Economics, University of Trento, Italia.

Mittone, L. and Patelli, P. (2000) Imitative behaviour in tax evasion, in *Economic Simulations in Swarm: Agent-Based Modelling and Object Oriented Programming* (eds B. Stefansson and F. Luna), Kluwer, Amsterdam, pp. 133–158.

Moore, E.F. (1956) Gedanken-experiments on sequential machines. *Automata Studies*, **34**, 129–153.

Muehlbacher, S. and Kirchler, E. (2016) Taxperiments. About the external validity of laboratory experiments in tax compliance research. *Die Betriebswirtschaft (DBW)*, **76**, 7–19.

Myles, G.D. and Naylor, R.A. (1996) A model of tax evasion with group conformity and social customs. *European Journal of Political Economy*, **12** (1), 49–66.

Parke, E.C. (2014) Experiments, simulations, and epistemic privilege. *Philosophy of Science*, **81** (4), 516–536.

Pickhardt, M. and Seibold, G. (2014) Income tax evasion dynamics: evidence from an agent-based econophysics model. *Journal of Economic Psychology*, **49**, 147–160.

Ping, X. and Tao, W. (2007) Cumulative Prospect Theory in Taxpayer Decision Making: A Theoretical Model for Withholding Phenomenon. 2007 International Conference on Management Science and Engineering, pp. 1680–1685.

Piolatto, A. and Rablen, M.D. (2014) Prospect theory and tax evasion: a reconsideration of the Yitzhaki Puzzle. IEB Working Paper.

Piolatto, A. and Trotin, G. (2016) Optimal income tax enforcement under prospect theory. *Journal of Public Economic Theory*, **18** (1), 29–41.

Quiggin, J. (1982) A theory of anticipated utility. *Journal of Economic Behavior & Organization*, **3** (4), 323–343.

Romera, M.E. (2000) Using finite automata to represent mental models. Unpublished Masters thesis. San Jose State University, San Jose, CA.

Rubinstein, A. (1986) Finite automata play the repeated prisoner's dilemma. *Journal of Economic Theory*, **39** (1), 83–96.

Sanchez, I. and Sobel, J. (1993) Hierarchical design and enforcement of income tax policies. *Journal of Public Economics*, **50**, 345–369.

Schepanski, A. and Shearer, T. (1995) A prospect theory account of the income tax withholding phenomenon. *Organizational Behavior and Human Decision Processes*, **63** (2), 174–186.

Scholz, J.T. and Pinney, N. (1995) Duty, fear, and tax compliance: the heuristic basis of citizenship behavior. *American Journal of Political Science*, **39** (2), 490–512.

Selten, R. (1998) Aspiration adaptation theory. *Journal of Mathematical Psychology*, **42** (2), 191–214.

Selten, R. (2001) What is bounded rationality, in *Bounded Rationality: The Adaptive Toolbox* (eds G. Gigerenzer and R. Selten), The MIT Press, Cambridge, MA, pp. 13–36.

Sheffrin, S.M. and Triest, R.K. (1992) Deterrence backfire? Perceptions and attitudes in taxpayer compliance, in *Why People Pay Taxes: Tax Compliance and Enforcement*, University of Michigan Press, Ann Arbor, MI, pp. 193–218.

Simon, H.A. (1955) A behavioral model of rational choice. *The Quarterly Journal of Economics*, **69** (1), 99–118.

Simon, H.A. (1956) Rational choice and the structure of the environment. *Psychological Review*, **63** (2), 129–138.

Sipser, M. (2006) *Introduction to the Theory of Computation*, 2nd edn, Thomson Course Technology, Boston, MA.

Snow, A. and Warren, R.S. (2005) Ambiguity about audit probability, tax compliance, and taxpayer welfare. *Economic Inquiry*, **43** (4), 865–871.

Spicer, M.W. and Becker, L.A. (1980) Fiscal inequity and tax evasion: an experimental approach. *National Tax Journal*, **33** (2), 171–175.

Spicer, M.W. and Hero, R.E. (1985) Tax evasion and heuristics: a research note. *Journal of Public Economics*, **26** (2), 263–267.

Spicer, M.W. and Thomas, J.E. (1982) Audit probabilities and the tax evasion decision: an experimental approach. *Journal of Economic Psychology*, **2** (3), 241–245.

Srinivasan, T.N. (1973) Tax evasion: a model. *Journal of Public Economics*, **2**, 339–346.

Taber, C.S. and Timpone, R.J. (1996) *Computational Modeling*, Quantitative Applications in the Social Sciences, Sage Publications, Inc., Thousand Oaks, CA.

Torgler, B. (2002) Speaking to theorists and searching for facts: tax morale and tax compliance in experiments. *Journal of Economic Surveys*, **16** (5), 657–683.

Trotin, G. (2010) Tax evasion decision under cumulative prospect theory. EQUIPPE, Université Charles-de-Gaulle Lille 3 and GREQAM-IDEP, Université de la Méditerranée working paper.

Tversky, A. and Kahneman, D. (1973) Availability: a heuristic for judging frequency and probability. *Cognitive psychology*, **5** (2), 207–232.

von Neumann, J. and Morgenstern, O. (1944) *Theory of Games and Economic Behavior*, vol. **60**, Princeton University Press, Princeton, NJ.

Watrin, C. and Ullmann, R. (2008) Comparing direct and indirect taxation: the influence of framing on tax compliance. *The European Journal of Comparative Economics*, **5** (1), 33–56.

Webley, P. (1991) *Tax Evasion: An Experimental Approach*, Cambridge University Press.

Winsberg, E. (2009) A tale of two methods. *Synthese*, **169** (3), 575–592.

Yitzhaki, S. (1974) Income tax evasion: a theoretical analysis. *Journal of Public Economics*, **3**, 201–202.

Part II

Agent-Based Tax Evasion Models

4

Using Agent-Based Modeling to Analyze Tax Compliance and Auditing

Nigar Hashimzade and Gareth Myles

4.1 Introduction

The economic analysis of tax compliance has the objectives of explaining and predicting compliance behavior. The achievement of these objectives is essential for the design of beneficial interventions that increase the level of compliance and raise revenue. Several different research methodologies can contribute to this program of research. Theoretical analysis can develop models that are evaluated by empirical studies and tested using laboratory and field experiments. The potential value of applying these methods is constrained by the limited detail that can be incorporated within a theoretical model if useful insights are to be obtained, the potential lack of external validity of laboratory experiments, and the legal (and cost) constraints on field experiments.

Agent-based modeling provides a way of circumventing these difficulties. It makes it possible to obtain insights from a much richer model than can be used for a purely theoretical analysis, and permits experiments to be conducted that would not be possible in the field. This does not imply that the other methods are not equally valuable. Without economic theory we would not know how to specify the agent-based model; and without empirical and experimental research we would not know how to calibrate the model. It is our belief that the real benefits of agent-based modeling will only be realized by combining the best parts

Agent-Based Modeling of Tax Evasion: Theoretical Aspects and Computational Simulations,
First Edition. Edited by Sascha Hokamp, László Gulyás, Matthew Koehler, and Sanith Wijesinghe.
© 2018 John Wiley & Sons Ltd. Published 2018 by John Wiley & Sons Ltd.

of economic theory with empirically based calibration. Together, the combination of methodologies will provide novel insights into the design of revenue service compliance strategy.

A successful application of agent-based modeling uses the best of economic theory to describe the behavior of agents with heterogeneous characteristics and allows for interaction among these agents in a rich environment. The components of economic theory on which we focus are recent behavioral advances in understanding the compliance decision, the effect of occupational choice in creating opportunities for noncompliance, and the role of social networks in the transmission of information. In brief, our model of the compliance decision and policy intervention combines *attitudes* toward compliance, *beliefs* about audit strategy, and *opportunities* for evasion. It also recognizes the *social setting* in which the compliance decision is made.

The chapter describes the theoretical background of the modeling and the numerical results from two nested agent-based models. The baseline model focuses on occupational choice and the distributional consequences of noncompliance. The extended model generalizes the baseline version by adding repeated social interaction and the transmission of attitudes and beliefs in a dynamic setting. The models demonstrate that noncompliance increases inequality and risk-taking in the economy and that different compliance behaviors can be established within occupational groups. It is also possible that taxpayers, on average, can systematically hold a belief about the probability of audit that remains consistently above the true rate. When audit strategies are compared we find that a strategy of auditing a fixed number of individuals within each occupational group delivers a higher level of revenue than strategies with randomness across groups or a systematic focus on a particular group or groups.

Section 4.2 provides a description of the version of an agent-based model used in our work, according to the ODD+D protocol (Müller *et al.*, 2013). Section 4.3 gives an overview of the literature on economic models of the individual compliance decision and outlines a theoretical model of social custom and epidemics of evasion. Section 4.4 describes the model of individual choice of a taxpayer in a setting where an opportunity for noncompliance plays a central role, along with earning opportunities, in the choice of occupation. The distributional outcomes are illustrated using an agent-based model where the Allingham and Sandmo (1972) and Yitzhaki (1974) framework is extended to include the choice of occupation. A further extension incorporates a behavioral approach to the individual choice, with specific assumptions about the preferences and the information set of individuals. In particular, we assume that each taxpayer forms subjective beliefs about the audit strategy of a tax authority and develops attitude to the social custom of paying tax; these *attitudes* and *beliefs* develop endogenously within a *social network*. Section 4.5 describes an agent-based model with the behavioral concepts of attitudes, beliefs, and social network effects (Hashimzade *et al.*, 2014, p. 135), and Section 4.6 illustrates the outcomes of different choices of audit strategy by a tax authority in this framework. The conclusions are summarized in Section 4.7. Mathematical syntax and parametrization used in model simulations are listed in the Appendix 4A.

4.2 Agent-Based Model for Tax Compliance and Audit Research

This section describes the agent-based model we have used to investigate tax compliance and the outcomes of alternative audit strategies of a revenue service. The results of the simulations in this setting are reported in Section 4.5. The version used to obtain the results reported in Section 4.4 is nested within this general model by removing the social interaction and adaptation. The description follows the ODD+D protocol outlined in Müller *et al.* (2013).

4.2.1 Overview

4.2.1.1 Purpose

The purpose of the model is to understand the observed differences in tax compliance behavior and attitudes in a society with social networks and to investigate the relative effectiveness of various tax audit strategies in that society. It can be used by social scientists interested in modeling the interaction of individuals in social networks leading to the emergence of social norms. The model can also be used by revenue services for testing alternative audit strategies, in order to compare their outcomes for tax yield and compliance rates.

4.2.1.2 Entities, State Variables, and Scales

There are two types of human agents: the individual taxpayers and a revenue service. In addition, there are three types of income sources: employment with certain income, and two types of self-employment with stochastic income. Human agents operate in an environment characterized by a tax law, according to which income is subject to tax, and a social custom of fully paying any tax due. Taxpayers can choose a source of income. Tax is withheld at source on employment income but the taxpayer is responsible for declaring income from self-employment. Each taxpayer is characterized by risk attitude, the earning ability for each of the three income sources, the subjective belief about the audit strategy of the revenue service, and the attitude to the social custom of tax compliance.

The revenue service is characterized by the audit strategy, that is, the rule used to select taxpayers for an audit. The revenue service selects and audits a set of tax declarations by self-employed, collects any unpaid tax, and imposes fines on evaders. Income net of tax and fines is consumed.

Incomes earned by individual taxpayers in a given period form a vector of exogenous state variables. Individual tax payments in a given period, individual subjective beliefs about audit strategies, and individual weights assigned to the utility from following social custom form the set of vectors of endogenous state variables.

4.2.1.3 Process Overview and Scheduling

In every period, representing a tax year, the following activities take place. Taxpayers choose an occupation, earn income, and make tax declarations. The revenue service selects a number of declarations for audit, uncovers any unpaid tax, and records each detected case of evasion. Evaders pay omitted tax and a fine. All taxpayers update their own beliefs about the audit strategy. After the round of audits each taxpayer may meet with another taxpayer to whom they are linked in the social network, and at a meeting the taxpayers may or may not exchange information. The taxpayers who exchange information update their beliefs and the attitude to the social custom of being compliant. The revenue service updates its database of recorded evasion cases and may or may not use it to update its audit strategy in the next period.

4.2.2   Design Concepts

4.2.2.1 Theoretical and Empirical Background

The model is designed to test the following two hypotheses. First, we hypothesize that different patterns in compliance behavior in a society for different groups of taxpayers can emerge endogenously through self-selection of taxpayers willing to take a risk of evading tax into occupations where these opportunities exist. Second, we hypothesize that subjective beliefs about the likelihood of being audited can remain persistently high, and substantially higher than the true audit frequency, through the information exchange between audited and unaudited taxpayers.

The model employs a combination of standard expected utility theory and behavioral theories of decision making under uncertainty. Specifically, a taxpayer maximizes expected utility, but instead of using the objective probabilities of different states of nature (audited or not audited) the taxpayer uses subjective beliefs. The objective probabilities are not known to the taxpayers. Furthermore, the model uses the behavioral concept of attitude to the social custom of tax compliance, or tax morale, whereby a taxpayer derives additional utility from paying tax in full. Finally, the model assumes a particular mechanism for updating of beliefs and attitudes, which takes place at the meetings of taxpayers belonging to the same social network.

The motivations for the specification of the model are the following. Firstly, the observed compliance rate is higher than predicted by the standard model for empirically plausible values of audit probabilities and fines. Secondly, the observed compliance rates differ for different occupational groups within a given economy, and the economy-wide compliance rates differ across countries. The model allows for such differences to emerge endogenously through self-selection of taxpayers into occupations based on their risk attitudes, and through the evolution of attitudes to the social custom.

The model was calibrated to reproduce the distribution of taxpayers across the three occupations that matches UK data.[1] The data used are at the aggregate (macroeconomic) level.

4.2.2.2 Individual Decision Making

Taxpayers solve an optimization problem to evaluate the maximal utility they can achieve in each occupation, taking into account evasion opportunities and utilizing their beliefs about the audit strategy of the tax authority. Each taxpayer chooses the occupation that delivers the highest expected utility. After income uncertainty is resolved self-employed taxpayers solve an optimization problem to calculate how much income to declare for tax purposes.

Income in employment is known with certainty. In contrast, each of the two types of self-employment is characterized by a distribution of possible realizations of income; the second type has higher mean and higher variance than the first type, and in this sense is "riskier." In the rest of the chapter we will refer to the employment with certain income as occupation 0, less risky self-employment as occupation 1, and riskier self-employment as occupation 2.

Taxpayers adapt their beliefs and attitudes based on their own experience of tax audits and on the experience of other taxpayers in their social network with whom they exchange information. A taxpayer adjusts his belief according to whether he was audited in the previous period and after an information exchange takes place with a network contact. The social norm is that of full compliance, and any taxpayer who paid tax in full receives additional utility. However, the attitudes, captured by individual's weights assigned to this extra utility, differ across taxpayers and are adapted according to the compliance behavior of others. Thus, the weight increases when an honest taxpayer is encountered at a meeting, and decreases when an evader is encountered.

The revenue service's decision to select tax declarations for audit has two components. An exogenous component is the number of declarations; this is either fixed in every period or varies each period around a given long-run average and is present in all audit strategies. An endogenous component is present in the optimized audit strategies as a decision rule. Two decision rules are

---

[1] Authors' calculations based on the U.K. Living Costs and Food Survey (LCFS) suggest 86% in employment, and 8% in "nonrisky" and 6% in "risky" self-employment. Occupations dominated by informal suppliers were classified as "risky," according to Alm and Erard (2016).

considered: the choice of the highest evaders and the choice of the most likely evaders. In either case the choice is made on the basis of data analytics (predictive analytics).

4.2.2.3   Learning

There is no learning in the model. All changes in the decision-making rules are classed as adaptations.

4.2.2.4   Individual Sensing

Each taxpayer observes and considers his attitude to risk and to the social custom of compliance, earning ability in each occupation, and the distribution of potential earnings in self-employment prior to the choice of occupation; he observes his earnings in the chosen self-employment occupation. The revenue service observes and considers tax declarations, taxpayers' occupations, and their past history of audits. The mechanism of sensing is a mixture of knowing and information exchange. Specifically, taxpayers exchange information on their beliefs about the audit strategies of tax authority and about their latest compliance behavior.

Information exchange between taxpayers is direct and intentional, it takes place only between taxpayers who are connected in the social network, and the probability of receiving information is higher when the taxpayers belong to the same occupation. There are no errors in sensing. The network is modeled as the Erdös-Rényi type, that is, each taxpayer has the same probability of being linked to any other taxpayer. There is no cost of cognition or gathering information. There is an implicit cost of audits for the revenue service reflected in the bounded number of audits.

4.2.2.5   Individual Prediction

Taxpayers use knowledge of the distribution of earnings and the subjective belief about the audit strategy, updated in every period, to predict expected utility in each of the available occupations. The internal model of a taxpayer is that of expected utility, modified to include subjective probabilities and a social custom. The prediction error is determined by the distribution of earnings and the deviation of the subjective belief from the objective probability. The tax authority uses declared income, taxpayer occupation, and taxpayer audit history to predict the level of evaded tax or the probability of evasion for each taxpayer. The internal model of the revenue service is a multivariate parametric regression model: a censored linear regression in the first scenario and a logistic regression in the second scenario.

The prediction error is determined by the sampling distribution of the estimated regression parameters.

4.2.2.6   Interaction

Taxpayers interact in a social network by exchanging information at pairwise meetings. Each information exchange leads to a change in the subjective beliefs about the audit strategy and in the weight assigned to the utility from following social custom for both parties. The revenue service interacts with taxpayers directly, by auditing selected tax declarations, and indirectly, by altering the beliefs and attitudes of taxpayers who were not audited but exchanged information with those audited.

4.2.2.7   Collectives

There are two independent types of social groups: occupations and networks. Taxpayers in the same occupation need not belong to the same social group, and taxpayers in the same social group do not necessarily belong to the same occupation. The network of social groups is specified at the outset by the modeler by drawing links randomly, whereas the occupational choices emerge from the simulations. Each taxpayer is equally likely to be connected or not connected to any other taxpayer.

4.2.2.8   Heterogeneity

Taxpayers have identical preferences and decision models but are heterogeneous in the parameters describing their preferences and productivities. Specifically, taxpayers differ in their attitude to risk, described by the constant relative risk aversion (CRRA) utility function with individual-specific CRRA coefficient, and in their earning abilities, or skill, in each occupation.

4.2.2.9   Stochasticity

The model includes a one-off randomization at the outset and recurring randomization in each period. Initial randomization involves assignment of the preference and productivity parameters to the individual taxpayers as independent random draws from specific distributions, and the network connections between taxpayers. Recurring randomization involves the realization of earnings of self-employed taxpayers after the choice of occupation is made; the pairwise meetings of taxpayers after the round of audits; and the exchange of information at the meetings.

The revenue service employs recurring randomization in one of its audit strategies by making independent random draws each period from the population of submitted tax declarations for an audit.

4.2.2.10   Observation and Emergence

The following model output is collected in every period in the simulation: the occupational choices, the realized incomes, the declared incomes, the identities of audited taxpayers, the amounts of paid taxes and fines, and the updated beliefs and attitudes.

4.2.3   Details

The programming language is Matlab. The model code was initialized for 6000 taxpayers, with randomly assigned skills in each of the three occupations, the risk aversion coefficients, the social custom attitudes, and the beliefs about audit strategy. The values for all parameters are given in the Appendix 4A. The interaction among agents and between the agents and the environment is contained within the model; there are no submodels.

4.3   Modeling Individual Compliance

This section provides a brief review of the economic theory that has been applied to modeling the individual compliance decision. We begin by describing the expected utility model and exploring its limitations. We then review attempts to address these limitations by incorporating ideas from behavioral economics into the modeling.

4.3.1   Expected Utility

The starting point for modeling the compliance decision is to consider an individual taxpayer with a given amount of income who has to choose how much of this income to declare. This is the situation studied in the seminal paper of Allingham and Sandmo (1972). It is assumed that the income is not observed by the revenue service. The taxpayer is assumed to know the objective probability of being subjected to an audit and to believe that if an audit is conducted then the true level of income is revealed with certainty. The discovery of undeclared income by an audit results in the payment of tax on the undeclared income plus an additional fine. This model provides a set of baseline predictions that can be assessed against evidence.

The actual level of income is Y, the level of income declared to the revenue service is X, and the amount of undeclared income is $E = Y - X$. The tax rate is constant at τ and the fine if audited is $f\tau E$.[2] The decision problem of the taxpayer is to choose the amount of evasion, E (or, equivalently, the declaration X).

After the compliance decision is made one of the two potential states of the world is realized. In the state of the world in which evasion is successful the taxpayer is left with disposable income Y^n, where n stands for "not caught,"

$$Y^n = Y - \tau X \tag{4.1}$$

The level of disposable income in the state of the world in which evasion is detected is denoted Y^c, where c stands for "caught." This is given by

$$Y^c = Y - \tau Y - f\tau E \tag{4.2}$$

These two states of the world occur with objective probabilities $1 - p$ and p, respectively. The taxpayer evaluates income levels in each state using a utility function $U(\cdot)$ satisfying the standard assumptions $U' > 0$ and $U'' < 0$. The optimization problem faced by the taxpayer can be written as

$$\max_{\{E\}} V = pU(Y[1 - \tau] - f\tau E) + [1 - p]U(Y[1 - \tau] + \tau E) \tag{4.3}$$

The first- and second-order conditions for an interior optimum are

$$-pfU'(Y^c) + [1 - p]U'(Y^n) = 0 \tag{4.4}$$

and

$$S \equiv pf^2U''(Y^c) + [1 - p]U''(Y^n) < 0 \tag{4.5}$$

A sufficient condition for tax evasion to take place is obtained by evaluating Eq. (4.4) at $E = 0$. Doing this shows that it is optimal to choose $E > 0$ if

$$f < \frac{1 - p}{p} \equiv \bar{f}(p) \tag{4.6}$$

It is important to observe that this sufficient condition is independent of preferences, so if one taxpayer chooses to evade then all should evade. Moreover, with the value of $f = 1$ (which is fairly representative of international tax systems) the model predicts that all taxpayers should evade if $p < 0.5$. Hence, a value of audit probability far in excess of what is observed is necessary to prevent all taxpayers choosing to conceal at least part of their incomes. The fact that the data on evasion (e.g., Slemrod and Yitzhaki, 2002) shows that a large proportion actually choose to declare all income presents the first challenge to the model. A detailed discussion of this framework is provided in Chapter 1.

We now turn to the comparative statics of an interior optimum. The objective function is strictly concave so if there is an interior optimum it must be unique.

---

[2] This assumption on fine is due to Yitzhaki (1974); in the original (Allingham and Sandmo, 1972) framework the fine is levied on concealed income rather than unpaid tax.

It can be shown that an increase in p or in f reduces E, so compliance is increased. The predicted effect of an increase in the tax rate on the level of evasion is more interesting. The effect of a tax increase is given by

$$\frac{dE}{d\tau} = -\frac{1}{S}\frac{\partial^2 V}{\partial E \partial \tau} \tag{4.7}$$

Using the Arrow–Pratt measure of absolute risk aversion

$$r(I) = -\frac{U''(I)}{U'(I)} \tag{4.8}$$

we have

$$\frac{\partial^2 V}{\partial E \partial \tau} = pfU'(Y^c)(Y[r(Y^n) - r(Y^c)] - E[r(Y^n) + r(Y^c)f]) \tag{4.9}$$

If absolute risk aversion decreases as income increases then $r(Y^n) < r(Y^c)$ so

$$\frac{\partial^2 V}{\partial E \partial \tau} < 0, \tag{4.10}$$

which implies that the level of evasion will *fall* as the tax rate rises. This result is counter to intuition and contradicts much of the empirical evidence. Intuition suggests that a high tax rate should provide an incentive to evade whereas the model predicts the opposite. The source of the results is that the fine is a multiple of the tax evaded: the "effective" punishment is $f\tau E$, so that the punishment becomes more severe as the tax rate increases. This discourages evasion. There is very little empirical evidence on which to base an assessment of this result but this has not prevented a presumption developing in the literature that the result is wrong, from which it follows that the model is inadequate and must be improved.

These two results – the sufficient condition for noncompliance and the comparative statics of the tax rate – have led many to criticize this model and to suggest alternatives that make different predictions.

It is worth noting, however, that the first reason cannot be used to dismiss the model. Indeed, if the value of p is estimated by dividing the number of audits conducted each year by the number of taxpayers, then the value obtained will certainly satisfy Eq. (4.6), and, thus, the model predicts that all taxpayers should underdeclare income – a claim that is inconsistent with the data for many countries.

Slemrod (2017) has argued that this process for constructing p is not correct because many taxpayers know that the income they earn will be reported by a third party to the revenue service. If this income is not declared on the tax return then an audit will be very likely since there will be an observable inconsistency. Knowing this the taxpayers will be compliant. To claim that the model fails to predict the behavior of these taxpayers is incorrect: if they are noncompliant the value of p they face is higher than the proportion of taxpayers audited. What matters is how the model predicts for the taxpayers who are not subject to third-party reporting so that income is unobserved by the revenue service. For this subset of taxpayers (or for the fraction of income that is unobserved) there is currently no substantive evidence that the model does not predict correctly.

4.3.2 Behavioral Models

The first aspect of the model to review is the assumption that the taxpayer makes a decision on the basis of the objective probability of audit. A more compelling description of the situation is that the taxpayer does not know the probability with which a declaration will be audited. If this is the case, the taxpayer will need to construct a *subjective* probability based on whatever evidence is available. It then becomes possible for the subjective probability to be significantly different from the objective probability. Two questions then arise: how is the subjective probability formed and how does it affect the compliance decision? We will describe a process for the formation of the subjective probability in the discussion of our agent-based model in Section 4.5. For the remainder of this discussion we will consider the consequences of moving to subjective probability.

Many alternative choice models have been proposed in the literature on behavioral economics.[3] These alternative models can be understood as relaxing the very restrictive assumptions of expected utility theory. The behavioral models can be viewed as special cases of the general formulation of value function

$$V = \omega_1(p)U(Y^c) + \omega_2(1-p)U(Y^n) \tag{4.11}$$

where $\omega_1(p)$ and $\omega_2(1-p)$ are *decision weights*. The interpretation is that the decision maker (the taxpayer for a compliance decision) transforms the true probabilities of the events p and $1-p$ into decision weights ω_1 and ω_2. The effect of the weighting on compliance choice can be seen by repeating the derivations that gave Eq. (4.6). A taxpayer will now be noncompliant if

$$f < \frac{\omega_2(1-p)}{\omega_1(p)} \equiv \bar{f}(\omega_1, \omega_2) \tag{4.12}$$

Since the transformation of probabilities can differ among individuals, different taxpayers facing the same objective probabilities can make different decisions if their weights differ.

Furthermore, the usual assumption in behavioral economics is that unlikely events are typically assigned a decision weight greater than the probability. So, if the chance of being audited is small, it is given much greater weight in the subjective assessment of the taxpayer. The individual logic underlying this is of the kind "I know only 4 % of people are audited but if I evade I will almost certainly

---

[3] See Hashimzade *et al.* (2013) for a comprehensive survey.

be audited." If this is the case, then

$$\omega_1(p) > p, \quad \omega_2(1-p) < 1-p \tag{4.13}$$

These two conditions together imply

$$\frac{\omega_2(1-p)}{\omega_1(p)} < \frac{1-p}{p} \tag{4.14}$$

or

$$\bar{f}(\omega_1, \omega_2) < \bar{f}(p)$$

Thus, the transformation of the probabilities into weights satisfying Eq. (4.13) results in the taxpayer choosing to be compliant at a lower fine rate than for the objective probabilities.

This application of behavioral economics can address the first claimed limitation of the expected utility model by modifying the sufficient condition for noncompliance to occur. What it cannot do is change how the tax rate affects the level of evasion. Repeating the comparative static analysis of changes in the tax rate for the choices arising from maximizing Eq. (4.11) just returns Eq. (4.9) with p replaced by $\omega_1(p)$. This change does not affect the conclusion that a higher tax rate will lead to less evasion.[4]

4.3.3   Psychic Costs and Social Customs

It has been assumed so far that the decision by any taxpayer to be compliant is independent of what the other taxpayers are doing. In practice we may expect that someone is more likely to be noncompliant when noncompliance is widespread compared to when it is confined to a small segment of the population. The reasoning behind this social interaction can be motivated along the following lines: "The amount of stigma or guilt I feel when noncompliant depends on what others do and think. If they underpay taxes then I will feel little guilt when I do not comply." This form of interdependence between taxpayers creates a social interdependency that can lead to multiple self-supporting equilibria.

The modeling can be motivated by considering the idea of a psychic cost of evasion introduced in Gordon (1989). The idea was that a noncompliant taxpayer would suffer feelings of guilt or anxiousness about being audited that reduced the level of utility. This can be captured by writing the taxpayer's payoff as

$$V = pU(Y[1-\tau] - f\tau E) + [1-p]U(Y[1-\tau] + \tau E) - \chi E \tag{4.15}$$

where χE is the utility cost of deviating from full compliance. A higher value of χ can be interpreted as a greater psychic cost from evasion. E will only be

---

[4] See Piolatto and Rablen (2017) for a detailed discussion of the role of various assumptions in the nonexpected utility framework in resolving these limitations.

positive when the marginal utility of evasion is greater than zero at a zero level of evasion. Formally, evasion will occur when $V_0 - \chi > 0$, where $V_0 \equiv [1 - p - pf]\tau U'(Y[1 - \tau])$. Otherwise, the taxpayer will choose to be compliant with $E = 0$. This modification to the expected utility model changes the sufficient condition for noncompliance to occur. The psychic cost can also be included in the model with subjective probability.

There is a second implication of introducing the psychic cost. Assume that the value of χ differs among taxpayers. Then taxpayers characterized by values of χ that satisfy $V_0 - \chi > 0$ will be noncompliant, and those with higher values of χ that give $V_0 - \chi \leq 0$ will be compliant. This allows the model to describe how a population of taxpayers can be separated into two groups with some compliant taxpayers and some noncompliant taxpayers.

An alternative assumption is that there is a utility benefit from abiding by a social custom of being compliant and that the benefit is lost when a taxpayer is noncompliant. Assuming that the benefit is an increasing function of the proportion of taxpayers who are compliant captures the fact that the loss of social prestige will be greater, the more out of step the taxpayer is with the remainder of society. This approach was used by Myles and Naylor (1996) to show that reputation effects can lead to multiple equilibria and epidemics of evasion.

Let the utility level for a taxpayer of income Y and facing the tax rate τ, who chooses not to evade, be given by

$$V^{NE} = U(Y[1 - \tau]) + bR(1 - \mu) + c \qquad (4.16)$$

with $b \geq 0$ and $c \geq 0$. The additional utility from adhering to the social custom of honest tax payment is $bR(1 - \mu) + c$, where μ is the proportion of population evading tax and $R' > 0$. The parameters b and c capture the attitude of the individual taxpayer toward the social custom and can be expected to be different across taxpayers.

When the decision to evade is taken, the resulting level of utility is

$$V^E = \max_{\{X\}} \{pU(Y[1 - \tau] - f\tau[Y - X]) + [1 - p]U(Y - \tau X)\} \qquad (4.17)$$

Equation (4.17) reflects the fact that the utility from following the social custom is lost when evasion is chosen. The declaration that maximizes V^E in Eq. (4.17) is denoted by X^* and the maximized level of utility by V^{E*}. The optimal level of reported income must satisfy the first-order condition

$$pfU'(Y[1 - \tau] - f\tau[Y - X^*]) - [1 - p]U'(Y - \tau X^*) = 0 \qquad (4.18)$$

From this condition it can be seen that for given Y, the choice of X^*, and hence the value of V^{E*}, are independent of b and c. Effectively, if evasion is chosen then the level of noncompliance is identical to that for the standard model. Hence, if individuals evade tax and deviate from the social custom the extent to which they are noncompliant does not depend on the importance attached to the social custom.

Whether an individual taxpayer is noncompliant or not depends upon the value of V^{NE} relative to V^{E*}. Noncompliance will only occur if $V^{E*} > V^{NE}$. The strict inequality captures the implicit assumption that the taxpayer prefers to follow the social custom if there is no utility loss from doing so. Writing this condition in detail, a taxpayer chooses to evade if

$$pU(Y[1-\tau] - f\tau[Y - X^*]) + [1-p]U(Y - \tau X^*)$$
$$> U(Y[1-\tau]) + bR(1-\mu) + c \qquad (4.19)$$

Note that separation into compliant and evading taxpayers in this model does not require an assumption of identical incomes: it occurs as long as Eq. (4.19) holds for some $\mu \in (0, 1)$.

The first point to note is that the model will generate no observations of taxpayers just evading a "little" tax. Taxpayers either do not evade or jump straight to evasion level $E = Y - X^*$. The second point to note is that the model, in common with the psychic cost model, can predict perfect compliance even when the expected financial gain from noncompliance is positive. This occurs when the gain from noncompliance is not sufficient to offset the loss from not following the social custom.

The key point of the model is that the choice of whether to evade or not depends on the proportion of the population who are noncompliant, μ. Hence, the compliance decision is not made by the taxpayer in isolation but is the outcome of a process of social interaction. The determination of which taxpayers are noncompliant is made by finding the proportion of noncompliance which is self-supporting. That is, in the equilibrium, precisely proportion μ of taxpayers will find it optimal to evade in response to that value of μ. Figure 4.1 depicts such an equilibrium.

In both the psychic cost and the social custom models a change in the tax rate has an effect on the intensive margin and the extensive margin. The effect on the

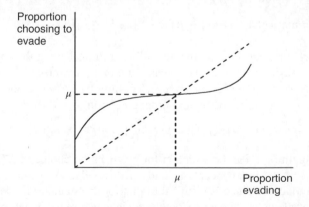

Figure 4.1   Social equilibrium

intensive margin – a taxpayer declaring $X^* < Y$ – is the same as for the standard model. The effect on the extensive margin depends on how the change in the tax rate affects the choice to be noncompliant. In the social custom model the extensive margin operates through changes in the self-supporting value of μ. It is through this effect that the model can generate a range of interesting outcomes. For example, a small change in the tax rate may cause a low-μ equilibrium to disappear so that the economy "jumps" to a high-μ equilibrium. Hence, an epidemic of evasion can occur in response to a minor change in the tax system or auditing process. Furthermore, a nonevasion equilibrium may exist under social pressure with low values of p and f.

Further analytical results can be obtained under the assumption of identical incomes. The effect on the extensive margin is found by taking given values of Y and τ and evaluating the critical proportion of evaders required for a taxpayer with characteristics $\{b, c\}$ to evade. If there exists μ^* satisfying

$$pU(Y^c) + [1 - p]U(Y^n) = U(Y[1 - \tau]) + bR(1 - \mu^*) + c \qquad (4.20)$$

then, since the right-hand side is monotonically increasing in μ, the taxpayer evades if $\mu > \mu^*$ and pays their full amount of tax if $\mu \leq \mu^*$. For a taxpayer for which there is a value μ^* satisfying Eq. (4.20) it can be calculated that

$$\frac{d\mu^*}{d\tau} = \frac{[pU'(Y^c) + [1 - p]U'(Y^n) - U'(Y[1 - \tau])]Y}{bR'}$$

The effect depends on the expected marginal utility from evasion compared to the marginal utility of correct declaration. This is not restricted by the model, so the effect may go in either direction. If $\frac{d\mu^*}{d\tau} < 0$ then an increase in the tax rate reduces the proportion of evaders required to encourage the (previously) marginal taxpayer to evade. The equilibrium number of evaders must therefore increase.

The social custom model emphasizes the importance of distinguishing between individual and aggregate effects of an increase in the tax rate. For each individual who is already evading an increase in the tax rate raises the level of income declared. However, an increase in the tax rate can increase the proportion of evaders. The net effect is determined by the resolution of these two effects. Consequently, aggregate data may show a positive relation between the tax rate and evasion even if the opposite is true at the individual level.

Traxler (2010) modified the analysis by assuming that the social custom is a loss of utility when evading. This gives utility

$$V^E = \mathcal{E}[U(X)] - E_i\theta_i c(\mu) \qquad (4.21)$$

where $\theta_i \geq 0$ is the individual-specific degree of norm internalization and $c(\mu)$ is the strength of the norm for a given fraction of evaders. It is assumed that $c'(\mu) \leq 0$ for $\mu \in [0, 1)$ so that an increase in the proportion of taxpayers deviating from the norm lowers the social custom cost of evasion. With this functional form the first-order condition for an interior solution is

$$-pf\tau U'(Y^c) + [1-p]U'(Y^n) = \theta_i c(\mu) \tag{4.22}$$

If $\theta_i = 0$ the first-order condition collapses to Eq. (4.4) so the standard necessary condition for evasion to occur will apply. If θ_i is sufficiently high, then there will be no evasion. Evaluating Eq. (4.22) at $X = Y$ gives the critical value of θ_i that separates nonevaders from evaders:

$$\hat{\theta}(\mu) \equiv \frac{(1-p[1+f])\tau U'([1-\tau]Y)}{c(\mu)} \tag{4.23}$$

The effect of an increase in the tax rate can be found from Eq. (4.22) as

$$\frac{\partial E_i^*}{\partial \tau} = \frac{1}{\mathcal{E}[U'']}[1-p]\tau U''(Y^n)[Y - E_i^*]$$

$$- \frac{1}{\mathcal{E}[U'']}\left(pf\tau U''(Y^c)[Y + f E_i^*] + \frac{\theta_i c(\mu)}{\tau}\right) \tag{4.24}$$

When $\theta_i = 0$ this collapses to the standard result described in Eq. (4.9), so that $\frac{\partial E_i^*}{\partial \tau} < 0$. When θ_i satisfies

$$\theta_i \geq \tilde{\theta}(\mu) \tag{4.25}$$

where $\tilde{\theta}(\mu) \equiv \frac{-\tau}{c(\mu)}(pf\tau U''(Y^c)[Y + f E_i^*] - [1-p]\tau U''(Y^n)[Y - E_i^*])$, then an increase in the tax rate raises the level of individual evasion. This is very similar to the result in Gordon (1989). What is different here is that the tax effect is dependent on the extent of evasion in the population through μ. A fall in the proportion of tax evaders reduces $\tilde{\theta}(\mu)$ so a positive tax effect applies for a larger range of values of θ_i.

The introduction of psychic costs and of social norms is capable of explaining some of the empirically observed features of tax evasion, which are not explained by the standard expected utility maximization hypothesis. It achieves this by modifying the form of preferences but the basic nature of the approach is unchanged. The social custom model can also be combined with subjective probabilities by replacing the probabilities with weighting functions.

4.4   Risk-Taking and Income Distribution

In this section, we illustrate the first application of our agent-based model, namely, the investigation of the effect of decisions involving risk-taking on the distribution of income, abstracting from the repeated interaction between taxpayers and from the adaptation of the audit strategy of the tax authority, and focusing on the occupational choice. The exposition follows the framework developed in Hashimzade *et al.* (2015).

A key element for understanding the compliance decision is the role the *opportunity* for noncompliance plays in the choice of occupation. Working as a paid employee either rules out noncompliance, if labor income is subject to a withholding tax (such as the PAYE system in the UK), or makes successful noncompliance very unlikely, if there is a system of third-party reporting. In contrast, choosing to be self-employed and accepting the responsibility for tax filing opens the opportunity for noncompliance. It is through this channel that occupational choice is interlinked with the compliance decision.

A second aspect of occupational choice is also linked to the compliance decision. Generally, the income generated in occupations 1 or 2 is uncertain, unlike the level of income received in occupation 0. This implies that choosing self-employment also involves accepting greater income risk and, therefore, all else being constant, agents in occupations 1 and 2 will have a lower degree of risk aversion than those in occupation 0. This directly determines the extent of noncompliance: the amount of income that is not declared increases as risk aversion decreases. In this way occupational choice self-selects those who will evade most into an occupation where they have the opportunity to evade.

Our first example of agent-based modeling incorporates occupational choice into a compliance model. This is achieved by extending the model of Allingham and Sandmo (1972) and Yitzhaki (1974) to permit each individual, first, to make an occupational choice and, second, to make an evasion decision based on the realization of income. The model can be seen as a generalization of the work of Pestieau and Possen (1991). The focus of the simulation is the effect that noncompliance has on the extent of risk-taking and income distribution in the economy.

The structure of the model is described in Section 4.2. In this simulation the occupational choice decision is made once. In the simulations of Sections 4.5 and 4.6 it is made at the start of every period because the choice may change as the information of the taxpayer evolves through interaction with others. Consequently, the results we report are the aggregate outcome of repeated static simulations rather than the outcome of a dynamic simulation. There is no social custom, or tax morale.

In what follows, we introduce the notations used in this and in the subsequent sections. Each taxpayer is randomly assigned a set of characteristics $\{w, \rho, s^1, s^2\}$, where w is the earning ability in occupation 0, or wage, ρ is the coefficient of (relative) risk aversion in a CRRA utility function, and s^α is the level of skill in occupation $\alpha \in \{1, 2\}$. The income earned in occupation α is $s^\alpha y^\alpha$ where y^α is drawn from a distribution with density function $g^\alpha(\cdot)$. In this simulation we used the beta distribution. The advantage of the beta distribution is in the flexible choice of parameters, allowing density functions with required skewness to be obtained, while the finite support ensures robust convergence of the numerical integration.

The variable y^α can be interpreted as local market conditions, so that income is determined jointly by individual skill and market conditions. The draw of y^α is unique for each taxpayer, so in a given round of simulation, a low-skill individual in occupation α may earn more than a high-skill individual if the former obtains a beneficial draw of y^α. It is assumed that $\mathcal{E}[y^1] = \mu^1 < \mu^2 = \mathcal{E}[y^2]$ and $\mathrm{Var}(y^1) = \sigma_1^2 < \sigma_2^2 = \mathrm{Var}(y^2)$, so that for a given skill level occupation 2 has a higher mean income but also a greater variance of income, or is riskier, than occupation 1. If a taxpayer has realized outcome $s^\alpha y^\alpha$ from occupation α the amount of income that is not declared, $E_i(y_i)$, is determined by

$$\max_{\{E^\alpha\}} U(E^\alpha; y^\alpha) = pU([1 - \tau]s^\alpha y^\alpha - f\tau E^\alpha) + (1 - p)U([1 - \tau]s^\alpha y^\alpha + \tau E^\alpha)$$

Taking account of the choice of E^α, the expected utility from self-employed occupation α is then

$$\mathcal{E}U^\alpha = \int U(E^\alpha(y^\alpha); y^\alpha)g(y^\alpha)\, dy^\alpha$$

The expected payoffs from the three occupations $\{U^0, \mathcal{E}U^1, \mathcal{E}U^2\}$ are compared, and the maximum payoff determines the chosen occupation. Once the occupation is chosen, incomes are realized and tax declarations are made. The tax authority conducts random audits and punishes any evasion that is detected.

The outcome is calculated for two different scenarios. The first scenario assumes that all income is honestly declared. This provides a baseline from which to judge the effect of noncompliance. The second scenario assumes that noncompliance may take place. Each simulation had 6000 individuals and was repeated 100 times. The data were pooled across the 100 rounds in order to smooth out the consequences of randomness and so the figures report the outcomes for a total of six million individuals. The following parameters were used for the illustrative example: the tax rate was 25 %, each self-employed taxpayer was audited with a probability of 0.05, and the fine rate was 150 % of evaded tax.

Our first two figures compare the distribution of occupational choices between the two scenarios. Figure 4.2 is a histogram of the distribution of taxpayers across the three occupations with honesty. The three occupations are on the horizontal axis and the vertical axis shows the number of taxpayers in each occupation. The corresponding histogram for when noncompliance was possible is shown in Figure 4.3. Comparing the figures shows that noncompliance causes the distribution of occupational choices to shift away from occupation 0 toward riskier occupations 1 and 2. As a consequence, there is more occupational risk-taking when noncompliance is possible. In addition to this increase in occupational risk-taking, there is a further increase in total risk-taking in the economy because some of the taxpayers choosing self-employment are also evading. Hence, the total amount of risk-taking in the economy is increased by the existence of tax evasion. This observation is interesting in view of the past discussion (Kanbur 1981; Black and de Meza 1997) on the efficiency of risk-taking in competitive economies.

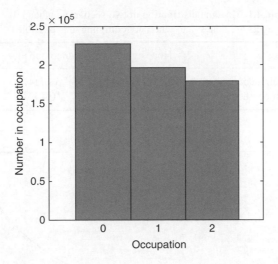

Figure 4.2   Occupational choice with honest tax payment

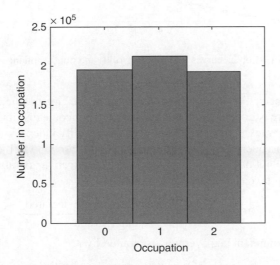

Figure 4.3   Occupational choice with noncompliance

The effect of noncompliance on income distribution is presented in two ways. Table 4.1 provides summary statistics of the income distributions with and without evasion, and Figure 4.4 plots the Lorenz curves for the two distributions. The effect of noncompliance is to increase the mean income level, where the mean is computed after both taxes and fines have been imposed. Noncompliance also

Table 4.1   Income distribution in two scenarios: when all
taxpayers are honest and when noncompliance is possible

	Honesty	Noncompliance
Mean income	28.211	33.424
Gini coefficient	0.464	0.492

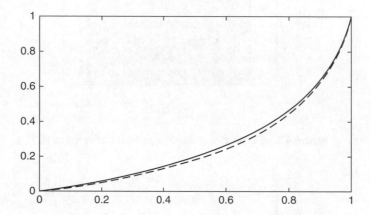

Figure 4.4   Lorenz curves for honesty (solid) and noncompliance (dashed)

increases the inequality of income as measured by the Gini coefficient. Figure 4.4
shows that there is Lorenz-curve dominance for the income distribution with honesty, and so the ranking is independent of the inequality index.

Another consequence of noncompliance is that those who fail to declare their
true income do not pay the statutory tax rate. The effective tax rate is defined for
a noncompliant taxpayer who is not audited by

$$ETR^{NA} = \frac{\text{Tax payment on income declared}}{\text{Actual income}} \qquad (4.26)$$

and for a noncompliant taxpayer who is audited by

$$ETR^{A} = \frac{\text{Correct tax payment plus fine}}{\text{Actual income}} \qquad (4.27)$$

ETR^{NA} will be below the statutory tax rate and ETR^{A} will be above the statutory
tax rate. The consequence of noncompliance by taxpayers is that the distribution
of effective tax rates is unrelated to income and does not correspond to the flat tax
intended by the government. This point is illustrated in Figure 4.5, which displays a
histogram of tax rates. This is trimodal, reflecting the three groups: noncompliant
taxpayers who are not audited (first two bars on the left), compliant taxpayers

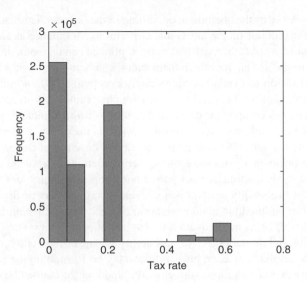

Figure 4.5 Histogram of effective tax rates

(the middle bar at 0.25 tax rate), and noncompliant taxpayers who are audited (the tail). Given the propensity for taxpayers to be noncompliant and the audit rate of 5%, the majority of taxpayers pay an effective tax rate below the statutory rate of 25%. The general observation is that noncompliance undermines the intended tax policy of the government.

These results illustrate some of the effects that noncompliance can have on the economy. The possibility of noncompliance encourages entry into risky occupations, while the consequence of noncompliance and auditing is increased inequality and a dispersion of the effective tax rate. The agent-based model reported in this section shows the importance of introducing opportunities, but there are more features of the compliance decision that need to be taken into account. The next section therefore reviews recent literature on the application of behavioral economics to the compliance decision.

4.5   Attitudes, Beliefs, and Network Effects

The empirical analysis of the determinants of tax evasion has demonstrated two important features. First, there is a strong evidence that the social setting influences the individual compliance decision. For example, individual perceptions of the justifiability of tax evasion in a country are positively associated with the measures of aggregate tax evasion in that country, according to the World Values Survey (Slemrod, 2007, p. 41). We refer to this effect as the *attitude to compliance*, or *attitude*. An aggregate measure of the individual attitudes to compliance across a society can also be viewed as the tax morale prevailing in that society. One can think about the effect of tax morale on the individual attitudes to compliance as an externality: an individual who holds the view that noncompliance can be (sometimes) justified contributes to the low tax morale in the society which, in turn, makes for that individual the decision to evade tax more easily acceptable. When individuals in a society belong to different social groups, with closer links within each group and looser links across groups, it is possible that social customs, including the attitude to compliance, are different across groups, and this difference persists in time. If the Myles and Naylor (1996) model is extended to have social customs formed by reference to behavior within subgroups, then epidemics of evasion can occur in one part of a society, while in other parts honest reporting continues to prevail.

Second, the probability of audit is not revealed to taxpayers by the revenue service. Therefore, in the individual evaluation of the expected benefit from evasion the probability of being audited and found to be noncompliant is subjective, rather than objective. While the objective probability is part of the audit strategy of the revenue service, the subjective probabilities may be formed on the basis of individual experience and available information, and can, of course, be different for different individuals. To distinguish between the objective and the subjective probabilities we refer to the latter as the *subjective belief*, or just *belief*.

If attitudes and beliefs are determined, among other factors, by experience and information, it is natural to assume that they can evolve and change for a given individual over time as he or she interacts with the revenue service (through accumulation of experience) and with other individuals (through accumulation and exchange of information). Furthermore, information exchange is more likely to occur (or more information is likely to be exchanged) if the individuals belong to the same occupational group. Individuals meet with their contacts in the social network, and meetings allow exchange of information on beliefs. In addition, at a meeting, individuals may observe each other's attitude to evasion. For example, individual 1 can infer something about individual 2's attitude simply by learning whether individual 2 has evaded tax previously. This will affect 1's own attitude and, through this channel, 1's future evasion decisions. The same may take place for individual 2.[5] This, in particular, can explain why social groups have different behavior with respect to tax evasion.

We have incorporated the dynamics of attitudes and beliefs into an agent-based model by adding to the individual compliance decision a process of *adaptation within a social network* according to the algorithm outlined below. It should be stressed that the behavior we model involves two dimensions of bounded rationality. First, the agents apply very simple adaptation rules and update information only from their contacts and not by observation of the wider world. Second, the agents are not forward looking and so do not make any strategic intertemporal decisions. The model therefore balances the pure rationality of much economic theory with the frequent empirical observation of bounded rationality.

---

[5] See Onu and Oats (2016) on empirical evidence of tax compliance information exchange among taxpayers in social networks.

4.5.1   Networks and Meetings

In an economy with N individuals the social network is described by a symmetric $N \times N$ matrix A with $A_{ij} = 1$ if individuals i and j are linked and $A_{ij} = 0$ otherwise. The links are bidirectional: if i "knows" j then j "knows" i.[6] In our simulations the network is fixed at the outset and does not change; one can also introduce random or endogenous changes in the network structure. Time is divided into discrete periods, and in every period each individual chooses an occupation, earns income, and decides how much of this income to declare. Declarations are audited (according to some randomizing device as described below), after which individuals linked in the network randomly meet and exchange information.

Here, we introduce two additional layers of randomness: not all individuals in the network meet in every period, and not every meeting results in an information exchange. This is implemented by introducing an $N \times N$ matrix C of zeros and ones, drawn randomly in each period; this matrix represents the probabilities of meetings between individuals. Thus, in each period a random selection of meetings occurs described by an element-by-element product of A and C: individuals i and j meet during a period if $A_{ij} C_{ij} = 1$ and do not meet otherwise.

Furthermore, at a meeting of i and j information is exchanged only with some probability. It is possible to consider various patterns in the probability of information exchange; one plausible assumption is that the probability depends on the occupational groups to which i and j belong. More specifically, we assume that the probability of information exchange between i and j is higher when i and j belong to the same occupational group, and that it does not depend on their individual characteristics or other model parameters. With three occupations, in general, six different probabilities can be introduced, denoted by $q^{\alpha\beta}$, where $\alpha, \beta \in \{0, 1, 2\}$, and $q^{\alpha\alpha} > q^{\alpha\beta}$ for all α and $\beta \neq \alpha$.

4.5.2   Formation of Beliefs

The choice of occupation in period $t + 1$ is made on the basis of the beliefs $\{p_{j,t}^1, p_{j,t}^2\}$ updated after the audits, that is, based on own experience, and after the information exchange, that is, based on the experience of others.

4.5.2.1   Audits and Beliefs

The updating process immediately after the audit is qualitatively similar to Bayesian updating, and is to assume that individuals feel marked as targets if they are audited, so that one audit is believed likely to be followed by another. We term this the *target effect*. In contrast, those not audited in a period believe that they

---

[6] A matrix that is not symmetric captures unidirectional links. This can be used to investigate the effect of a "celebrity."

are less likely to be audited in the next period. Formally, if audited in period t, an individual's belief about being audited in the next period is raised to probability P; otherwise, it decays. The updating rule for the subjective probability is therefore

$$\begin{cases} \tilde{p}_{j,t+1}^{\alpha} = A_{j,t}P + (1 - A_{j,t})\delta p_{j,t}^{\alpha}, \ \delta \in [0,1], \ P \in [0,1] \\ \tilde{p}_{j,t+1}^{\beta} = p_{j,t+1}^{\beta}, \ \beta \neq \alpha \end{cases} \tag{4.28}$$

where $A_{j,t} = 1$ if taxpayer j was audited in period t and $A_{j,t} = 0$ otherwise. This can also be written as

$$\tilde{p}_{j,t+1}^{\alpha} = \begin{cases} P \in [0,1] & \text{if audited at } t, \\ \delta p_{j,t+1}^{\alpha}, \ d \in [0,1] & \text{otherwise} \end{cases} \tag{4.29}$$

We refer to the case of $P = 1$ as the maximal target effect.

Under an alternative assumption of the *bomb crater effect* (e.g., Guala and Mittone, 2005, p. 505), immediately after an audit the belief falls, possibly to zero, and then rises. That is, if j is audited and caught in period t he believes that he is less likely to be audited again (similar to the belief that a bomb is unlikely to hit a crater made by the previous bomb), but subsequently worries that his turn to be audited again is approaching. The results of the simulations under this assumption are qualitatively similar. Empirical findings in Advani *et al.* (2015) seem to support the target effect: immediately after, the audit compliance increases and gradually declines after several periods.

4.5.2.2 Information Exchange and Beliefs

After the audit process is completed the taxpayer may meet with a contact. The information that may (or may not) be exchanged at a meeting includes the subjective probabilities and whether or not the agents were audited. If taxpayer j in occupation α meets individual i who works in occupation β the subjective probability is updated according to the rule

$$p_{j,t+1}^{\alpha,\beta} = \begin{cases} \mu \tilde{p}_{j,t}^{\alpha,\beta} + (1 - \mu)\tilde{p}_{i,t}^{\alpha,\beta} & \text{with probability } q^{\alpha\beta} \\ \tilde{p}_{j,t}^{\alpha,\beta} & \text{with probability } 1 - q^{\alpha\beta} \end{cases} \tag{4.30}$$

4.5.2.3 Formation of Attitudes

The importance assigned to the social custom is also determined by interaction in the social network. The weight, $\chi_{j,t}$, is updated in period t if information exchange occurs between j and some other taxpayer in that period. Assume that individual j meets individual i at time t and information exchange takes place. If i evaded ($\tilde{E}_i^{\alpha} > 0$), j's weight on honesty is adjusted downwards. Conversely, if i was honest, j's weight on honesty is adjusted upwards. The magnitude of adjustment is assumed to be larger for the intermediate weights ($\chi_{j,t}$ close to 0.5) and smaller for low (close to 0) and high (close to 1) weights. The updating process is described by

$$
\chi_{j,t+1} = \begin{cases} \chi_{j,t} + \lambda(2 \times \mathbf{1}_{[\tilde{E}_i^\alpha > 0]} - 1)(\chi_{j,t} + \mathbf{1}_{[\chi_{j,t} > 0.5]}(1 - 2\chi_{j,t})) \\ \text{with probability } q^{\alpha\beta}; \\ \\ \chi_{j,t} \\ \text{with probability } 1 - q^{\alpha\beta}. \end{cases}
$$

where $\tilde{E}_i^\alpha$ is the level of evasion of i, $\mathbf{1}_{[\tilde{E}_i^\alpha > 0]}$ is an indicator function with value of 1 if $\tilde{E}_i^\alpha > 0$ and 0 otherwise, $\mathbf{1}_{[\chi_{j,t} > 0.5]}$ is an indicator function with value of 1 if $\chi_{j,t} > 0.5$ and 0 otherwise, and $\lambda \in [0, 1]$ is the rate of adjustment. Hence, $\chi_{j,t+1} > \chi_{j,t}$ if information is exchanged with a compliant taxpayer, $\chi_{j,t+1} < \chi_{j,t}$ if information is exchanged with an evader, and $\chi_{j,t+1} = \chi_{j,t}$ if no information exchange has taken place. This is similar in spirit to the model of Myles and Naylor (1996), described in Section 4.3.

4.6 Equilibrium with Random and Targeted Audits

As the benchmark case, we first assume a simple random probability of audit: each self-employed individual is audited with the same constant probability; those in paid employment are not audited.[7] The tax authority is assumed to know that in paid employment income tax is fully deducted at source, and there is no opportunity for earning additional income that could be concealed. This assumption could be modified in a more general model to allow an additional income for individuals in employment and a possibility to evade tax on that income. As an alternative strategy we consider targeted audits, or predictive analytics, where the tax authority uses information from the past audits and from the submitted declarations to select taxpayers for audit. Here, we focus on the strategy whereby only the latest declaration is audited; another interesting extension would be to consider back audits, which would require introducing intertemporary optimization in taxpayer's problem.

We assume that earnings in occupation α, where $\alpha \in \{0, 1, 2\}$, are drawn from lognormal distribution, $\log \mathcal{N}(\mu_\alpha, \sigma_\alpha^2)$, and that skills in self-employment are drawn from $\frac{1}{1-\gamma U}$, where U is a uniform $[0, 1]$ random variable, and $\gamma \in (0, 1)$ is a constant parameter. Each individual knows their wage in occupation 0, $y_{j,t}^0 = w_j$, as well as the skill, s_j^α, and the distribution of outcomes, $g^\alpha(\cdot)$, in the occupations $\alpha \in \{1, 2\}$. At time 0 each individual is randomly assigned a vector of subjective beliefs, $\{p_{j,0}^1, p_{j,0}^2\}$, the level of importance of social custom, $\chi_{j,0}$, both drawn independently from a uniform $[0, 1]$ distribution, and "age" (years of work) drawn

---

[7] The results presented in this section draw on Hashimzade and Myles (2017).

from a uniform $[0, \overline{T}]$ distribution. Taxpayer j works until the "age" of $T_j^{\max}$ before being replaced by a new taxpayer with a randomly drawn set of characteristics; the "retirement age," $T_j^{\max}$, is drawn from a uniform $[\underline{T}^R, \overline{T}^R]$. The first 70 periods of the simulation are conducted with random audits of self-employed taxpayers. The probability of a random audit for all taxpayers in occupations 1 and 2 is 0.1; those in occupation 0 are not audited. Data from audits are recorded from period 21 onwards to avoid the initial condition effects. Audits based on predictive analytics are implemented from period 71 onwards.

The simulation involves the random realization of earnings in every period for every self-employed taxpayer. As a consequence, the outputs of the simulation (compliance levels, audit outcomes, tax revenues, etc.) are realizations of random variables. For this reason the level of tax and fine revenues are contrasted in Figure 4.6 by plotting the (empirical) cumulative distribution functions for random auditing and for audits targeted using predictive analytics where revenue is on the horizontal axis. The cumulative distribution function for targeted audits exhibits first-order stochastic dominance over that for random audits. In fact, the worst outcome for targeted is better than the best outcome for random. This comparison shows clearly that predictive analytics successfully target audits and achieve increased revenue (in this example, by about 10 %).

The change in compliance behavior that lies behind the increase in revenue from targeting is shown in Figure 4.7. The plot shows proportions of income reported by

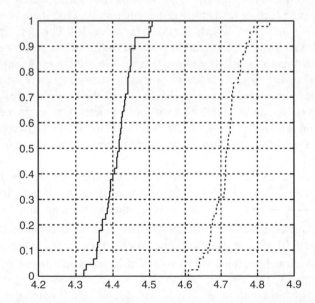

Figure 4.6 Cumulative distribution functions for tax and fine revenues (random: solid; targeted: dashed)

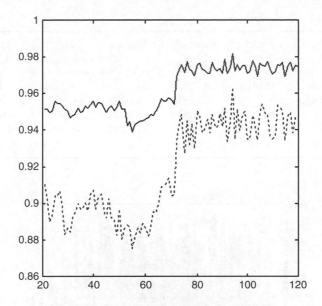

Figure 4.7 Compliance levels (occupation 1: solid; occupation 2: dashed

the self-employed for each of the periods of the simulation. The compliance levels show the link between risk aversion and compliance: occupation 2 is riskier and the compliance level is lower. The change in compliance after the implementation of predictive analytics after period 70 is very clear. Furthermore, predictive analytics reduce the compliance gap between the two occupations.

The predicted level of evasion for each taxpayer is estimated using a Tobit model; the regression equation is re-estimated in every period using the information from the new audits. Table 4.2 shows the output of the Tobit regression estimated in the last period of simulations. The dependent variable for the model is the amount of evasion. *Declaration* is the income declared on the tax return. *Audit* is a dummy with value 1 if audited in the previous period. *Occupation* is a dummy with value 1 if the taxpayer is working in occupation 1. ME refers to the estimated marginal effect, calculated either at the average of all data points or as the average of the marginal effects at each data point. All estimated coefficients have the expected sign, so that predicted noncompliance is greater when declared income is lower, the taxpayer was not audited in the previous period, the taxpayer is young, and is engaged in occupation 2. Only the coefficient on the age variable is not statistically significant.

The consequences of following the audit recommendation of predictive analytics can be understood by plotting the characteristics of the audited taxpayers in each percentile of the declared income distribution. Figure 4.8 shows the proportion of audited taxpayers in occupation 1 in the population of the audited.

Table 4.2 Tobit regression

Variable	Coefficient	Standard error	z-Statistic	ME (avg. data)	ME (ind. avg)
Constant	20.420	0.250	81.652	4.7299	12.8898
Declaration	−3.9722	0.1297	−30.6299	−0.920	−2.5074
Audit	−6.447	1.5053	−4.2829	−1.493	−4.0697
Age	−0.4686	0.4346	−1.0783	−0.108	−0.2958
Occupation	−0.4815	0.0927	−5.195	−0.1115	−0.3039

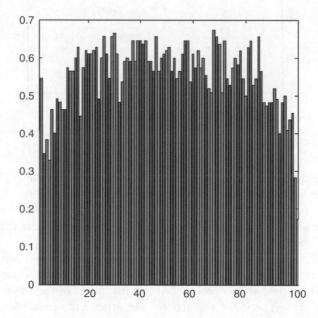

Figure 4.8 Proportion of audited in occupation 1 by percentile of declaration

Approximately half of audited taxpayers are in occupation 1, but this group has about twice the numbers as occupation 2. Consequently, a greater proportion of the taxpayers in occupation 2 are audited.

Theoretical characterizations of optimal audit rules have been developed in models in which declared income is the only distinguishing characteristic of taxpayers. When the tax authority can commit to an audit function (probability of audit as a function of income) then the optimal rule is to operate a cut-off (Reinganum and Wilde, 1985). Without commitment, the rule is a decreasing function of declaration (Reinganum and Wilde, 1986). Figure 4.9 plots the audit probability implied by predictive analytics for each percentile of the declaration

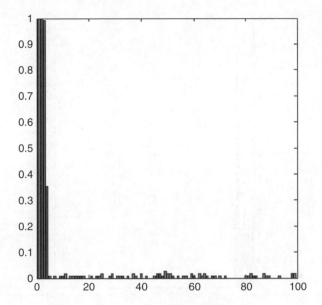

Figure 4.9   Proportion of audited in occupation 1 by declaration percentile

distribution. While there is some probability throughout the range of declaration, the probability function is very close to being a cut-off rule. Almost all taxpayers are audited in the lowest third percentile; after that the probability falls away rapidly.

The effect of predictive analytics is made clear in the figures of the previous section. However, these figures reflect only one particular choice of parameter values and realization of the random variables. To provide some support for the argument that they indicate general features a robustness check is now reported. This check varies the audit rate and the social custom.

The first revision is to reduce the audit rate from 0.1 to 0.05. The second is to double the utility benefit of compliance. Figure 4.10 shows that this does not affect the relationship between audit probability and declared income or the impact on compliance.

4.7   Conclusions

The compliance decision combines a range of economic, psychological, and social elements. Included among these are perceptions of risk and attitudes toward risk-taking, the importance of social standing and conformity to group norms, and the transmission of information through social contacts. A compelling model of the compliance decision requires these components to be combined and embedded within a taxpayer equilibrium.

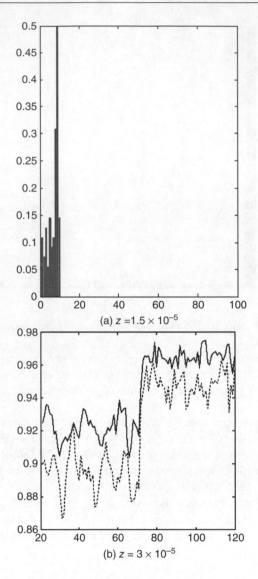

Figure 4.10   Probability of audit and compliance levels

Agent-based modeling provides the ideal methodology for bringing disparate elements into a cohesive whole. The combination of the agent-based model with the structure of a social network to govern interaction provides a rich environment in which to explore compliance. A particular strength of agent-based modeling is that it has the potential to accommodate complex optimization and learning processes.

The models that we have presented in this chapter emphasize the importance of opportunities for noncompliance, and the link that this creates between occupational choice and risk attitude. Risky occupations will be chosen by those who are most willing to accept risk and to exploit most fully the available opportunities for noncompliance. As a consequence, compliance behavior can vary significantly across occupational groups. Our simulations suggest that the possibility of noncompliance encourages entry into risky occupations, thereby increasing the amount of risk-taking in a society. At the same time, noncompliance and auditing lead to increased inequality and a dispersion of the effective tax rate.

The methodology is very flexible and is, therefore, able to incorporate recent advances in the theory of compliance. Our work emphasizes the role of attitudes, beliefs, and opportunities, and draws ideas from advances in behavioral economics. A further advantage of an agent-based model of tax compliance is that it can incorporate a variety of different intervention strategies by the revenue service. Here we explored the effect of *predictive analytics*, or the use of past information on taxpayers for predicting their future compliance behavior, on audit outcomes.

The process for the selection of audit targets is important for a revenue service to maximize the effectiveness of audit resources. The tools of predictive analytics have been adopted by many revenue services including HMRC in the United Kingdom and the IRS in the United States – although in practice the results of predictive analytics serve predominantly only as a guide rather than a rule, and an "auditor judgment" retains an important role in selecting audit targets. We have investigated the consequences of predictive analytics in a rich network model that employs ideas from behavioral economics to model the compliance decision of taxpayers. The behavioral aspects of the model are the formation of a subjective belief about the probability of audit and a social custom toward compliant tax payment through experience and social interaction. Since attitudes and beliefs emerge endogenously they can differ across occupational groups in equilibrium. Given this fact, there is likely to be a benefit to the revenue service of selecting audit targets on the basis of occupation as well an declared income level. The existing theoretical models of optimal auditing have not been able to explore this issue because they do not identify different occupations or sources of income.

The results of the simulation show that predictive analytics are effective in raising revenue when compared to a strategy of random audits. A basic feature of the outcome of the simulation model is that, conditional on attitudes and beliefs, taxpayers self-select into occupations on the basis of risk aversion. It is therefore no surprise that predictive analytics focus audits disproportionately upon taxpayers in the riskier occupation. Predictive analytics reduce the difference in compliance rates between occupations but do not eliminate it entirely. Interestingly, the outcome of predictive analytics has the form of a cut-off rule: almost all audit effort is focused on declarations of low income.

Acknowledgments

Thanks are due to the ESRC for financial support under grant RES-194-23-0002. We thank the editors and two anonymous referees for helpful comments and suggestions.

References

Advani, A., Elming, W., and Shaw, J. (2015) How long lasting are the effects of audits? TARC Discussion Paper, 011-15.

Allingham, M.G. and Sandmo, A. (1972) Income tax evasion: a theoretical analysis. *Journal of Public Economics*, **1**, 323–338.

Alm, J. and Erard, B. (2016) Using public information to estimate self-employment earnings of informal suppliers. *Public Budgeting and Finance*, **36**, 22–46.

Black, J. and de Meza, D. (1997) Everyone may benefit from subsidising entry to risky occupations. *Journal of Public Economics*, **66**, 409–424.

Gordon, J.P.P. (1989) Individual morality and reputation costs as deterrents to tax evasion. *European Economic Review*, **33**, 797–805.

Guala, F. and Mittone, L. (2005) Experiments in economics: external validity and the robustness of phenomena. *Journal of Economic Methods*, **12**, 495–515.

Hashimzade, N. and Myles, G.D. (2017) Risk-based audits in a behavioural model. *Public Finance Review*, **45**, 140–165.

Hashimzade, N., Myles, G.D., Page, F., and Rablen, M.D. (2014) Social networks and occupational choice: the edogenous formation of attitudes and beliefs about tax compliance. *Journal of Economic Psychology*, **40**, 134–146.

Hashimzade, N., Myles, G.D., Page, F., and Rablen, M.D. (2015) The use of agent-based modelling to investigate tax compliance. *Economics of Governance*, **16**, 143–164.

Hashimzade, N., Myles, G.D., and Tran-Nam, B. (2013) Applications of behavioural economics to tax evasion. *Journal of Economic Surveys*, **27**, 941–977.

Kanbur, S.M. (1981) Risk taking and taxation: an alternative perspective. *Journal of Public Economics*, **15**, 163–184.

Müller, B., Bohn, F., Dressler, G., Groeneveld, J., Klassert, C., Martin, R., Schlüter, M., Schulze, J., Weise, H., and Schwarz, N. (2013) Describing human decision in agent-based models – ODD+D extension of the ODD protocol. *Environmental Modelling & Software*, **48**, 37–48.

Myles, G.D. and Naylor, R.A. (1996) A model of tax evasion with group conformity and social customs. *European Journal of Political Economy*, **12**, 49–66.

Onu, D. and Oats, L. (2016) Tax talk: an exploration of online discussions among taxpayers. *Journal of Business Ethics*. Available at https://doi.org/10.1007/s10551-016-3032-y (accessed 18 October 2017).

Pestieau, P. and Possen, U.M. (1991) Tax evasion and occupational choice. *Journal of Public Economics*, **45**, 107–125.

Piolatto, A. and Rablen, M.D. (2017) Prospect theory and tax evasion: a reconsideration of the Yitzhaki Puzzle. *Theory and Decision*, **82**, 543–565.

Reinganum, J.F. and Wilde, L.L. (1985) Income tax compliance in a principal-agent framework. *Journal of Public Economics*, **26**, 1–18.

Reinganum, J.F. and Wilde, L.L. (1986) The tax compliance game: toward an interactive theory of law enforcement. *Journal of Law, Economics and Organization*, **2**, 1–32.

Slemrod, J. (2007) Cheating ourselves: the economics of tax evasion. *Journal of Economic Perspectives*, **21**, 25–48.

Slemrod, J. (2017) Tax compliance and enforcement: new research and its policy implications. In: Auerbach, A.J. and Smetters, K. (Eds), *The Economics of Tax Policy*. Oxford: Oxford University Press. Available at Oxford Scholarship Online. DOI:10.1093/acprof:oso/9780190619725.003.0006.

Slemrod, J. and Yitzhaki, S. (2002) Tax avoidance, evasion, and administration, in *Handbook of Public Economics*, vol. **III** (eds A.J. Auerbach and M. Feldstein), Elsevier Science B.V, Amsterdam, pp. 1423–1470.

Traxler, C. (2010) Social norms and conditional cooperative taxpayers. *European Journal of Political Economy*, **26**, 89–103.

Yitzhaki, S. (1974) A note on income tax evasion: a theoretical analysis. *Journal of Public Economics*, **3**, 201–202.

Appendix 4A

Probability of being linked in the network: $v = 0.5$;

Probability of audit: $p = \{0.05; 0.1\}$;

Tax rate: $\tau = 0.25$;

Fine rate: $f = 1.5$;

Weight in information exchange: $\mu = 0.75$;

Probability of meeting a network contact: $\phi = 0.25$;

Probability of information exchange

Focussed information transmission: $q^{\alpha\beta} = 0$, $q^{\alpha\alpha} = 1$ for $\alpha \neq \beta$, $\alpha, \beta = \{0, 1, 2\}$;

Diffused information transmission: $q^{\alpha\beta} = 0.15$, $q^{\alpha\alpha} = 0.75$ for $\alpha \neq \beta$, $\alpha, \beta = \{0, 1, 2\}$;

Number of agents: $N = 6000$.

Time horizon: $T = 90$.

Probability distributions ($\mathcal{N}$ = normal; $\mathcal{U}$ = continuous uniform)

Wage in occupation 0: $\log w \sim \mathcal{N}(\mu_0, \sigma_0^2)$;

$\quad E[w] = 13.0425$; $\text{Var}(w) = 4$;

Skill in occupation 1:

$\quad s_j^1 = \frac{1}{1-0.5\tilde{x}}$, $\tilde{x} = \mathcal{U}[0, 1]$;

$\quad E[s_j^1] = 1.387$; $\text{Var}(s_j^1) = 0.078$;

Income in occupation 1:

$\quad s_j^1 y^1$, $\log y^1 \sim \mathcal{N}(\mu_1, \sigma_1^2)$;

$\quad E[y^1] = 8.01$; $\text{Var}(y^1) = 12.26$;

Skill in occupation 2:

$\quad s_j^2 = \frac{1}{1-0.5\tilde{x}}$, $\tilde{x} = \mathcal{U}[0, 1]$;

$\quad E[s_j^2] = 1.387$; $\text{Var}(s_j^2) = 0.078$;

Income in occupation 2:

$\quad s_j^2 y^2$, $\log y^2 \sim \mathcal{N}(\mu_2, \sigma_2^2)$;

$\quad E[y^2] = 8.3$; $\text{Var}(y^1) = 22.8$;

Risk-aversion: $\rho \sim \mathcal{U}[0.1, 5.1]$;

Initial belief on audit probability: $p_0 \sim \mathcal{U}[0, 1]$;

Initial weight on the payoff to following the social custom: $\chi_0 \sim \mathcal{U}[0, 1]$;

Value of social custom: $z \sim \{3; 1.5\} \times 10^{-5} \times \mathcal{U}[0, 1]$;

Age (years in work) at the outset: $\mathcal{U}[0, \overline{T}]$, $\overline{T} = 19$;

Retirement age (total years in work): $\mathcal{U}[\underline{T}^R, \overline{T}^R]$, $\underline{T}^R = 20$, $\overline{T}^R = 29$.

5

SIMULFIS: A Simulation Tool to Explore Tax Compliance Behavior

Toni Llacer, Francisco J. Miguel Quesada, José A. Noguera and
Eduardo Tapia Tejada

5.1 Introduction

Tax evasion, usually defined as the voluntary reduction of the tax burden by illegal
means (Elffers *et al.*, 1987), is a problem of huge social relevance at present times.
This is so, first, because tax evasion reduces the volume of resources available
for the public sector. This reduction is especially damaging in the Spanish case:
Arrazola *et al.* (2011) estimates that Spanish shadow economy represents approximately 17% of GNP, while (GESTHA, 2011) estimates a higher level of 23.3%.
Second, since tax evasion behavior is not equally distributed among taxpayers, it
violates the principles of fairness, equality, and progressivity that the tax system
ought to satisfy (Alvarez and Herrera, 2004). Academic researchers who aim to
explain tax evasion and tax compliance are increasingly acknowledging the need
to include psychological, social, and cultural factors in their explanatory models;
traditional explanations were too often linked to the strict assumptions of rational
choice theory and the *Homo Oeconomicus* model (Allingham and Sandmo, 1972)
(see Chapter 1 for a detailed discussion). Instead, new studies focus, for example,
on taxpayers' tax morale (their tolerance toward tax fraud (Torgler, 2007)).
Recently, it has been shown that there is a causal link between aggregated tax
morale and the volume of the shadow economy at the national level (Halla, 2010);

Agent-Based Modeling of Tax Evasion: Theoretical Aspects and Computational Simulations,
First Edition. Edited by Sascha Hokamp, László Gulyás, Matthew Koehler, and Sanith Wijesinghe.
© 2018 John Wiley & Sons Ltd. Published 2018 by John Wiley & Sons Ltd.

at the individual level, some contributions have proved the relationship between individual tax morale and self-declared levels of tax evasion (Torgler *et al.*, 2008, 2010; Cummings *et al.*, 2009) as well as tax noncompliance behavior in the laboratory (Kirchler and Wahl, 2010). Besides tax morale, factors such as social norms, social influence (e.g., see Chapters 6–8), fairness concerns, and perceptions of the distributive outcomes of the tax system are increasingly taken as likely determinants of tax compliance (Alm *et al.*, 2012; Braithwaite and Wenzel, 2008; Hofmann *et al.*, 2008; Kirchler, 2007; Kirchler *et al.*, 2010; Meder *et al.*, 2012). There is little doubt that the understanding of the causes and determinant factors of the variations of tax evasion levels across time and space is a pressing need in the present context of economic crisis and scarcity of public resources. Such an understanding would help design robust institutional strategies and policies in order to tackle tax evasion and, thereby improve the efficacy and fairness of the tax system. Besides, reducing tax evasion allows increasing public resources without the need to raise tax rates. This is especially interesting when one looks at the difficulties that governments face today in order to achieve public budget equilibrium and fund welfare programs. The SIMULFIS project is conceived as the first stone of a research strategy to fill three gaps in contemporary research on tax evasion: (i) most of the studies on tax compliance deal separately with different factors that hypothetically affect tax behavior; (ii) standard economic models of tax evasion find it difficult to simulate complex social dynamics in a realistic way while keeping at the same time mathematical tractability; and (iii) most models are not properly calibrated and validated for a specific case, so they are not likely to become a useful tool in order to assess existing tax policies as well as their possible reforms, by providing virtual outcomes on their direct and indirect behavioral and economic effects. In the next section, we present in detail an ABM that allows creating virtual societies of agents, which have specific individual and relational characteristics and which take decisions about tax compliance. In the last section, we provide some results of experimental simulations by using the model (Llacer *et al.*, 2013; Noguera *et al.*, 2012, 2013).[1]

5.2 Model Description

This section includes a description of the SIMULFIS model following the Overview, Design, Details, and Decision (ODD) plus Decision-Making protocol proposed by Müller *et al.* (2013). The in-bracket codes matches with the ODD+D protocol sections.

---

[1] We refrain from describing the academic dispute (Feige, 2016; Schneider, 2016) on how useful the estimates for the size of the shadow economy are and what the caveats are, since these issues are addressed in Chapter 9.

5.2.1   Purpose

SIMULFIS aims to test the consistency and acceptability of different theoretical hypotheses proposed in the literature on tax compliance [I.i.a]. The purpose of the research line is to fill a number of gaps in previous contemporary research on tax evasion in order to increase the understanding of the causes for the variations of tax evasion levels across time and space. Contrary to most of the studies on tax compliance that test separately different factors that hypothetically affect tax behavior, our model seeks to study the integration and interaction of different mechanisms that have been tested in isolation by previous research (SIMULFIS theoretically oriented aim). The use of an ABM simulation allows to overcome one of the most important shortcomings of standard economic models on tax evasion: to build more realistic models with heterogeneous individuals and social interaction mechanisms properly modeled and to simulate complex social dynamics in a realistic way (SIMULFIS methodologically oriented aim). Once the model is properly calibrated and validated against the Spanish empirical case, it is likely to become a useful tool to assess existing tax policies as well as their possible reforms (SIMULFIS politically oriented aim). The following are the questions that drive the research and development of the ABM tax model:

- To what extent is rational choice theory enough to explain estimated levels of tax compliance?
- What is the effect of normative commitments on tax compliance?
- What is the effect of social influence on tax compliance?
- Which social scenario optimizes tax compliance?
- Which combination of tax policies (deterrence scenario) is able to reduce tax evasion?
- How is an agent-based model calibrated empirically for the study of tax compliance?

[I.ii.b] While SIMULFIS has been mainly developed as a tool for theory development, it also aims to allow quantitative predictions to be contrasted with different empirical cases as external validation strategy. SIMULFIS has been designed to be so open and flexible as to benefit both tax policies researchers and decision makers.

5.2.2   Entities, State Variables, and Scales

[I.ii.a] In SIMULFIS there are two kinds of interacting behavioral entities: humans and institutions. Humans are modeled in different subtypes, while the tax central authority is modeled as an individual institution. Each human agent represents an individual taxpayer with some economic income facing a decision-making process about how much of that income she will declare to be taxed in a compulsory tax system with progressive rate brackets, and some audit and fine procedures with redistributive aims. Humans are embedded in a social network with a small-world-like topology and a homophily structure corresponding to six subtypes as a result of their occupational status and income level.

The space dimension is abstract and meaningless, without spatial units, although the social network topology is relevant for defining the agents' information scope. There is no spacial environment but endowment, institutional, and social environment: the agent constrictions (the decision context) could be considered as composed by (i) a set of income original distribution (ii) a set of normative taxing rules – the tax rates by income brackets, the audit procedure, and the institutional fines; and (iii) a set of neighbors' behaviors.

[I.ii.b] The main attributes or state variables that characterize the tax agents in SIMULFIS are[2] identity code, gross income (Υ_i), income level category (high, medium, low), declared income (χ_i), class (wage earners, self-employers), homophily group (income level $\times$ class), fraud opportunity use rate (FOUR; see [II.ii.c] for details), behavioral type (unconditional evader, unconditional taxpayer, conditional taxpayer using the decision filters), perceived sanction risk, perceived personal tax balance, perceived neighborhood tax balance, amount of tax due, amount of tax paid, amount received from social benefits (Z_i), social influence coefficient (ω), support to progressivity, perception of system progressivity, normative satisfaction, period tax behavior (tax abider, tax evader), memory of previous fraud behavior, memory of previous audits, and memory of previous sanctions. Agents' neighborhood is defined as a list of linked agents. Other auxiliary attributes are used for intermediate computations and procedures. The TCA institutional agent (Tax Central Authority), acting as the tax system, has no attributes and no memory other than aggregated period by period records about gross tax revenue – as line chart. All economic attributes use currency as unit (Euro, in the case of Spain).

[I.ii.c] The exogenous parameters of the model include the following, with no change along the simulation run. In the up-to-date publications it is set for the Spanish case at year 2011, but SIMULFIS model can be set to other sociohistorical contexts with ease.

(1) Income (Υ_i) brackets and tax rates (τ_{Υ_i}), thus offering the possibility of empirical calibration for specific existing tax systems or even counterfactual tax policies. In case of Spain (2008) the corresponding tax rates for annual income brackets, in Euro, are as follows: 0% for less than 5050; 24% for 5051–17,360; 37% for 17,361–32,360; 43% for 32,361–120,000; 44% for 120,001–175,000, and 45% for more than 175,000 after the Personal Income Tax (IRPF) normative.

(2) Occupational status, determined by the percentage of wage earners in the population – the rest being self-employed. In case of Spain the corresponding ratio is 81.9% after Social Security affiliation (March 2012).

(3) Support for progressivity in the tax system, set as the percentage of agents that support the principle of progressivity in the tax system, so that it may be

---

[2] See the nomenclature of parameters and functions in Tables 5.1 and 5.2.

Table 5.1   Parameters: no change = Greek

Notation	Description
n	Total number of agents
α	Audition probability
θ	Amount of fine for tax evasion
Υ_i	Total (gross) income of agent i
χ_i	Income declared by agent i
τ_{Υ_i}	Tax rates (over real income)
τ_{χ_i}	Tax rates (over declared income)
ρ	Number of previous tax periods
ω	Social influence coefficient
$\mathcal{U}\mathcal{E}_i(\chi_i)$	Expected utility of declaring χ_i for agent i
M	Minimum income threshold for receiving the social benefit
Z_i^χ	Social benefit received by agent i when $\chi_i(1 - \tau_{X_i}) < M$
Z_i^Υ	Social benefit received by agent i when $\Upsilon_i(1 - \tau_{\Upsilon_i}) < M$

Table 5.2   Parameters: functions = roman

Notation	Description
$UE_i(\mathcal{X}_i)$	Expected utility
TB_i	Tax balance (personal and in the neighborhood)
PT_i	Perceived progressivity of the tax benefit system
SP_i	Support for the progressivity principle
PS_t	Perception that the system is progressive
P_i	Perceived probability for agent i of being caught if she evades
Z_i	Social benefit received by agent i
A_i^I	Audits to agent i in previous periods
A_i^N	Audits in agent's neighborhood (including i) in the preceding period
N_i	Agent's number of neighbors
F_i^t	Fraud opportunity use rate (FOUR) for agent i in period t
F_N^*	Median of FOUR in agents' neighborhood in the preceding periods

empirically calibrated with data from attitudinal survey studies. In case of Spain, SIMULFIS use a ratio of 80.0% after a 2010 study (Noguera *et al.*, 2011).

(4) Income threshold for receiving social benefits (μ), which determines eligibility for a means-tested public cash benefit, so that if an agent's net income after tax is below the threshold, it is topped up to reach that level. The marginal withdrawal rate of the benefit is 100%, so that each eligible agent only receives the difference between his declared income and the minimum income determined by the threshold. In case of Spain, this parameter is set to 8700 Euro, 50% of the median income (poverty threshold), after the Spanish Household Budget Survey, 2008 (INE, 2016).

(5) Audit rate, modeled as the probability of any agent being randomly audited. In our Spanish SIMULFIS experiments the rate has been set at different arbitrary values of 0.03, after data from the Spanish Tax Agency for 2008 (ATE, 2008), and 0.25, 0.50, and 0.75. SIMULFIS is aimed, at the present stage, at simulating taxpayer behavior and not tax inspector or tax administration behavior. Since the cost of audits is to be taken into account by the tax administration, it is not included as a parameter of the model. The different values of audit probability are tested hypothetically in order to explore how they affect taxpayers' behavior.

As in the traditional model of Allinghman and Sandmo, audits are random, that is, there is a random selection of the audited taxpayers. However, alternative auditing strategies (Bloomquist, 2012; Hashimzade, 2015, 2016) should be considered in future developments of SIMULFIS. Particularly, we could include audits that target specific groups of taxpayers: for instance, the most likely or largest tax evaders. Such strategy would improve the realism of the model since real-world tax agencies mostly perform risk-based preselected audits, which have proved to increase tax compliance and tax yield, and decrease audit costs.

(6) Amount of fine (θ), modeled as a percentage over the evaded taxes charged if an audit succeeded in finding an agent with tax-evading behavior. In the present version, there is no chance for tax evaders to be audited but not sanctioned: when tax evaders are randomly audited, then they pay the fine. In our Spanish SIMULFIS experiments the amount has been set at different rates from 1.5 (after the Spanish 2008 regulation (BOE, 2003)), 3.0, 4.5, and 6.0. In SIMULFIS, taxpayers can end a round with negative income, but the distribution of income starts again from the initial one in each round. However, agents have memory of their past audits. The model aims to test in this sense, how taxpayers adapt and change behavior as their past audit record evolves.

The combination of audit rates and fines charged will result in a two-dimensional ordered scale of deterrence taxing scenarios.

(7) Behavioral or decision filters, which can be activated and deactivated as initial parameters in any SIMULFIS simulation run. Since they are necessary for the agents to make a decision, by default the opportunities to evade (O) and rational choice (RC) are activated (as the base model or zero model for testing). On the contrary, the other two filters, the normative commitments (N) and the social influence (SI), can be activated or deactivated for experimental purposes. Further explanation on decision filters will be done at Section 5.2.5.

(8) Social influence coefficient (ω), as the strength of social influence from her close network neighbors over an agent, modeled as a numerical value from 0 (no social influence) to 1 (full social influence). In our Spanish SIMULFIS experiments three social influence scenarios are considered, with $\omega = 0.25$, $\omega = 0.50$, and $\omega = 0.75$.

[I.ii.d] Space is not included in SIMULFIS model. [I.ii.e] One time step represents one tax period and the simulations were typically run for 100 years, without changing tax system parameters. That was unrealistic, but allows us to explore if outcomes show a stable or convergent tendency.

5.2.3   Process Overview and Scheduling

[I.iii.a] The general model dynamics is as follows: when SIMULFIS is initialized, agents randomly receive a salary and a number of neighbors in the way described at the initialization section [III.ii]. Then they go through the decision algorithm and end up making a decision about how much of their income they declare, as a result of the activated behavioral filters. Then incomes are taxed and random audits and fines are executed. The benefits are paid to those who are entitled, and endogenous parameters are updated for the next time period.

Pseudo-code of SIMULFIS (v.11, February 2013):

- **User** sets the initial parameters – using interface sliders and monitors,
- Procedure Setup,
- **System** generate agents, set individual attributes, set network links between agents.
- For each tax period (sequential order, asynchronous update):
- Procedure Decision about FOUR:
- For each agent (random order, asynchronous update)
  - **Agent** run Opportunity filter (O). Update FOUR,
  - **Agent** run Normative filter (N). Update FOUR,
  - **Agent** run Rational Choice filter (RC). Update FOUR,
  - **Agent** run Social influence filter (SI). Update FOUR,
  - Next agent.
- **Central Authority** Collect Taxes,
- **Central Authority** Run random Audits and Execute Fines,
- **Central Authority** Pay Social Benefits,
- **System** Compute the Outcome Variables,
- **System** Update Plotting display and Print outcomes,
- If period = 100 stop running,
- Otherwise Next tax period.

There is no endogenous stop condition for the simulation. The experimental plan has been performed with a limitation of 100 tax periods, after checking that it is enough to assure that a stable state has been achieved by the system.

5.2.4   Theoretical and Empirical Background

[II.i.a] Specifically, the model includes the possibility of complex interactions between the three main types of mechanisms that have been considered by the literature as likely determinants of tax evasion decisions: fairness concerns, rational choice or utility maximization, and social influence. SIMULFIS design relies on the assumption that independently none of the single mechanisms mentioned allows for an explanation of the aggregate level pattern on tax evasion. By means of (i) the ABM agents' interaction and (ii) the quasi-experimental manipulation of the mechanisms triggered in the decision process, SIMULFIS aims to study the complexity of a socioeconomic system with tax evasion decisions.

[II.i.b] The agents' decision model is based on a set of assumptions supported by both (i) established theories about microeconomic decision models, as bounded rationality with cognitive and with social network informational constrictions, and (ii) observational mechanistic explanations, as social influence. In line with an analytical, sequential, and procedural decision-making approach, after Jon Elster's view of decision process (Elster, 1984, p. 76) (Elster, 1989, p. 13), agents decide how much tax they evade after going through four successive "filters" that affect their decision: (i) opportunity, (ii) normative commitments, (iii) rational choice, and (iv) social influence.

[II.i.c] This approach aims to capture recent developments in behavioral social science and cognitive decision theory, which disfavor the usual option of balancing all determinants of decision in a single individual utility function (Bicchieri, 2006; Elster, 2007; Gigerenzer *et al.*, 2011).

[II.i.d] Some submodels used in SIMULFIS are based on sets of empirical data. The reference value for some exogenous parameters uses household surveys, statistical census, and field or laboratory experiments. Specifically, the income tax rates and brackets and the income threshold for receiving social benefits come from the Spanish regulations, and the occupational status distribution data are taken from Social Security Affiliation official records, while support for progressivity in the tax system data are from a specific study (Noguera *et al.*, 2011).

Some laboratory experiments provide data about the ambivalent effect of social influence (ω) (Muchnik *et al.*, 2013), but they were considered inconclusive at the time SIMULFIS experiments were developed, so instead of using a reference value, we explore a discrete and reduced space of the parameter. There might be specific reasons and tests designed to choose optimal values for the social influence coefficient (Lorscheid *et al.*, 2016). However, since our aim here is just to give a description of the model, we do not intend to choose an optimal parameter setting but just to explore sufficiently different value combinations along the parametric space.

[II.i.e] The data mentioned were available at subpopulation aggregation level, so they were useful in calibrating SIMULFIS for the case of Spain.

5.2.5 Individual Decision Making

[II.ii.a] The subjects of decision making are the agents and the tax central authority. The tax central authority makes system-level top-down decisions about whom to audit and then about applying the corresponding fines if necessary and redistributing taxes by means of social benefits. Each taxable agent makes her individual-level decision about the amount of income economic units she declares as subject to taxation, that is, the percentage of her income she will conceal. The use of sequential filters approach for modeling the final decision making could be understood as an aggregation of multiple levels of decision, powered by different situational logics.

[II.ii.b] In SIMUFIS, the conditional agents are modeled so that they pursue *an explicit set of objectives*: to conceal as much amount of their income as possible given their occupational social context, and keeping a general normative fairness perception, while avoiding a sucker feeling given their close social network. Individual agents use a number of success criteria to generate a final decision along a sequence of four decisional filters. Each decision outcome from a filter is taken as an input for the next, so that different basic rationalities are behind each of the decision-making stages. The first filter (opportunity) is modeled after *a routine-based rationality*, without considering any objective, as a maximum fraud opportunity correlated with the level-class category. The second filter (normative) is modeled following a *satisficing* approach, *as bounded rationality* based on personal normative perception and comparison with the social environment. The next filter (rational choice) is modeled as a cognitive procedure where agents *maximize a utility function* to decide. The last filter (social influence) is modeled as *a reference group convergence function* where the agent adjusts and moderates her previous decision about what percentage of her income she will conceal.

[II.ii.c] In SIMULFIS, agents decide what *portion of their income* they will conceal, beyond binary or ternary choice, which is typical in previous models. As an example, if two taxpayers – say, A and B – both have a gross income of 100 units and they both decide to hide 10 units, but taxpayer A had the objective chance to hide 50 while taxpayer B could only hide 20, although in absolute terms they comply the same, in relative terms taxpayer B is making use of 50% of his opportunity to evade tax while taxpayer A only makes use of 20%. Agents make their decisions about tax compliance as produced by a specific sequence or combination of different mechanisms or "filters," most of which may be activated or deactivated in order to run controlled experiments. The decision process sequence includes the opportunity filter (O), the normative filter (N), the rational choice filter (RC), and finally the social influence filter (SI). If a filter is triggered then the agent modifies the concealed part of her income that results from the previous filter, so that the decision process is modeled as a complex interaction between this kind of cognitive mechanisms and the perceived socioeconomic environment. It is worth noting that filters have different logics or rules to generate the corresponding outcome: the O filter is structural deterministic, the N filter is relative and satisfaction based, the RC filter is utilitarian and optimization based, while the SI filter is ruled by convergence to the social environment. See detailed information about each filter in Section 5.2.17.

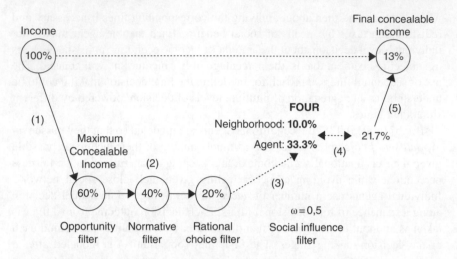

Figure 5.1   An example of the decision algorithm

As an example of the FOUR decision process, Figure 5.1 displays a numerical case of how agents' decision algorithm operates. The encircled numbers express percentages of a hypothetical agent's concealed income, and how they change when the agent goes through the consecutive behavioral filters; as explained above, the opportunity filter results in a maximum percentage of concealable income, while the normative, the rational choice, and the social influence filters lead the agent to a decision as to what percentage of her income she will conceal. The bold numbers at the middle level express the equivalents to these two latter percentages in terms of FOUR once the agent has gone through the rational choice filter, the resulting percentage is transformed in terms of FOUR (doted ascending arrow) this FOUR is compared with the agent's neighborhood's FOUR; then, the application of the social influence formula results in a different FOUR (bidirectional horizontal arrow), which is again transformed into a percentage of income to be concealed by the agent (doted ascending arrow).

Specifically, the example works as follows:

(1) Let us assume that in a given period of a simulation, agent i after the O filter may conceal 60% of her income.

(2) Let us assume that she is satisfied only with respect to her tax balance but not with respect to the progressivity of the system; then, she reduces by one third her opportunity to conceal so that the maximum she can conceal after applying the N filter is 40% of her income.

(3) Let us assume that after applying the RC filter, this percentage is further reduced to 20% of her income. Therefore, her FOUR ($\mathcal{F}_i^t$) will be 33.3%, resulting from 20/60.

(4) Let us also assume that in the previous period her neighborhood's median FOUR (F_N^*) was 10%. If the social influence coefficient (ω) is set to 0.5, the agent's FOUR after the SI filter will be: $F_i^t + \omega(F_N^* - F_i^t)) = 33.3 + 0.5 \ (10-33.3) = 21.7\%$.

(5) The resulting FOUR of 21.7% is equivalent to concealing 13% of the agent's income, since she is making use of 21.7% of her initial opportunities, which were 60%, so: $21.7\% \times 60\% = 13\%$.

In this example, the SI filter affects the agent's compliance by reducing the percentage of income she decides to conceal from 20% to 13%.

[II.ii.d] The agents adapt their behavior to changing endogenous and exogenous state variables. For instance, the perceived audit probability (P_i) of agents at each period of the simulation depend on agents' decisions in the previous periods and on information gathered from their close social network. Other parameters that co-evolve as a function of previous decisions of the agent and other agents are eligibility for benefits, tax balance (personal and in the neighborhood), perceived progressivity of the tax benefit system, or neighborhood's compliance rate.

[II.ii.e] Social norms or cultural values play a role in the decision-making process, because agents' normative commitments and fairness concerns toward taxation are explicitly modeled. The decision model takes into account both factual and normative beliefs, and also relative deprivation feelings (Manzo, 2009, 2011). The agents' support for the progressivity principle in the tax system models their normative beliefs on progressivity, and is used in behavioral filter N to moderate agents' decision about the maximum amount of concealable income. In that way, sociocultural differences could be introduced in specific SIMULFIS settings.

Spatial aspects do not play any role in the decision process [II.ii.f], because the space in SIMULFIS is modeled as an abstraction. On the contrary, temporal aspects play a relevant role in the decision about the concealed income [II.ii.g]. Specifically, in the expected utility function $(U\mathcal{E}_i(\chi_i))$ that rules the rational choice filter, (i) individuals use a discounting function for taking into account the risk of being audited and sanctioned – the cube root included in the aforementioned function – and also (ii) individuals keep a temporal memory to update their perceived probability of being sanctioned if they evade (P_i), as a function of their own audit record in all previous periods and the audit rate in their neighborhood in the immediately previous period. In that way, uncertainty is included in the agents' decision rules [II.ii.h] by means of their explicit consideration of uncertain situations or risk while applying the rational choice decision filter.

5.2.6 Learning

[II.iii.a] In the strong sense of the concept, individual learning is not included in the decision process modeled in SIMULFIS, because individuals never change their decision rules over time as consequence of their experience. In a weaker sense, agents can change aspirational levels depending on a combination of past experiences and present perceptions – in normative and social influence filters – and also can change reference values for risk aversion – in rational choice filter – so that their behavior evolves in time as they update the memory records. No collective learning is implemented in the model [II.iii.b], beyond the convergence decisional mechanism implied in the social influence filter.

5.2.7 Individual Sensing

[II.iv.a] Agents are assumed to sense and consider in their decisions both endogenous and exogenous state variables. These perception processes are not subject to noise or errors, but are constrained by the structural position of agents in their close social network.

[II.iv.b] In the opportunity filter, agents are aware of their own occupation status and income level. In the normative filter, agents perceive (i) their own normative beliefs on progressivity (exogenous); (ii) the income after taxes and benefits for all the population – to compute if the income ratio between the richest 10% and the poorest 10% is reduced by more than 30% after each tax period; (iii) their own personal previous tax burden ($\chi_i t_{x-1}$) and their own previous received benefits ($Z_i t_{x-1}$) – to compute whether they are net contributors or net recipients; and (iv) their neighbors' personal previous tax burden ($\chi_v t_{x-1}$) and previous received benefits ($Z_v t_{x-1}$) – to compute whether the majority in the neighborhood are net contributors or net recipients.

In the rational choice filter, agents perceive (i) their own audits in all previous periods (A_i^I), the audits in their neighborhood in the preceding period (A_i^N), the number of neighbors (ρ) – to update their own audit risk (P_i); (ii) their own gross income; and (iii) some relevant system parameters, such as the applicable tax rates, the fines for evading taxes, and the expected social benefit.

In the social influence filter, agents are aware of (i) the exogenous system parameter of social strength (ω), (ii) their own FOUR after the RC filter, and (iii) the median FOUR in their neighborhood (F_N^*) – to compute their final FOUR as a convergence to the group median pondered by (ω).

There is diversity in the spatial scale of sensing in SIMULFIS [II.iv.c]. Although some global or aggregate variables are perceived from the whole model space, most of the information used by individuals is available from their local network. On the contrary, the central tax authority is assumed to perceive any internal attribute for individuals subject to audit processes. No mechanism by which agents obtain information is modeled explicitly [II.iv.d], and no costs for cognition and gathering information are included [II.iv.e].

5.2.8 Individual Prediction

[II.v.a] Individual agents use a number of sources to predict future conditions and to make inference about tendencies. Extrapolation from experience is explicitly modeled by implementing a memory of past events – audits – that have affected both the agents and their neighborhood. While the space is an abstraction in SIMULFIS, these predictions are in fact based on close network observations.

[II.v.b] Rational choice filter (RC) is the only procedure that models how agents are assumed to estimate the consequences of their decisions: By using a temporal discount modeling function, individuals compute their expected utility under uncertainty about the probability of being audited, and make a decision about how much to conceal by optimizing this function.

The strictly local nature of memory data in SIMULFIS model includes the possibility for agents to be erroneous in the prediction process [II.v.c].

5.2.9   Interaction

In SIMULFIS, interactions among agents and entities are assumed as direct [II.vi.a] and depending on social distance, or network proximity, configured by the type of agents – occupational status and income level [II.vi.b]. No explicit representation of the communication involved in social interactions is modeled [II.vi.c]. Since agents take into account the behavior of their direct ties, there is a direct network effect on agents' decisions [II.vi.d]. Agents are randomly linked with a number of neighbors under some constraints: each agent has a minimum of neighbors, and a subset of each agent's neighbors are similar to him or her in terms of occupational status and income level. The generated social network type is similar to the so called "small world," and does not change over time.

5.2.10   Collectives

[II.vii.a] In SIMULFIS, individuals form aggregations that affect, and are affected by, other individuals. These aggregations are imposed by the modeler: A social structure in terms of local constrictions to interaction are imposed at each initialization, without any change or relinking during the simulation run.

[II.vii.b] These collectives are represented by generating link-agents between pairs of individuals, not as a separate kind of entity with its own state variables and traits but just a set of related agents.

To generate the homophily network, agents are randomly linked with a number of neighbors under some constraints: each agent has a minimum of neighbors, and a percentage of each agent's neighbors that are similar to him or her in terms of occupational status and income level. Both parameters – minimum and percentage – can be adjusted in the graphic user interface. Agents are also randomly assigned an occupational status (wage earners or self-employed) and an income level (high, the top decile of the income distribution; low, the three lowest deciles; and middle, the six deciles left in between). Both classes are calibrated using the Spanish case (Llacer et al., 2013). Each agent is assigned one of six subtypes of occupational status crossed by income level following Table 5.3.

Table 5.3   Agent subtypes (occupational status by income level) and homofily network building criteria

Agent type	Similar agents
HW (High income, wage earner)	HW, HS, MW
HS (High income, self-employer)	HS, HW, MS
MW (Medium income, wage earner)	HW, MW, MS, LW
MS (Medium income, self-employer)	HS, MW, MS, LS
LW (Low income, wage earner)	MW, LW, LS
LS (Low income, self-employer)	MS, LW, LS

5.2.11   Heterogeneity

[II.viii.a] Agents are heterogeneous in terms of market position. Different degrees of opportunity to conceal part of their income are assigned to different categories of agents, as a function of their income level and occupational status. This is theoretically justified because the FOUR is a much better indicator of the intensity of agents' tax fraud efforts than the amount of money evaded or the percentage of their income they conceal.

In SIMULFIS there is no heterogeneity in the decision making for conditional agents [II.viii.b]. Once a simulation is set in the initializing procedure – a subset of decisional filters are activated by the user – all individuals with conditional strategy will follow the selected decisional stages. The possibility of including subpopulations of unconditional agents is open as an option in the initial setting: these agents do not decide using the activated filters, but behave as unconditional tax evaders or unconditional tax compliants.

5.2.12   Stochasticity

[II.ix.a] In the initialization process, each individual agent is randomly assigned an occupational class (wage earners or self-employers) – after the ratio established in the initial settings for the Spain case. A random uniform algorithm assigns an income decile to each agent. A mean income by deciles by occupational class is then assigned to agents – using a decile range table for the Spain case – so that the income distribution follows a uniform distribution for each range of income decile (Llacer *et al.*, 2013, Table 1). The final income distribution is modified so that 10% of the highest deciles agents get a top level income – in the range (109,116–2,000,000) for wage earners, and (144,000–2,000,000) for self-employers – with a uniform distribution.

In the initialization process, a random uniform distribution in the range (0.0–1.0) is used to initialize the individual's perceived risk of audits. Agents are also assigned some binary attribute values with a probability of 0.5: perception of the progressivity of the system, personal tax balance, neighborhood tax balance, and behavior in a previous period.

In the auditing process, the central tax authority agent samples the population using a simple random sampling without replacement with the probability set at initialization (α).

5.2.13   Observation

[II.x.a] Since our main focus is behavioral, SIMULFIS outcome data go beyond traditional indicators for compliance – such as the amount of their personal income agents conceal – toward determining how much relative advantage agents take of their opportunities to conceal income, or (FOUR). The outcome indicators collected from the ABM are as follows:

(1) The mean FOUR of all agents – typically 1000 – computed at each tax period of a simulation

(2) The mean FOUR of all agents in all periods of a single simulation run – typically computed after 100 periods.

When agents' FOUR is converted into the equivalent amounts of evaded tax (in Euro), we may have also outcomes such as the following:

(1) Mean percentage of gross income concealed by each agent in all periods of a simulation

(2) Mean percentage of gross income concealed by all agents in each period of a simulation

(3) For each agent, the absolute amount of tax evaded, which is the result of $\Upsilon_i \tau_{\Upsilon_i} - \chi_i \tau_{\chi_i}$. Note that progressive tax rates applied to different income brackets are not necessarily the same for total gross income as for declared income, so this calculation is necessary.

From these individual outcomes it is possible to compute aggregated outcomes for the system, such as (i) aggregated concealed income as a percentage of total income in the system, (ii) aggregated fiscal pressure as a percentage of total income in the system, (iii) aggregated absolute amount of tax evaded, and (4) aggregated tax evaded as a percentage of total tax due, or "aggregated tax gap," which is the result of

$$\frac{\sum_{k=1}^{n} \left[\left(\frac{\Upsilon_i \tau_{\Upsilon_i} - \chi_i \tau_{\chi_i}}{\Upsilon_i \tau_{\Upsilon_i}} \right) 100 \right]}{n} \tag{5.1}$$

Finally, all these results may be cross-tabulated by different initial settings, categories of agents, values of a parameter, and so on, thus allowing SIMULFIS users to run controlled virtual experiments on tax compliance behavior.

The aforementioned aggregated results are an outcome from decisions at the individual level [II.x.b]. The behavioral filters make each decision locally dependent on the agent position, her previous events history, and her neighborhood behavior, and while a random initial state is generated by each simulation run, the system aggregated results could be understood as emergent properties of the ABM.

5.2.14   Implementation Details

[III.i.a] The model was originally implemented using the Netlogo modeling environment version 4.1.3. Several versions and enhancements of SIMULFIS were developed from October 2011 (v0.1) until November 2013 (v12.0.1). Most of the simulations executed for the published experimental outcomes (Llacer et al., 2013; Noguera et al., 2012, 2013) were run in the stable version v11, using Netlogo v5.0.4, on a quad-core PC with Windows 7 OS. A typical simulation runtime – with 1000 agents, 100 tax periods, and all filters activated – takes around 10 minutes (626.57 seconds).

Any simulation is run in 10 different "worlds" – different random initial assignations of neighbors and income – and the outcome means are used to contrast the corresponding hypothesis.

In the present state SIMULFIS model is not public [III.i.b], but authors can distribute versions of the code to interested researchers on demand.

5.2.15   Initialization

SIMULFIS can be easily set for a diversity of initial states [III.ii.a] of the virtual society at $t = 0$, providing a tool for experimenting in different socioeconomic contexts and tax systems. In the Spain case, the following initial parameters were used: 1000 tax agents, 81.9% of wage earners in the population, income is distributed following a data-based decile range table (Llacer et al., 2013, Table 1), each agent has a minimum of 10 neighbors with 80% neighbors of the same category, all individual agents use conditional strategies, the estimate risk of audition is randomly distributed between 0 and 1, and the tax rates are set empirically after the rate brackets for the case.

Some initial parameters vary among simulations [III.ii.b], so that SIMULFIS allows to assign different values to initial parameters (e.g., the weight to social influence in agents' decision algorithm). This variability is useful in the design of systematic controlled experiments, where the space of values of a parameter – or combination of state variables – is explored in discrete steps with the corresponding simulations.

Our published simulation outcomes (Llacer et al., 2013; Noguera et al., 2012, 2013) explore, for the case of Spain, a set of scenarios that involve the following initial settings:

- Activation of normative filter,
- Activation of social influence filter,[3]
- Audit probability rates of 15%, 30%, 45%, and 60%,
- Fines of 1.5, 3, 4.5, and 6 as multipliers of the amount of evaded tax,[4]
- Social influence weight of 0.25, 0.50, 0.75.

The initial values are in part chosen arbitrarily and in part based on data [III.ii.c]. A number of parameters are established in terms of the experimentation plan with clear hypothesis to be tested.

Income distribution: Each run agents are assigned an amount of annual income following an exponential-shape distribution. In the Spanish case setting, SIMULFIS is calibrated empirically for an income decile distribution of the Spanish. The top of the distribution has also been adjusted for including a small percentage of big fortunes. After the distribution has been completed, the model assigns each agent to one of three income levels: "high" (the top decile of the distribution), "low" (the three lowest deciles of the distribution), and "middle" (the six deciles left in between). This distribution also may be differentiated for self-employed workers and wage earners in order to calibrate the model in a more realistic way, after data from the Spanish Household Budget Survey (INE, 2016).

Social network: Each run agents are randomly linked with a number of neighbors under some constraints: each agent has a minimum of neighbors (10 by default), and a percentage of each agent's neighbors are similar to him or her in terms of occupational status and income level (80% by default).

5.2.16   Input Data

[III.iii.a] SIMULFIS does not use any input from external sources to represent processes nor exogenous variables that change over time.

5.2.17   Submodels

[III.iv.a] The submodels that represent the processes listed in "Process overview and scheduling" are the following:

Opportunity filter (O): The outcome is a function of the agent income level and occupational status. The maximum concealable income as a result of the O filter will be used as the denominator for computing the agents' FOUR ($\mathcal{F}_i^t$).

Normative filter (N): The outcome is a reduction of the agents' concealable income resulting from their previous opportunity filter, using a three-way conditional

---

[3] Recall that the opportunity filter O is always activated and agents always decide using the rational choice filter.

[4] Audits and fines combined in 16 different levels of deterrence scenarios.

function: (i) If an agent is satisfied in both "progressivity" ($\mathcal{PT}_i$) AND "tax balance" ($\mathcal{TB}_i$) she reduces by two thirds (66%) the proportion of income she may conceal; (ii) if an agent is satisfied only in "progressivity" ($\mathcal{PT}_i$) OR "tax balance" ($\mathcal{TB}_i$), the reduction is by one third (33%); and (iii) if an agent is NOT satisfied in any sense, she may conceal the full percentage of income resulting from the opportunity filter.

Rational choice filter (RC): Over the percentage of their gross income resulting from previous filters O and N, agents rationally maximize their net income by estimating the expected utility of concealing under uncertainty conditions ($\mathcal{UE}_i(\chi_i)$). An agent's decision to declare income (χ_i) is a function of tax rates (τ), fines (θ), her real income level (Υ_i), and the probability of being caught if she evades ($\mathcal{P}_i$). Homogeneous risk aversion is assumed by using cube roots in both addends of the equation, and the expected social benefit, which agents compute by determining their eligibility in the immediately preceding period of the simulation, is also included. The usual square root is avoided to prevent applying it to eventual negative numbers in the second addend – which could be the case, for example, if agents declare a low percentage of their income and fines are very hard.[5, 6]

$$\mathcal{UE}_i(\chi_i) = (1 - \mathcal{P}_i)\sqrt[3]{(\Upsilon_i - \chi_i\tau_{\chi_i} + \mathcal{Z}_i^\chi)}$$

$$+ \mathcal{P}_i\sqrt[3]{(\Upsilon_i - \Upsilon_i\tau_{\Upsilon_i} - \theta(\Upsilon_i\tau_{\Upsilon_i} - \chi_i\tau_{\chi_i}) + \mathcal{Z}_i^\Upsilon)} \qquad (5.2)$$

where $\mathcal{Z}_i^\chi = M - \chi_i(1 - \tau_{\chi_i})$; $\mathcal{Z}_i^\Upsilon = M - \Upsilon_i(1 - \tau_{\Upsilon_i})$; $\mathcal{Z}_i^\chi$, $\mathcal{Z}_i^\Upsilon \geq 0$

Note that when $\chi_i(1 - \tau_{\chi_i}) < M$ but $\Upsilon_i(1 - \tau_{\Upsilon_i}) \geq M$, that is, when the agent is not legally eligible for benefits but she evades enough to be so, then $\mathcal{Z}_i^\Upsilon = 0$. Similarly, when $\chi_i(1 - \tau_{\chi_i}) \geq M$, that is, when the agent is not eligible whether she evades or not, then $\mathcal{Z}_i^\chi = 0$ and $\mathcal{Z}_i^\Upsilon = 0$.

Social influence filter (SI): After applying the previous RC filter, agents' FOUR ($\mathcal{F}_i^t$) converge to the median in their neighborhood ($\mathcal{F}_N^*$), according to $\mathcal{F}_i^t + \omega(\mathcal{F}_N^* - \mathcal{F}_i^t)$. The result of this calculation is the agents' final FOUR, which is expressible in terms of the percentage of the agents' income that is declared or concealed, and in terms of absolute amount of income concealed.

The model parameters, dimensions, and reference values [III.iv.b] are different in a diversity of experimental conditions simulated by running SIMULFIS.

The reference values for opportunities to conceal (O filter) – in percentage of agents' gross income are 80% for high income level self-employed, 60% for the

---

[5] SIMULFIS does not include at this stage the possibility of taxpayers' errors or heuristic behavior. This is something to be taken into account in future developments of the model (see, e.g., Hokamp and Pickhardt (2016)).

[6] The nonconcavity of these utility function is not a problem – although no global maximum is found numerically – because we assume here some sort of bounded rationality: an agent simply uses a rule of thumb by comparing expected utilities from concealing 100%, 90%, and so on, to 0% of income.

rest of self-employed, 30% for high income level wage earners, 10% or 20% for medium income level wage earners, and 10% or 20% for low income level wage earners.

Normative filter (N) has two dimensions of satisfaction: (i) "progressivity" of the tax system (PT_i), and (ii) personal "tax balance" (TB_i).

(1) Satisfaction with the progressivity of the tax system (PTi = PT_i) is a function of two parameters: If agents support progressivity (SPi = SP_i) = 1) and perceive that the system is progressive (PSt = PS_t) = 1), they are "satisfied"; otherwise, they are not.

(1a) The agents' support for the progressivity principle in the tax system (SP) is an exogenous parameter, available through survey research. Agents are randomly assigned normative beliefs about progressivity (SP_i) according to the aggregated percentage (support = 1, do not support = 0).

(1b) The agents' perception of "real" progressivity in the tax system is updated after each tax period. The system is perceived as progressive if the mean income ratio – after taxes and benefits – gap between the richest 10% subpopulation and the poorest 10% subpopulation is reduced by more than 30%. Perfect information is assumed for agents' estimations.

(2) Satisfaction with the tax balance (TB_i) is a function of two parameters: If agents are net contributors while the majority of their neighbors are net recipients they are "unsatisfied"; otherwise, they are satisfied.

(2a) Agents' personal tax balance is computed by comparing their personal tax burden ($\chi_i t_x$) with the benefits they eventually receive (Z_i): agent i is a "net contributor" if $\chi_i t_x > Z_i$, or a "net recipient" if $\chi_i t_x \leq Z_i$.

(2b) To compute their neighborhood's tax balance, agents observe whether the majority (more than 50%) of agents in their neighborhood (including themselves) are net contributors or net recipients. Perfect information is assumed.

Rational choice filter (RC): Agents' perceived probability of being sanctioned if they evade tax by concealing income (P_i) is randomly distributed in the initial period of the simulation and afterwards is updated endogenously as a function of the agents' audit records in all previous periods and the audit rate in the agents' neighborhood in the immediately previous period, according to the following equation.

$$P_i = \left(\frac{A_i^I}{\rho} - \frac{A_i^N}{N_i} \right) \Big/ 2 \tag{5.3}$$

Social influence filter (SI): Agents' decision about declared income converge to that of their neighborhood, and SIMULFIS models the strength of social influence by an exogenous parameter ω equal for all agents, ranging (0, 1) from no social influence to full social influence. Note that, when $\omega = 1$ (full social influence), the individual effect of the RC filter is cancelled; and when $\omega = 0$ (no social influence), the agents keep their original FOUR resulting from the RC filter. The reference value for the social influence coefficient is set according to three social influence scenarios, with $\omega = 0.25$, $\omega = 0.5$, and $\omega = 0.75$.

[III.iv.c] Submodels are designed and parametrized with specific assumptions or justifications. In the opportunity filter (O), in order to determine values for different agents' opportunities, we adopt some simple – and arguably realistic – assumptions: (i) wage earners have fewer opportunities to conceal than self-employed workers, because their income is typically withheld at origin in a more substantial proportion. (ii) Agents with high income – the top decile of income distribution – have more opportunities to conceal than the rest, because they receive income from many different sources or they have the resources, techniques, and abilities needed to successfully conceal a higher proportion of it. (iii) For similar reasons, agents with middle income have more opportunities to conceal than agents with low income – although in other set of experiments low-income workers have higher probability to participate in the shadow economy (Noguera *et al.*, 2013). (iv) For all agents there is some percentage of their income that they cannot conceal, since the government always has some information on at least a minimum proportion of every agent's income.

Normative filter (N) models agents' normative attitudes and satisfaction feelings toward the tax system's design and performance in terms of its fairness. This filter combines two elements in order to determine the outcome level of satisfaction of each agent: (i) agents' satisfaction with the progressivity of the tax system, and (ii) agents' satisfaction with their tax balance when compared with that of their neighborhood. The agents' support for the progressivity principle in the tax system models their normative beliefs on progressivity, while agents' satisfaction with their tax balance is determined by comparing their personal tax balance with that of their neighborhood. This method tries to capture the well-known findings of the literature on relative deprivation, which show that people's feelings of satisfaction with their endowments depend more on the comparison with their reference group than on the amount enjoyed in absolute terms (Manzo, 2011).

Rational choice filter (RC) models the agents' cognitive process of maximizing expected utility according to an adaptation of the classical tax fraud expected utility function (Allingham and Sandmo, 1972). SIMULFIS uses a discrete computational approach to determine a sequence of 11 expected outcomes in terms of agents' net income, which results from reducing concealed income by intervals of 10% – from evading 100–0% – of agents' concealable income. Since the expected utility formula we are using introduces substantive complications in relation to Allingham and Sandmo – such as a progressive tax rate and a social benefit – this way of formalizing expected utility makes the computational working of the model easier, while ensuring the consistency of agents' decisions. Despite other filters and mechanisms affecting agents' compliance, rational choice has always some weight in the final decision – except under full social influence. We take this as a realistic assumption, since an economic decision such as tax compliance is almost always rationally considered by taxpayers, and, in most cases, assessed by experts or professionals.

Social influence filter (SI) models the extent to which agents' decision about declared income (in terms of FOUR) converge to that of their neighborhood as a result of any kind of social influence (Bruch and Mare, 2006; Centola and Macy, 2007; Durlauf, 2001, 2006; Rolfe, 2009; Salganik and Watts, 2009; Watts and Dodds, 2009). In SIMULFIS, the strength of social influence is determined by an exogenous parameter (ω) equal for all agents, ranging (0, 1) from no social influence to full social influence. Note that, when $\omega = 1$ (full social influence), the individual effect of the RC filter is cancelled; and when $\omega = 0$ (no social influence), the agent keeps his original FOUR resulting from the RC filter. The operation of social influence mechanisms affecting tax evasion behavior has been recently questioned by some scholars (Hedström and Ibarra, 2010) on the basis of the "privacy objection": since tax compliance is taken to be private and unobservable by peers, no social influence could take place. However, as survey studies repeatedly show (IEF, 2012; CIS, 2011), citizens usually have an approximate idea on the tax compliance level in their country, occupational category, or economic sector, its evolution in time, and its main causes. These ideas may be formed from information received through mass media, personal interaction, or indirect inference (e.g., shared social characteristics, when compared with economic lifestyles, may be proxies for inferences about neighbors' and peers' tax compliance). Additionally, in countries where there is low tax morale and high social tolerance toward tax evasion – such as Spain – it is usual to have access to public "street knowledge" about personal tax compliance, and to give and receive advice between neighbors and peers on how to evade. Finally, there is some literature modeling social influence mechanisms triggered by estimated or revealed information on criminal and dishonest behavior (Diekmann *et al.*, 2011; Gino at al., 2009; Groeber and Rauhut, 2010).

5.3 Some Experimental Results and Conclusions

The main conclusions to be drawn from the analysis of results from experimental simulations running SIMULFIS (Llacer *et al.*, 2013; Noguera *et al.*, 2012, 2013) could be summarized as follows:

(1) As suggested by theoretical literature on tax compliance, strict rational agents would produce much less compliance than is usually estimated, except with unrealistically high deterrence levels. The simulated results for the arguably most realistic conditions in terms of deterrence, which is 3% for audit probability and 1.5 for fine multiplier for the Spanish case, generates a behavioral outcome where – in the RC (rational choice) scenario – all agents end up with a FOUR of 100%. That means they are taking full advantage of their opportunities to evade – the higher the FOUR, the less the compliance.

It has been often demonstrated that there is a trade-off between severity and probability of punishment (Kahan, 1997, p. 377ss), since raising that probability – what economists of crime call the "certainty of conviction" – is often more costly than raising the intensity of punishment and even inefficient if the resources and efforts needed for effective surveillance are considerable, as is the case with tax supervision. With the aid of an agent-based model, it is possible to assess this trade-off: SIMULFIS results show that different combinations of audits and fines produce equivalent compliance levels under the same or different behavioral scenarios. For instance, under the RC scenario, a combination of a fine multiplier of 5 and an audit rate of 25% is very similar to another of a fine multiplier of 1.5 and an audit rate of 50%; both produce compliance levels that are close to the ones obtained under the F+RC+SC scenario with fine multipliers of 1.5 or 2.5 and an audit rate of 25%. Other results (Noguera *et al.*, 2013, Table 2) shows that higher audits and fines always improve compliance – always decrease mean FOUR – but increasing audits is proportionally more effective than raising fines – except when we pass from a fine multiplier of 2.5–5 of the evaded tax under an audit rate of 25%, where a "phase transition" seems to take place. Interestingly, under the most realistic audit rate of 3%, increasing fines does not have an effect on compliance in any behavioral scenario. Conversely, under realistic values for fines – multipliers of 1.5 and 2.5 of evaded income – increasing the audit rate has a more substantial effect under all behavioral scenarios. This may challenge the standard conception that increasing audits is inefficient because of its high cost: as (Kahan, 1997) notes, when individuals do not decide in an isolated context, high-certainty/low-severity strategies may be better than the reverse owing to the signal that individuals get that they will be most likely caught if they cheat. In our model, this is captured by the local way in which agents estimate the probability of being audited and punished.

In order to confirm this trend statistically, a linear regression analysis was performed taking as dependent variable the mean FOUR in the last iteration of each simulation (Noguera *et al.*, 2013, Table 3). The results show that audits have a strong effect in decreasing FOUR, which triplicates the effect of fines in all behavioral scenarios. Similarly, in their meta-analysis of laboratory experiments in this area, Alm and Jacobson (2007) find an elasticity of 0.1–0.2 for declared income/audits, but below 0.1 for declared income/fines. The regression models also included the intensity of social contagion as an independent variable in the two behavioral scenarios where it is activated, and the results show that it also has a negative effect on FOUR, which is higher than the effect of fines but lower than that of the audit rate. Interaction effects between social contagion and audits/fines were also tested separately and the results show that deterrence increases its negative effect on FOUR when the intensity of social contagion decreases – that is, social contagion acts as a "brake" of the negative effect of deterrence on FOUR.

These results strongly suggest that rational choice theory is not enough on its own to generate empirically estimated compliance levels through simulations and that other normative and social mechanisms are therefore necessary in any plausible model of tax compliance behavior.

(2) Contrary to what is assumed by other agent-based models of tax evasion, social influence does not always improve compliance. In particular, it has been shown that when deterrence is strong, RC (rational choice) and RC + N (rational choice plus normative commitments) fare better in terms of compliance. The reason for this ambivalent effect of social influence is that its presence, by making agents' decisions dependent on those of their peers, makes tax compliance level less sensitive to increased deterrence levels. So, contrary to what Korobow *et al.* (2007) seem to assume, social influence does not have the same directional effect on compliance independently of deterrence level. The *social influence conception of deterrence* developed by Kahan, (1997, p. 351) is close to our interpretation of this result. According to him, the effect of deterrence must be considered in relation with the power exerted by social influence: from a given point, the marginal gain in compliance achieved by higher levels of deterrence may be lower in a situation with social influence than in a situation with isolated rational agents – and the higher the intensity of social influence, the lower this relative gain will be. In a similar way, our model shows that social contagion attenuates the net effect of deterrence on raising compliance when deterrence is high and increases it when it is low. Consistently, in Noguera *et al.* (2013) we also showed that with higher tax rates the same transition starts from a lower deterrence level. This social influence effect, as well as its foundations at the micro level, would be difficult to observe and analyze without the aid of an agent-based model such as SIMULFIS.

(3) Similarly to most experimental studies (Alm and Jacobson, 2007; Franzoni, 2007), we find that audits are comparatively more effective than fines in order to improve tax compliance. A key factor to explain this may be the link between being audited and being fined. Further experiments performed with SIMULFIS may try to disentangle both facts in order to test whether this trend is confirmed, but the implication so far seems clear that policies to tackle tax evasion should rely more on improving the efficacy of audits, as well as their number and scope, than on raising penalties.

The policy implications of the results at this stage of model development seem clear: first, policies to tackle tax evasion should rely more on improving the efficacy of audits, as well as their number and scope, than on raising penalties. Second, smart use of public information on tax compliance levels may be a forceful weapon to induce taxpayers to comply more. In any case, we would like to emphasize the utility of agent-based models for understanding compliance patterns and, therefore, for assessing public decisions along the many trade-offs involved in tax policy. Agent-based models are flexible tools that offer many possibilities to improve social-scientific knowledge of tax behavior, a research field that is still in its infancy, but as promising as the present compilation shows.

Acknowledgments

This chapter is based on a research project funded by the Institute for Fiscal Studies of the Spanish Ministry of Economy, and has also benefited from financial support by the National Plan for R+D+I of the Spanish Ministry of Science and Innovation and the Spanish Ministry of Economy and Competitiveness through Grants CSO2012-31401 and CSD2010-00034 (CONSOLIDER-INGENIO). Toni Llacer has enjoyed financial support by the CUR-DIUE of the Generalitat de Catalunya and the European Social Fund.

References

Agencia Tributaria Española (Spanish Tax Bureau) (2008) *Memoria 2008 Agencia Tributaria (2008 Tax Report)*, Agencia Tributaria Española, Madrid.

Allingham, M.G. and Sandmo, A. (1972) Income tax evasion: a theoretical analysis. *Journal of Public Economics*, **1**, 323–338.

Alm, J. and Jacobson, S. (2007) Using laboratory experimentsin public economics. *National Tax Journal*, **60**, 129–152.

Alm, J., Kirchler, E., and Muehlbacher, S. (2012) Combining psychology and economics in the analysis of compliance: from enforcement to cooperation. *Economic Analysis and Policy (EAP)*, **42**, 133–152.

Alvarez, S. and Herrera, P.M. (2004) Etica Fiscal. Technical Report 10/04, Instituto de Estudios Fiscales, Madrid, http://www.ief.es/contadorDocumentos.aspx?URLDocumento=/documentos/recursos/publicaciones/documentos_trabajo/2004_10.pdf (Last accessed 09 November 2016).

Arrazola, M., de Hevia, J., Mauleon, I., and Sanchez, R. (2011) Estimación del volumen de economía sumergida en españa. *Cuadernos de Información económica*, **220**, 81–88.

Bicchieri, C. (2006) *The Grammar of Society: The Nature and Dynamics of Social Norms*, Cambridge University Press, New York.

Bloomquist, K.M. (2012) Agent-based simulation of tax reporting compliance. Doctoral dissertation in Computational Social Science. George Mason University, Fairfax, VA, http://digilib.gmu.edu/dspace/bitstream/1920/7927/1/Bloomquist_dissertation_2012.pdf (Last accessed 09 November 2016).

Boletin Oficial de Estado (Spanish Official State Gazette) (2003) Ley 58/2003, de 17 de diciembre, general tributaria. BOE 302, artículo 191.

Braithwaite, V. and Wenzel, M. (2008) Integrating explanations of tax evasion and avoidance, in *The Cambridge Handbook of Psychology and Economic Behaviour*, Cambridge Handbooks in Psychology (ed. A. Lewis), Cambridge University Press, pp. 304–331.

Bruch, E.E. and Mare, R.D. (2006) Neighborhood choice and neighborhood change. *American Journal of Sociology*, **112**, 667–709.

Centola, D. and Macy, M. (2007) Complex contagion and the weakness of long ties. *American Journal of Sociology*, **113**, 702–734.

CIS, Centro de Investigaciones Sociologicas (Centre for Sociological Research) (2011) Encuesta sobre opinión pública y política Fiscal (xxviii). Technical report, CIS (Centro de Investigaciones Sociológicas), Madrid.

Cummings, R.G., Martinez-Vazquez, J., McKee, M., and Torgler, B. (2009) Tax morale affects tax compliance: evidence from surveys and an artefactual field experiment. *Journal of Economic Behavior and Organization*, **70**, 447–457.

Diekmann, A., Przepiorka, W., and Rauhut, H. (2011) Lifting the veil of ignorance: an experiment on the contagiousness of norm violations. Discussion Papers 2011004, University of Oxford, Nuffield College, http://ideas.repec.org/p/cex/dpaper/2011004.html (Last accessed 09 November 2016).

Durlauf, S.N. (2001) A framework for the study of individual behavior and social interactions. *Sociological Methodology*, **31**, 47–87.

Durlauf, S.N. (2006) Groups, social influences, and inequality: a memberships theory perspective on poverty traps, in *Poverty Traps* (eds S. Bowles, S.N. Durlauf, and K.R. Hoff), Russell Sage Foundation, Princeton University Press, New York, Princeton, NJ, pp. 141–175.

Elffers, H., Weigel, R.H., and Hessing, D.J. (1987) The consequences of different strategies for measuring tax evasion behavior. *Journal of Economic Psychology*, **8**, 311–337.

Elster, J. (1984/1984) *Ulysses and the Sirens: Studies in Rationality and Irrationality*, Cambridge University Press, Cambridge.

Elster, J. (1989) *Nuts and Bolts for the Social Sciences*, Cambridge University Press, Cambridge.

Elster, J. (2007) *Explaining Social Behaviour: More Nuts and Bolts for the Social Sciences*, Cambridge University Press, Cambridge.

Feige, E.L. (2016) Reflections on the meaning and measurement of Unobserved Economies: what do we really know about the "Shadow Economy"? *Journal of Tax Administration*, **2** (1), 5–41.

Franzoni, L. (2007) A tax evasion and tax compliance, in *Encyclopaedia of Law and Economics*, Edward Elgar, Cheltenham.

GESTHA, Sindicato de Tecnicos del Ministerio de Hacienda (Spanish Technicians Syndicate at the Finance Ministry) (2011) III informe de la lucha contra el fraude fiscal en la agencia tributaria. Technical report, GESTHA (Sindicato de Técnicos del Ministerio de Hacienda), Madrid.

Gigerenzer, G., Hertwig, R., and Pachur, T. (2011) *Heuristics: The Foundations of Adaptive Behavior*, Oxford University Press, Oxford, New York.

Gino, F., Ayal, S., and Ariely, D. (2009) Contagion and differentiation in unethical behavior: the effect of one bad apple on the barrel. *Psychological Science*, **20**, 393–398.

Groeber, P. and Rauhut, H. (2010) Does ignorance promote norm compliance? *Computational and Mathematical Organization Theory*, **16**, 1–28.

Halla, M. (2010) Tax morale and compliance behavior: first evidence on a causal link. IZA Discussion Papers 4918, Institute for the Study of Labor (IZA), http://ideas.repec.org/p/iza/izadps/dp4918.html (Last accessed 09 November 2016).

Hashimzade, N., Myles, G.D., Page, F., and Rablen, M.D. (2015) The use of agent-based modelling to investigate tax compliance. *Economics of Governance*, **16** (2), 143–164.

Hashimzade, N., Myles, G.D., and Rablen, M.D. (2016) Predictive analytics and the targeting of audits. *Journal of Economic Behavior & Organization*, **124**, 130–145.

Hedström, P. and Ibarra, R. (2010) On the contagiousness of non-contagious behavior: the case of tax avoidance and tax evasion, in *The Benefit of Broad Horizons: Intellectual and Institutional Preconditions for a Global Social Science* (eds H. Joas, B. Wittrock, and B.S. Klein), Brill Publishers, Leiden [etc.].

Hofmann, E., Hoelzl, E., and Kirchler, E. (2008) Preconditions of voluntary tax compliance: knowledge and evaluation of taxation, norms, fairness, and motivation to cooperate. *Journal of Psychology*, **216**, 209–217.

Hokamp, S. and Pickhardt, M. (2016) Income tax evasion in a society of heterogeneous agents – evidence from an agent-based model. *International Economic Journal*, **24** (4), 541–553.

IEF, Instituto de Estudios Fiscales (Spanish Institute of Fiscal Studies) (2012) Opiniones y actitudes fiscales de los españoles en 2011: Documento de trabajo del IEF n.19/2012. Technical report, IEF (Instituto de Estudios Fiscales), Madrid.

INE, Instituto Nacional de Estadistica (Spanish National Bureau of Statistics) (2016) *Household Budget Survey. Base 2006 – Year 2015*, National Bureau of Statistics, Madrid, http://www.ine

.es/dyngs/INEbase/en/operacion.htm?c=Estadistica_C&cid=1254736176806&menu=ultiDatos& idp=1254735976608 (Last accessed 09 November 2016).

Kahan, D.M. (1997) Social influence, social meaning, and deterrence. *Virginia Law Review*, **83** (3), 349–395.

Kirchler, E. (2007) *The Economic Psychology of Tax Behaviour*, Cambridge University Press, New York.

Kirchler, E., Muehlbacher, S., Kastlunger, B., and Wahl, I. (2010) Why pay taxes? A review of tax compliance decisions, in *Developing Alternative Frameworks for Explaining Tax Compliance*, Routledge International Studies in Money and Banking, vol. **59** (eds J. Alm, J. Martinez-Vazquez, and B. Torgler), Routledge, London and New York.

Kirchler, E. and Wahl, I. (2010) Tax compliance inventory: TAX-I voluntary tax compliance, enforced tax compliance, tax avoidance, and tax evasion. *Journal of Economic Psychology*, **31**, 331–346.

Korobow, A., Johnson, C., and Axtell, R.L. (2007) An agent-based model of tax compliance with social networks. *National Tax Journal*, **60**, 589–610.

Llacer, T., Miguel, F.J., Noguera, J.A., and Tapia, E. (2013) An agent-based model of tax compliance: an application to the Spanish case. *Advances in Complex Systems*, **16** (04n05), 1350007.

Lorscheid, I., Heine, B., and Meyer, M. (2016) Opening the 'black box' of simulations: increased transparency and effective communication through the systematic design of experiments. *Comput Math Organ Theory*, **18**, 22–62.

Manzo, G. (2009) *La Spirale des inégalités: choix scolaires en France et en Italie au XXe siècle*, PUPS. Presses de l'Université Paris-Sorbonne, Paris.

Manzo, G. (2011) Relative deprivation in silico: agent-based models and causality in analytical sociology, in *Analytical Sociology and Social Mechanisms* (ed. P. Demeulenaere), Cambridge University Press, Cambridge and New York, pp. 266–308.

Meder, Z.Z., Simonovits, A., and Vinczeb, J. (2012) Tax morale and tax evasion: social preferences and bounded rationality. *Economic Analysis and Policy (EAP)*, **42**, 171–188.

Muchnik, L., Aral, S., and Taylor, S.J. (2013) Social influence bias: a randomized experiment. *Science*, **341** (6146), 647–651.

Müller, B., Bohn, F., Dressler, G., Groeneveld, J., Klassert, C., Martin, R., Schluter, M., Schulze, J., Weise, H., and Schwarz, N. (2013) Describing human decisions in agent-based models ODD + D, an extension of the ODD protocol. *Environmental Modelling and Software*, **48**, 37–48, http://www .sciencedirect.com/science/article/pii/S1364815213001394 (Last accessed 09 November 2016).

Noguera, J.A., Guijarro, X., Leon, F., Llacer, T., Miguel, F.J., Tapia, E., Tena, J., and Vinagre, M. (2011) Valors i actituds sobre justícia distributiva: prestacions socials i fiscalitat. Technical report, Centre d'estudis d'Opinió (CEO), Generalitat de Catalunya, Barcelona, http://ceo.gencat.cat/ceop/ AppJava/export/sites/CEOPortal/estudis/monografies/contingut/justiciadistributiva01.pdf (Last accessed 09 November 2016).

Noguera, J.A., Llacer, T., Miguel, F.J., and Tapia, E. (2012) Exploring tax compliance: an agent-based simulation, in *Shaping Reality through Simulation* (eds K.G. Troitzsch, M. Moehring, and U. Lotzmann), European Council for Modelling and Simulation, Koblenz.

Noguera, J.A., Miguel, F.J., Tapia, E., and Llacer, T. (2013) Tax compliance, rational choice, and social influence: an agent-based model. *Revue francaise de Sociologie*, **55** (4), 765–804.

Rolfe, M. (2009) Conditional choice, in *The Oxford Handbook of Analytical Sociology* (eds P. Hedstrom and P.S. Bearman), Oxford University Press, Oxford and New York, pp. 419–447.

Salganik, M.J. and Watts, D.J. (2009) Social influence: the puzzling nature of success in cultural markets, in *The Oxford Handbook of Analytical Sociology* (eds P. Hedstrom and P.S. Bearman), Oxford University Press, Oxford and New York, pp. 315–341.

Schneider, F. (2016) Comment on Feige's paper "Reflections on the meaning and measurement of unobserved economies: what do we really know about the "Shadow Economy"?" *Journal of Tax Administration*, **1** (2), 82–92.

Torgler, B. (2007) *Tax Compliance and Tax Morale: A Theoretical and Empirical Analysis*, Edward Elgar, Cheltenham and Northampton, MA.

Torgler, B., Alm, J., and Martinez-Vazquez, J. (eds) (2010) *Developing Alternative Frameworks for Explaining Tax Compliance*, Routledge International Studies in Money and Banking, Taylor & Francis.

Torgler, B., Demir, I.C., Macintyre, A., and Schaffner, M. (2008) Causes and consequences of tax morale: an empirical investigation. *Economic Analysis and Policy (EAP)*, **38**, 313–339.

Watts, D.J. and Dodds, P.S. (2009) Threshold models of social influence, in *The Oxford Handbook of Analytical Sociology* (eds P. Hedstrom and P.S. Bearman), Oxford University Press, Oxford and New York, pp. 475–497.

6

TAXSIM: A Generative Model to Study the Emerging Levels of Tax Compliance in a Single Market Sector

László Gulyás, Tamás Máhr and István J. Tóth

6.1 Introduction

Taxation is a crucial means to operate the complex machinery of society. Its web of complex patterns of interaction and transfers are not only based on law, but trust, social pressure, enforcement, and economic motivations contribute to its emerging delicate balance alike.

This chapter describes TAXSIM[1], a generative approach to study tax evasion. TAXSIM is an agent-based model, concerned with the operations of a single market sector. The economic well-being of agents depends on the employment contracts they make among each other and on their tax content. In this model tax evasion is a technique to reduce costs (and to raise wages); therefore, tax evasion is a matter of degree rather than a simple binary or ternary choice (e.g., complier/evader, or complier/evader/sceptic) (see Chapter 11). TAXSIM was inspired

---

[1] The first versions of TAXSIM were created in cooperation with Attila Szabó. His contributions are gratefully acknowledged.

Agent-Based Modeling of Tax Evasion: Theoretical Aspects and Computational Simulations, First Edition. Edited by Sascha Hokamp, László Gulyás, Matthew Koehler, and Sanith Wijesinghe. © 2018 John Wiley & Sons Ltd. Published 2018 by John Wiley & Sons Ltd.

by the Hungarian tax system, where the burden of the employee's personal income tax is split between employers and employees. Other taxes, such as corporate income tax and value-added tax are not considered.

In TAXSIM, agents have no perfect information; they estimate the various system-level parameters (such as the frequency and accuracy of audits). These estimations are based on the agents' individual experiences, as well as on experiences of others. Because of the latter, the social network of agents plays an important role in disseminating information.

The final core component of TAXSIM is the assumption that both employees and employers interact regularly with the government and take advantage of the public services offered. The quality of these services (e.g., observed efficiency or level of corruption encountered) determines the agents' underlying motivation to pay their dues. If the experiences with public services are bad, the agents become more prone to accepting the economic incentives of tax evasion (provided that they can rationally do so, given the expected costs derived from the estimated system-level variables).

Earlier works on TAXSIM discussed its basic capabilities (Szabó *et al.*, 2008, 2009a, 2010) and provided an overview of the various emergent responses it can provide (Szabó *et al.*, 2009b; Gulyás *et al.*, 2015). Moreover, Szabó *et al.* (2011) discussed explorations with various types of social networks Erdős and Rényi (1959), Watts and Strogatz (1998), Barabási and Albert (1999). In Kurowski *et al.* (2009) computational experiments with realistic population sizes (10^5–10^6) were considered. (As a comparison, Hungary, the home of the authors, has about 10^7 inhabitants.)

The main goal of this chapter is to provide a consolidated description of the TAXSIM model, providing a detailed specification according to the ODD+D protocol (Müller *et al.*, 2013). In addition, the chapter also provides a demonstration of the main capabilities of the model. This is first done through two *what–if* scenarios: one studying the effect of extended efforts of the government to improve its services, and the other exploring the consequences of the entry of a few tax-preferred, tax-compliant multinational companies. Our results show that the increasing quality of public services helps make the market more tax compliant, but the effect size depends on the current state of the economy. On the other hand, the entry of tax-preferred firms increases the number of hidden contracts.

A detailed study of the model's rich parameter space is also provided, identifying the main driving factors and then systematically studying their effects. Our demonstration is completed by analyzing two "virtual policy experiments": one studying the potential consequences of adopting a policy of minimum wages and the other exploring an audit strategy that selects audited companies from the social network of previously discovered evaders. We identify audit probability and audit frequency as the most important factors across the entire parameter space, with a phase transition between hidden and legal economy across both axes. However, for realistic combinations of these parameters, values of other parameters become important. Here the quality of government services stands out again, having a greater effect on employers than on employees.

The rest of the chapter is structured as follows. Section 6.2 provides a detailed, ODD+D compliant description of TAXSIM and its variants. This is followed by discussions of various computational results generated by TAXSIM. Section 6.4 concludes the chapter.

6.2   Model Description

TAXSIM is a family of models that has been created between 2007 and 2015 (Szabó *et al.*, 2008, 2009a,b, 2010, 2011; Kurowski *et al.*, 2009; Gulyás *et al.*, 2015)

This section provides a consolidated description, focusing on the base model and its properties, but briefly discussing the most important extensions separately as well.

6.2.1   Overview

Szabó *et al.* (2010, p. 17) introduces TAXSIM as follows:

> The TAXSIM model is concerned with the operations of a single market sector, where there are four kinds of agents involved: employee, employer, (tax) authority, and government. The economic well-being of employees depend on their net wages, while that of the employers' is a function of the market demand and the level of gross wages they are forced to pay. The rate of tax evasion is an agreement between an employer and an employee that is made when the employee occupies a new job. As the agreed employment type [see Table 6.1] determines the income of the employee and the (producing) costs of the employer, both participating agents have a motivation to evade.
>
> The government and the tax authority have service providing and regulatory roles, respectively.

In TAXSIM market demand is generated by a submodel and thus is exogenous to the four types of agents concerned. This market sector submodel is minimalistic. Companies (employers) producing the cheapest goods sell first. When demand is less than the total productivity, companies producing expensively will not be able to sell. Since the product is a perishable good, unsold items will be lost, causing financial losses to their producers. Employers (producers) and employees (workers) are assumed to be homogeneous in technological and productive ability: each employment contract between a pair of an employee and an employer agent produces exactly one unit of good in every production (procurement) cycle. The homogeneity also means that employers have the same production costs, produce

Table 6.1 The 23 employment types considered in TAXSIM

No.	Reported wage	Fringe benefits	Ad hoc engagement agreement[a]	Unreported wage	Payment in kind	Class
1	✓					Legal
2	✓	✓				Legal
3	✓		✓			Mixed
4	✓			✓		Mixed
5	✓				✓	Mixed
6	✓	✓	✓			Mixed
7	✓	✓		✓		Mixed
8	✓	✓			✓	Mixed
9	✓		✓	✓		Mixed
10	✓		✓		✓	Mixed
11	✓			✓	✓	Mixed
12	✓	✓	✓	✓		Mixed
13	✓	✓	✓		✓	Mixed
14	✓	✓		✓	✓	Mixed
15	✓		✓	✓	✓	Mixed
16	✓	✓	✓	✓	✓	Mixed
17			✓			Hidden
18			✓	✓		Hidden
19			✓		✓	Hidden
20			✓	✓	✓	Hidden
21				✓		Hidden
22				✓	✓	Hidden
23					✓	Hidden

Illicit practices are marked with gray background.
[a] "Ad hoc engagement agreement" means permanent employment reported as ad hoc agreement.

the same quality of goods, and have the same profit rate. Thus, every employer offers the same gross wage to every employee. As a consequence, the level of income for the employees and the competitiveness of the employers are determined by their respective approach to taxes. Therefore, in TAXSIM, tax evasion is a technique to reduce costs (and to raise wages): the gains from taxes evaded are split equally between the employee and the employer involved. Importantly, this utilitarian approach means that the agents' decision about tax compliance or tax evasion is not a binary choice: it is a decision about the *level of taxes* to be paid.

TAXSIM considers five types of payments, those found empirically most common in Hungary – both legal and illegal ones (Tóth and Semjén, 2008). Among these, two are perfectly legal (but pay different levels of taxes), while three are considered as illegal practices. An *employment type* chosen by TAXSIM agents is a combination of these five payment types: reported wage, fringe benefits, ad hoc engagement agreement, unreported wage, and payment in kind.[2] TAXSIM uses 23 combinations of these five types of payments (see Table 6.1; for more discussion see Section 6.2.4.3). Note that fringe benefits have no meaning without a reported wage, so all combinations that include the first without the other (eight items) are omitted. Furthermore, the employment types are grouped so that when there's no reported wage it is termed illegal (or hidden). On the other hand, we call combinations with only reported wage and/or fringe benefits legal. The remaining combinations belong to the group of mixed employments.

Employers incurring financial losses (from unsold goods) need to increase their competitiveness by reducing their costs. They can achieve this only by renewing their pool of employment contracts to include employment types with lower tax content. A similar situation may be faced by the employees. If their contracts are deemed too costly by their current employer, they become unemployed. After a period of unemployment they gradually decrease their job expectations (i.e., the amount of taxes paid) and will eventually accept any job offer. (The employee's period of patience depends on its reserves that, in turn, depend on the length of the agent's previous employment.)

As discussed, lowering the amount of taxes paid increases the competitiveness of both employers and employees. However, it comes with the risk of being eventually discovered and with fines expected to be paid. The level of this risk and the amount of the fines depend on the efficiency of the tax authority.

The agents, however, have no perfect information. Thus employees and employers have to learn about the effectiveness and accuracy of the tax authority. They learn from their own previous experiences, but they also share and gather information via their respective social networks. Similarly, they also gather information about their neighbors' tax preferences. This is crucial for optimizing their competitiveness. The agents use their collected knowledge when they negotiate the terms of a new employment contract.

---

[2] Fringe benefit is an extra benefit supplementing the employee's wage (company car, private healthcare, etc.). Ad hoc engagement is a special set of employment rules with lower taxes, designed to help serving short term and occasional demand. Payment in kind is the use of a good or service instead of wage (a computer, food, etc.). None of these options are illegal *per se*. However, they are often much less documented than wages and their excessive use may point to noncompliant intentions. For a period in Hungary payment in kind and ad hoc engagement were overused. This is why we consider them as illegal for our purposes here. Nonetheless, the exact labeling of these options is secondary: they simply represent illicit options.

Employees live forever: there is no aging or any fluctuation in the population of the employees. The financial status of the employee has no effect on its work abilities, but it shortens the period it looks for a desirable job. Employers, on the other hand, may go bankrupt, in which case they are replaced with new market players.

In TAXSIM agents are assumed to regularly utilize some public services provided by the government (a company benefiting from transport infrastructure or a person registering its new car, etc.). Satisfaction with the level of these services (the experienced effectiveness or perceived corruption, etc.) motivates the agents to pay their taxes, providing a potential balance to their economic incentives to evade.

TAXSIM has a rich structure that enables it to produce very different market equilibria with a wide variety of employment type distributions. This will be demonstrated in Section 6.3.

6.2.1.1 Purpose

The TAXSIM model was originally created in 2007 as an exercise to understand the applicability of the method of agent-based simulation to the study of tax compliance and tax evasion. Since its first version, TAXSIM was extended significantly in a series of steps, but the general scope and focus of the model remains. The purpose of TAXSIM is to perform abstract, theoretical studies in the domain of tax evasion and shadow economy in the form of computational thought exercises. In particular, TAXSIM is capable of demonstrating complex, nontrivial potential consequences of changes in regulations or government policies (Szabó *et al.*, 2008).

The model was designed for researchers interested in the theoretical questions of the economics of taxes. It can be used to perform *what–if* scenario analyses or to study the complex dependence between the model's behavioral or regulational parameters and the emerging system level tax compliance, as well as the potential consequences of different control strategies by tax authorities.

6.2.1.2 Entities, State Variables, Scales

The main concern of TAXSIM is the behavior of two kinds of agents: employees and employers. This is complemented by two agents: the (tax) authority and the government. (For our purposes, the tax authority is *not* the government. The former is responsible for enforcing the collection of taxes, while the latter represents *the rest of* the government and provides services to other agents.) The authority and government are singletons: there is only a single existing copy of them in the model. In contrast, there is a collection of both employees and employers.

There is a fifth separate component to the model: the demand function (over time) of the market in question. This is modeled as an external function (submodel), which, in the interest of simplicity, is implemented as a constant function. For further details see Section 6.2.4.3.

Employees

The most important state variable links the agent to its current employer (no parallel employment is possible) or shows that the agent is unemployed. Employees store their current *employment type* (see Section 6.2.4.3) and *salary* (net and gross) as well.

Employees have a *desired tax rate* based on their satisfaction with the government. They also estimate the chance of their misbehavior getting discovered (*chance of discovery*) by the authorities. The estimation is partially based on the actual percentage experienced by the given agent. Thus, each of them keeps tabs on the total length of the time periods it was illegally employed and on the number of times it was discovered.

Employees also have a *reserve* that they fill up when employed and deplete when unemployed (in the current version both happen linearly and the *reserve* is bound from above by a parameter). When reserves are depleted, the employee will accept any job offer, independent of its own tax preferences.

Employees are also embedded in a social network. (About the different types of social networks used, see Section 6.2.4.3.) At every couple of time steps, employees exchange their current preferred tax rate and their estimation of the chance of getting discovered with their neighbors in the social network.

Employers

Employers maintain a list of their current employees and have a certain amount of financial resources (*checking balance*). They keep their books, administrating an update on their total salary cap (*total salary cost*), their budget per time step (*budget per month*, based on the *checking balance* and the value of the parameter governing the frequency of sales), the net salary and employment type of their last offered employment, as well as the average effective tax content of their employment contracts.

Employers also have a *desired tax rate* based on their satisfaction with the government and on their perception of the average tax compliance of others. They also estimate the chance of their misbehavior getting discovered (*chance of discovery*) during an audit and the likelihood of getting audited by authorities (*chance of audit*). These estimations are based on their actual experiences as well as on information from their social networks. (About the different types of social networks used, see Section 6.2.4.3.)

Similarly to employees, employers also regularly exchange information with their social network neighbors. At every couple of time steps, they share their current preferred tax rate and their current estimation of the chance of getting discovered.

There are also a number of technical variables maintained by employers that are used during the negotiations of new employment contracts. These keep track of the feedbacks received about the current job offer (the salary was too low, the employment type did not satisfy the employee, etc.)

The Authority Agent
The authority agent has no internal variables in the original TAXSIM model. Its behavior is driven entirely by three system-level parameters: *audit probability*, *audit accuracy*, and the *fine* (penalty, a percentage of the tax evaded) to be paid by evaders.

One of the extensions of TAXSIM implements a tax authority with an adaptive strategy. This strategy prefers to select the targets of audits from the social network of employers previously discovered with noncompliant behavior. In this extension the tax authority also maintains a list of previously discovered wrongdoers (employers).

The Government
The Government has a single role in TAXSIM: to provide services for both employees and employers. In the interest of simplicity, these are implemented as probabilistic draws of quality from two separate uniform distributions. The draws are controlled by two global parameters: *quality of government services to employees* and *quality of government services to employers*.

Scales
The model has no spatial component and thus it does not have a spatial scale either. (Some versions of the social networks for employees or employers do rely on an assumed geographical distance between the nodes; see Section 6.2.4.3. Yet, this assumed geographical distance is abstract without any grounding in any empirical scale.)

Since TAXSIM is intended for abstract thought experiments, its temporal scale is also abstract and arbitrary. Sometimes, it helps to think of time steps as months, but in most cases we simply talk about abstract units of time: time steps.

6.2.1.3   Process Overview and Scheduling

In TAXSIM, none of the agents have perfect information. Thus, the tax authority does not know about possible tax evasion until control audits are executed and even then there is uncertainty about discovering any potential wrongdoing. Similarly, employees and employers do not know about the behavior of other agents: they do not know the tax adherence of others (except for the cases of their own contracts with each other), about the authorities' audit strategy, frequency, or accuracy, or about the quality of governmental services. On the other hand, employees and employers share information in their respective social networks and make estimates based on these communications, in addition to their own experiences related to audits and governmental services.

The employees and employers use this collected knowledge and estimates during the negotiations of their new employment contracts. During this procedure, the agents' expectations depend on their respective satisfaction with the government and on the estimated costs and benefits of evasion. Previous interactions with the authority agent (audits) and information derived from the social network determine cost and benefit estimations.

Activities of Employees

In every time step, every employee is activated (in a randomized order) to perform the following activities:

(1) If the agent is employed, then it collects its wages and increases its reserves. With probability *employee job search probability* it looks for a new job (but accepts only jobs superior to its current employment). If unemployed, the agent decreases its reserves and looks for a new job.
(2) If looking for job, the agent selects the best job offered.
(3) With probability *chance of illness*, the agent gets sick. If sick, the agent receives public medical services[3] and updates its *desired tax rate (d)* based on the quality of services. (See Section 6.2.2.3 for details.)
(4) The agent updates its estimation of the chance of getting sick. (See Section 6.2.2.3 for details.)
(5) The agent exchanges information with its social network neighbors in every *social network update frequencyth* step and adjusts its estimations. (See Section 6.2.2.3 for details.)

When an agent is caught evading, it is forced to quit its current employment and updates its estimation of the chances of getting discovered. (See Section 6.2.2.3 for details.)

Activities of Employers

In every time step, every employer is activated (in a randomized order) to perform the following activities:

(1) The agent pays the salaries of its employees.
(2) The agent updates its estimation of the audit probability in every *Estimations update frequencyth* time step. See Section 6.2.2.3 for details.
(3) Employers maintain their budget (income) and the total salary costs of their enterprise. Given the frequency of market sales (a parameter), they calculate

---

[3] Medical costs in Hungary, our motivating example, are covered by a state insurance system that is perceived by most people as an additional form of tax. While free riding is relatively easy, uninsured workers take a significant financial risk.

their salary budget per time step. If this budget has room for an additional employee at the cost of the last employee, the employer attempts to hire a new person. That is, if the employer has enough money and did not reach the maximum size allowed (parameter *maximum size of companies*), then the employer posts an advertisement on the job market.

- If it did not have an open position in the previous time step, then it generates a new one copying its last offer.
- If it did have an open position, then the agent updates it. (Updates happen with an agility proportional to the number of employees: small companies adapt fast, while large ones are more slow.)
- If the last offer was not accepted, the agent's updates are based on the feedback received when candidates turned down the offer. See Section 6.2.2.3 for details.
- If the last offer was accepted, then the employer updates its offer based on its own preferences. See Section 6.2.2.3 for details.[4]

(4) The next round of negotiation takes place. If the agent's offer was accepted, it hires the new employee. If the new hire puts the employer over budget, then it fires its most costly employee (the one with the highest net salary) immediately (only one).

(5) Bankrupt employers are removed from the system. (An agent is bankrupt if it does not have any employees left and it does not have enough money to pay one month's worth of salary of an employee.)

(6) If there were removed employers in the previous time steps (their number is below parameter *number of employers*), then a new employer enters the market with probability *new employers probability*, initialized as the employers that entered at the start of the simulation. The entry is probabilistic in order to "smoothen out" the changes.

(7) With a certain probability (parameter *employer frequency of requesting government services*) the employer obtains services from the government and updates its *desired tax rate* based on the experience. See Section 6.2.2.3 for details.

(8) The agent produces its products and sells them (maybe partially) if contacted by the market (as one of the cheapest sellers). Unsold items are disposed of.

(9) If the agent did not make enough money to cover the salary costs of the next round, it fires the most costly employees until the total salary costs become covered or all employees are fired.

(10) The agent exchanges information with its social network neighbors in every *social network update frequency*[th] step and adjusts its *desired tax rate*, plus its estimation of *chance of discovery*. See Section 6.2.2.3 for details.

---

[4] The offer is always updated by a number of steps up or down on the employment type table (see Table 6.1) if that is still possible. The number of steps is governed by the *preference updating delta* parameter (defaulting to 1).

When an employer is discovered to have illegal employment contracts,

- it pays a penalty of *fine* (parameter) times the taxes evaded (in the given time step);
- all its illegal employment contracts get removed (the employees get fired);
- the employer updates its estimation of audit accuracy (discovery). See Section 6.2.2.3 for details.

Market Activities

The market is not a real agent, but we discuss the activities of the demand submodel here. Market demand (for the products of the employers) is a constant number of units (governed by the *market demand* parameter) purchased at regular intervals (according to the *market update frequency* parameter).

The market buys from the cheapest producer (employer) first. If it has more units on sale than the current demand, only the amount satisfying the demand will be sold. If the demand is larger than the inventory of the cheapest producer, then the remaining amount is bought from the second cheapest producer, and so on.

The Negotiation Process

In TAXSIM employers and employees have to agree on the terms of employment. By assumption all employees possess the same skills and productivity, and all employers offer the same salary. Thus, the negotiation is concerned with the employment type only. The amount of taxes not paid is divided equally between the employee and the employer. This provides an economic incentive to both parties to hide (parts of) the contract. However, a deal must be reached that is compatible with the risk estimations and desired tax rates of both sides. This is done through the negotiation process that may span several time steps. This is detailed now.

- Employers wanting to hire new employees post their open positions (one at a time). The posts are accessible to all employees that seek jobs in the given time step.
  - If the employer did not have an open position in the previous time step, then it generates a new one copying its last offer.
  - If the employer did have an open position then it updates it based on feedback received (see below for details). Updates happen with an agility proportional to the number of employees: small companies adapt fast, while large ones are more slow.
- Employees looking for a new job pick a fixed number (2 in the current versions) of open positions at random in every time step.
- Employees evaluate the selected open positions:
  $\omega(o) = salary_o - \widehat{fine} - \widehat{medcost}$, where o is the offer in question, $salary_o$ is the net wage offered, while the expected fine and expected medical costs are

calculated as follows, respectively:

$$\widehat{fine} = \widehat{discovery} \cdot salary_o \cdot tax_{evaded} \cdot fine$$

$$\widehat{medcost} = \widehat{illness} \cdot medcost \cdot insurance_{evaded}, \text{ where}$$

$$tax_{evaded} = tax_{rate} - tax_{content}(o) \text{ and}$$

$$insurance_{evaded} = tax_{evaded}/tax_{rate}.$$

Here tax_{rate}, $medcost$, and $fine$ stand for the system level values of *tax rate*, *medical costs*, and *fine*, respectively. $tax_{content}(o)$ is the rate of tax to be paid according to the current offer and $\widehat{discovery}$ and $\widehat{illness}$ are the agent's current estimates of the *chances of discovery* and of the *chances of illness*, respectively.

– If the best evaluated offer is superior to the agent's current job (if it has any), the agent accepts the offer and takes the contract. Note that several agents may be evaluating the same offer. Thus, the (randomized) order of their actions is important. If their selected best offer is not available anymore, agents work with their second best option, and so on.

– For any offer evaluated that proved inferior to another offer in the same time step, the employee sends a feedback to the employer posting the offer. If $tax_{content}(o) < d$ then a signal is sent that the *employment type was unacceptable*. Otherwise if $\omega(o) < \omega(job)$ then the signal says that the *net wage was too low*. Here d stands for the *desired tax rate*, while $\omega(job)$ denotes the valuation of the employee's current job.

• The employers update their offers. If the last offer was not accepted, the updates are based on the feedback received when candidates turned down the offer. See Section 6.2.2.3 for details.

– If the majority of the rejecting candidates said that the offered employment type was not compliant enough for them, then the employer makes its offer more legal.

– If the majority of the rejecting candidates said that the offered amount in the hands was too low, the employer makes its offer more illicit.

• If the last offer was accepted, then the employer updates[5] its offer based on its own preferences. See Section 6.2.2.3 for details.

Activities of the Tax Authority

The basic behavior of the tax authority is rather simple. In every time step, it selects every employer for audit with a fixed probability (parameter *audit probability*, also called *audit frequency*).

For each audited company, the tax authority iterates through all employment contracts and discovers their true nature with probability *audit accuracy* (or *accuracy* for short). If the authority fails to observe the true type of the contract, it is always mistaken to think it is legal.

---

[5] The offer is always updated by a number of steps up or down on the employment type table (see Table 6.1) if that is still possible. The number of steps is governed by the *preference updating delta* parameter (defaulting to 1).

If the observed true type of the contract is illicit, the authority removes the contract in question (the employee gets fired) and fines the employee and the employer. The fine to be paid is the taxes not paid (in the given time step), multiplied by the *fine* parameter.

In Szabó *et al.* (2011) and Gulyás *et al.* (2015) an extension of TAXSIM was introduced that includes a simple adaptive strategy on the part of the tax authority. In this version, the expected number of audits per time step is still *audit probability* · *number of employers* as in the base version. However, the targets of some of these audits are selected differently. With a certain probability (parameter *ratio of adaptive audits*), the targets are selected from the social network neighborhood of previous tax offenders. The remaining targets (or if there are not enough discovered offenders yet) are selected at random. The audits are carried out in the same way as in the base model.

Activities of the Government
The government is a passive agent. It does not initiate any actions; it only reacts to requests by employees or employers. Upon request, it draws the quality of the given service from one of the two separate uniform distributions. With probability *quality of government services to employees* and *quality of government services to employers*, respectively, the service is of satisfactory quality.

Schedule: The Ordering of Activities
After initialization, the process flow of TAXSIM is uniform: agents perform the same activities in the same order as described above. (Agents in the same collective perform the same actions in a randomized order, reshuffled at the beginning of each time step.)

Audits happen after employers had a chance to hire new employees and before bankrupt employers are replaced by new entries. Market demand is satisfied after agents interact with the government. This is followed by the exchange of information over the social network.

6.2.2   Design Concepts

This section details the theoretical considerations behind the TAXSIM model, as well as provides a detailed description of the agents' actions and behavior.

6.2.2.1   Theoretical and Empirical Background

In the study of tax compliance and evasion (Cowell, 1985; Alm, 1988), as discussed in Chapter 1, there is a history of the utilitarian approach, as started by Allingham and Sandmo (1972), operating with an explicit utility function (Sandmo, 1981). Agent-based versions usually incorporate heterogeneity by using an ensemble of various such utilities (Mittone and Patelli, 2000; Davis *et al.*, 2003). Interactions and a social network among the agents play an important role in other works (Balsa *et al.*, 2006; Bloomquist, 2006; Korobow *et al.*, 2007). The effect of social networks is also considered in Chapters 5, 7, and 8.

TAXSIM joins this volume of literature by considering tax evasion as an economic choice by agents, where the choice is finer grained than a simple binary or ternary choice (see Table 6.1 and Section 6.2.4.3) (Mittone and Patelli (2000), Davis *et al.* (2003)). In addition, in TAXSIM the utilitarian choice driven by competition among market players is colored by ethical preferences and by satisfaction with public services. Social networks also play an important role in sharing and disseminating information in TAXSIM. The design of the model is based on empirical observations (see Tóth and Semjén (2008)), but not on actual data.

The key theoretical component in TAXSIM is comprised of the respective decision-making processes of employees and employers during job negotiations. This is the point when agents try to balance their economic incentives (i.e., the difference between the amount of evaded taxes and the expected costs of discovery) and their individual beliefs (i.e., desired rate to pay, based on own and social neighbors' experiences with public services).

During this procedure the agents' expectations depend on their respective satisfaction with the government and on the estimated costs and benefits of evasion. Previous interactions with the authority agent (audits) and information derived from the social network determine cost and benefit estimations. It is assumed that all agents utilize some services provided by the government (e.g., a company wants to register a trademark, or a person wants to get a passport). These interactions (the experienced effectiveness, corruption, etc.) determine the contentment level of the agent.

From the perspective of the employer (who needs to make a job offer) the economic incentives are simple; the following function needs to be maximized:

$$O(c, tax_{content}) = c \cdot tax_{evaded} - p(audit) \cdot accuracy \cdot fine \cdot c \cdot tax_{evaded} \quad (6.1)$$

where c is the *salary cost* (constant during the simulation), $O(c, tax_{content})$ is the offer made by the employer, and $p(audit)$, *accuracy*, and *fine* are the *audit probability*, *authority accuracy*, and *fine* for unpaid taxes, respectively (all constants during the simulation), while tax_{evaded} is the percentage of taxes not paid according to the given offer:

$$tax_{evaded} = tax_{rate} - tax_{content}$$

The employer estimates the values of $p(audit)$ and *accuracy*. The following formula is equivalent to Eq. (6.1):

$$O(c, tax_{content}) = c \cdot tax_{evaded} \cdot (1 - p(audit) \cdot accuracy \cdot fine) \quad (6.2)$$

The latter formula implies that $tax_{content}$ is indifferent if
$\widehat{costs} = p(audit) \cdot accuracy \cdot fine = 1$.

It also implies that if the estimated costs of evasion ($\widehat{costs}$) are greater than 1 non-compliance produces deficit, and that when it is below 1 tax evasion is profitable.

There is a very similar decision faced by employees, making illicit behavior profitable if the gains from avoided taxes exceed the expected costs of discovery and illness.

TAXSIM complements the above utility functions with another dimension of tax compliance. It is assumed that both employers and employees regularly use public services and that their level of satisfaction acts as a motivating force to comply. This component is valuable because it provides a kind of counter-balance to the economic incentives (purely pointing toward evasion, except for the potential costs incurred).

Employers will not make offers below their current level of *desired tax rate*, which depends on their satisfaction with public services. (The latter value is estimated from own experiences and from information via their social network.) Similarly, employees will not accept job offers with less tax content than they feel desirable. (They also develop this notion from their experiences with government services and through interactions in their social network.) There is a single exception: employees having been unemployed for long enough to deplete their reserves will accept any job offer to make a living.

6.2.2.2 Individual Decision-Making

The dynamics of the model are driven by the decisions made by employees and employers and by the controlling behavior of the tax authority. These decision-making algorithms are described in detail in Sections 6.2.1.3 and 6.2.2.3. Similarly, to many agent-based models, the agents of TAXSIM are bounded rational (Mansury and Gulyás, 2007).

In particular, on the part of the employees and employers the most important decision-making occurs during the negotiations of the terms (employment type, tax content) of a new contract. This is discussed in Section 6.2.1.3.

During negotiations agents try to balance their economic incentives (i.e., the difference between the amount of evaded taxes and the expected costs of discovery) with their individual beliefs (i.e., desired rate to pay, based on own and social neighbors' experiences with public services). The background of this fundamental topic was discussed in Sections 6.2.1 and 6.2.2.1.

On the part of employers, there is another important decision-making that ensures their budget balance. This relatively simple rule makes sure they fire their most costly employees if they earn less than expected (from their current sales) and that they hire new employees if they have the budget to grow. These rules are described in Section 6.2.1.3.

Social norms are not modeled explicitly, although both employees and employers maintain a *desired tax rate*, which bounds their behavior and which is partially dependent on observed behavior in their respective social networks. (Employers never offer jobs with "darker conditions" than this value.)

TAXSIM is a nonspatial model. Therefore, decisions do not have a spatial component either. They also lack the temporal dimension, except for the agents' adapting their internal decision-making parameters. On the other hand, several aspects of the agents' decision-making process are stochastic. These aspects are summarized in Section 6.2.2.9.

6.2.2.3 Learning

Müller *et al.* (2013) differentiates between adaptation and learning by stating that *"decision rules are prone to adaptation, where the information used by the rules to generate a decision changes and prone to learning where the rules themselves change over time."* Given this definition, agents in TAXSIM do not learn. Employees and employers do adapt, however, by adjusting their *desired tax rate* (i.e., target employment type) and their estimations of relevant system-level properties based on their own experiences and on information gathered through their social network about, for example, the frequency and accuracy of audits by the tax authority or about the tax adherence of other employees and employers.

In the base version of the model, the tax authority implements a static audit strategy and thus neither learns nor adapts. In some of the extended versions of the model the tax authority implements an adaptive audit strategy, preferring to probe employers in the social neighborhood of employers who have previously been found evading.

Adaptation plays an important role in the behavior of both employees and employers and this is at the heart of their negotiations about the terms of new contracts as well (see Section 6.2.1.3). These will be detailed next.

Employees
(1) When an employee gets sick and receives public medical services, it updates its *desired tax rate (d)* based on the experience (i.e., on the quality of services received):
$d = \min(\text{tax}_{\text{rate}}, \max(0, d + \Delta))$ where
$\Delta = 0.05 \cdot \text{tax}_{\text{rate}}$ if the agent is satisfied and $\Delta = -0.05 \cdot \text{tax}_{\text{rate}}$ otherwise.
(2) At the same time, the employee also updates its estimation of the chance of getting sick:
$\widehat{illness} = 0.9 \cdot \widehat{illness} + \delta$, where $\widehat{illness}$ stands for the *chance of illness* and $\delta = 0.1$ if the agent was sick and 0 otherwise.
(3) Adaptation also occurs, when an employee exchanges information with its social network neighbors (in every *social network update frequency*[th] step). During the exchange, the agent adjusts its estimations:
$\widehat{discovery} = (3 \cdot \widehat{discovery} + info_{<\widehat{discovery}>})/4$ updates the *chance of getting discovered* based on the average of the neighbors' estimation, and

$d = (3 \cdot d + info_{<d_i>})/4$, updates the *desired tax rate*, where $< d_i >$ stands for the average of the neighbors' d (*desired tax rate*).

(4) When an employee is caught evading, it updates its estimation of the chances ($\widehat{discovery}$) of getting discovered:

$\widehat{discovery} = 0.9 \cdot \widehat{discovery} + 0.1 \cdot (\overline{discovered/illegal})$, where $\overline{illegal}$ and $\overline{discovered}$ stand for the number of times the agent had an illegal job and when it had been discovered, respectively.

Employers

(1) The agent updates its estimation of the audit probability in every *Estimations update frequency*[th] time step: $\widehat{audit} = 0.9 \cdot \widehat{audit} + \Delta$, where $\Delta = 0.1$ if the agent was subject to an audit since the last update and 0 otherwise.

(2) During the negotiation process unsuccessful employers update their open positions. If the last offer was not accepted, the agents' updates are based on the feedback received when candidates turned down the offer:

   - If the majority of the rejecting candidates said that the offered employment type was not compliant enough for them, then the employer makes its offer more legal.
   - If the majority of the rejecting candidates said that the offered amount in the hands was too low, the employer makes its offer more illicit.

(3) If the last offer of the employer was accepted during negotiations, then the employer updates[6] its offer based on its own preferences:

   - If the expected value of fines to be paid renders evasion not profitable ($\star$) (see Section 6.2.2.1) or if the current *employment profile* of the company (the average tax content of its current contracts) is below its *desired tax rate* *(d)*, then it makes its offer more legal.
     ($\star$) $\widehat{audit} \cdot \widehat{discovery} \cdot fine > 1$
   - Otherwise, if the expected value of fines to be paid makes evasion profitable ($\star\star$) (see Section 6.2.2.1), then it makes its offer more illegal.
     ($\star\star$) $\widehat{audit} \cdot \widehat{discovery} \cdot fine < 1$

(4) After obtaining services from the government employers update their *desired tax rate (d)* based on the experience:

$d = min(tax_{rate}, max(0, d + \Delta))$, where

$\Delta = 0.05 \cdot tax_{rate}$ if the agent is satisfied and $\Delta = -0.05 \cdot tax_{rate}$ otherwise.

(5) Adaptation also occurs when employers exchange information with their social network neighbors and adjust their estimations:

$\widehat{discovery} = (3 \cdot \widehat{discovery} + info_{<\widehat{discovery}>})/4$ updates the *chance of getting discovered* based on the average of the neighbors' estimation, and

---

[6] The offer is always updated by a number of steps up or down on the employment type table (see Table 6.1) if that is still possible. The number of steps is governed by the *preference updating delta* parameter (defaulting to one).

$d = (3 \cdot d + info_{<d_i>})/4$, updates the *desired tax rate*, where $< d_i >$ stands for the average of the neighbors' d (*desired tax rate*).

(6) When an employer is caught evading, it updates its estimation of audit accuracy (discovery):

$\widehat{discovery} = (\widehat{discovery} + \rho)/2$, where ρ is the ratio of discovered illegal contracts to the number of illegal contracts the agent actually had.

6.2.2.4 Individual Sensing

In the traditional sense of the term, agents in TAXSIM do not sense their environment. However, they do collect information about

- the *desired tax rate* of others through the average value of neighbors in the agent's social network (employees, employers);
- the *chance of discovery* by the tax authority (employees, employers) through personal experiences and through the average of the neighbors' estimations in the agent's social network;
- the *chance of audit* by the tax authority (employer) through personal experiences;
- level of governmental services (employees and employers) through personal experiences;
- tax adherence of employees and employers (authority) through audits.

The government is a static, reactive agent. It does not collect information. It provides a static, stochastic response to queries by other agents (employees and employers).

None of the agents have perfect information. Thus, the tax authority does not know about possible tax evasion until control audits are executed and even then there is uncertainty about discovering any potential wrongdoing. Similarly, employees and employers do not know about the behavior of other agents: they do not know the tax adherence of others (except for the cases of their own contracts with each other), the authorities' audit strategy, frequency, or accuracy, or about the quality of governmental services. On the other hand, employees and employers share information in their respective social networks and make estimates based on these communications, in addition to their own experiences related to audits and governmental services.

Errors or noise are not modeled. All information collected in the above interactions is perfect (but often based on imperfect estimations of others) and free. The only exception is during the tax audits, when the authority discovers the true nature of every illicit contract only with a certain probability (governed by the *audit accuracy* parameter).

6.2.2.5   Individual Prediction

Agents in TAXSIM do not make predictions. Employees and employers base their tax-evading decisions on the expected value of fines, based on their individual estimations of the probability of discovery.

6.2.2.6   Interaction

Employees and employers both occasionally request services from the government. This is the only type of direct interaction between the government and agents of other types. During such interactions employees and employers contact the government directly. The tax authority regularly audits employers (and their employees). This is the single interaction that the tax authority participates in.

Both employees and employers have their respective (separate) social networks. The agents exchange information about their estimations and preferences through these contacts. In the social networks the agents interact only with their network neighbors. (The structure of the social networks used is discussed in Section 6.2.4.3.) Interaction in the social networks is implemented in a minimalistic way: average values of neighbors' relevant variables are passed directly to the updating agent.

Interaction between employees and employers takes place during salary payment via direct money transfer and during negotiations via posted job offers and feedbacks about them.

6.2.2.7   Collectives

There are two collectives in the model: one containing all the employees and another grouping all the employers.[7]

6.2.2.8   Heterogeneity

Employee and employer agents are initially homogeneous within their respective groups, except for their location in their respective social networks. Differences among the agents, that is, varying estimations, emerge due to different experiences during the simulation. An important part of the emerging agent heterogeneity is induced by interaction among the agents over their respective social networks (Dugundji and Gulyás, 2008).

---

[7] In one large-scale experiment with TAXSIM, carried out with its implementation on the QosCosGrid platform (Kurowski *et al.*, 2011), the social networks of both the employees and the employers were implemented as a network of networks, as a proxy for local communities.

6.2.2.9 Stochasticity

There are several sources of stochasticity in TAXSIM.

The social networks of employees and employers are generated using stochastic models of social networks (see Section 6.2.4.3). Apart from this, randomization plays no further role in the initialization of the model.

However, stochasticity is at the heart of the dynamics of TAXSIM. Agents of the same type always execute the same action (e.g., paying salaries, hiring, or exchanging information in their social networks) in a randomized order that is reshuffled before each time step so as to avoid artifacts and lock-step behavior.

Furthermore, each agent type has some random component in its activity set:

- **Employees** pick a random selection of job advertisements to evaluate when looking for a new job. Also, when employed, they look for a better job governed by a random variable. In addition, employees get sick and seek public medical services according to a random variable.
- **Employers** request services from the government according to a random variable. In addition, bankrupt employers are replaced with a certain probability (which avoids immediate replacement).
- **The tax authority** selects its audit targets randomly. (Stochastic elements are present even when using the adaptive audit strategy.) Moreover, illicit employment contracts are discovered stochastically (c.f., limited information or noise).
- **The government** services requests by employers and employees at random. More precisely, whether these services are satisfactory or not is determined by a random variable.

In addition to the agents' actions, the market submodel also has a random component. When fulfilling purchase orders, products are bought from the cheapest producer first, but ties are resolved randomly.

6.2.3 Observation and Emergence

The following emergent model output is collected at the end of every time step of the simulation: employment rate (the total number of active employment relationships), number of legal, mixed, and hidden employment relationships, the taxes collected by the tax authority.

There are a number of interesting emergent behaviors generated by the model. One important observation is that changes in the composition of employment types (and thus in the taxes collected) are often nonlinear. This means that the actual response to parameter or policy changes may be very dependent on the actual state of the economy. In particular, sharp transitions occur in several parameters (e.g., audit frequency and accuracy), meaning that minor changes in parameter values result in drastically different outcomes in certain parameter ranges.

6.2.4   Details

TAXSIM was originally implemented using Repast J (version 3.1). North *et al.* (2006) One version was re-implemented in the QosCosGrid framework (Kurowski *et al.*, 2011) to increase scalability and to perform large-scale experiments (Kurowski *et al.*, 2009). Later, the results were confirmed by a NetLogo implementation (Wilensky, 1999).

6.2.4.1   Initialization

TAXSIM is initialized by creating the agents and the social networks of the employers and employees, respectively. The initialization is driven by the parameters that are not directly anchored in empirical data. Initialization with a given set of parameters always yields the same initial configuration, except for the stochasticity in the generating models underlying the social networks. (This randomness, however, can be controlled for by the appropriate seed parameter to the pseudo-random generator used for network generation purposes.)

The agents' internal variables are initialized assuming no prior knowledge. That is, they expect no audits and, not having prior experience with the government either, their *desired tax rate* is zero.

6.2.4.2   Input Data

TAXSIM, in its current form, is not an empirically grounded, data-driven model. Thus, it does not use input from external sources.

6.2.4.3   Submodels

There are no real submodels in TAXSIM - at least, not in the sense of compact, self-containing complex models. However, there are four parts of TAXSIM that were designed to be independent, easy-to-change-and-replace components, even though their baseline implementations are relatively simple. We discuss these in this section.

The Market Sector

The market sector is responsible for generating the demand for the products produced through the collaboration of employers and employees. It is also responsible for regularly balancing the economy. The exact algorithm for the purchases is dependent on the submodel.

In the baseline implementation, the market sector purchases a fixed amount of products (controlled by the *market demand* parameter) at every *market update frequency* time steps. The products are bought from the cheapest producers first, and then in increasing price order.

The Employment Type

The employment type system is a model of the complexities of illicit employment practices. The shadow economy is a resourceful place when it comes to implementing laws creatively or coming up with new types of (illicit) compensations. (Tóth and Semjén, 2008) TAXSIM considers 23 combinations of 5 types of benefits (see Table 6.1). The employment-type system submodel is responsible for converting the discrete employment type (combination of compensations) into a scalar tax content.

In the baseline implementation a simple linear function is used; the tax content (percentage) for the ith combination (row in the table) is given by

$$tax_{content}(i) = \frac{23 - i}{22} tax_{rate} \tag{6.3}$$

As discussed in Section 6.3.4, one of the extensions of TAXSIM includes the possibility of minimum wage policies. The impact of such policies on the tax system is that it forces a *minimum tax content* on every employment contract that has at least one legal component. Technically, this means a modification of Eq. (6.3) to guarantee a minimum value for all mixed and legal employment mixes.

Social Networks

Social networks in TAXSIM play the role of disseminating and sharing information among market players, providing the bases for the agents' estimations. There are two separate networks: one for the employees and one for the employers.

The baseline version of TAXSIM uses *random networks* (*Erdős–Rényi networks* Erdős and Rényi (1959)) for these components. The single parameter of these networks is *density* (or the uniform probability of any edge to be present in the network). In later versions, Watts–Strogatz type (small-world) networks (Watts and Strogatz, 1998) were also experimented with. Here, two parameters regulate the networks: the size of the neighborhood within which geographically close nodes are all connected and the probability of rewiring (or shortcuts), that is, the amount of probabilistic (long-range) connections laid upon the geography-driven regular structure.

Experiments showed that there is little difference between TAXSIM's emergent outcomes in case of Erdős–Rényi or Watts–Strogatz networks, provided that the average path length is low. (This is the case for all connected Erdős–Rényi networks and for most Watts–Strogatz networks.) However, regular, two-dimensional lattices violate this constraint, which exhibit more shadowy economies as information spread is slow and thus agents have lower chances to correct their estimations. (Szabó *et al.*, 2011) Scale-free networks (see Barabási and Albert (1999)) were also experimented with, but were not included in the consolidated version, as having short average path lengths, they provided no significant contributions to the emergent results.

During a series of experiments with realistic population sizes (several millions of agents) a special network structure was implemented. (Kurowski *et al.*, 2009) This structure classified employees and employers in groups (a form of submarkets), so that employers could only contract employees from the same group. Each group had its own social network for both employees and employers (Watt–Strogatz type of networks) for information sharing. In order to make information exchange among groups possible, group-level social networks were connected loosely in a ring.

Services by the Government

Services by the government are regularly requested by both employees and employers. It is only the quality of these services that is of concern in TAXSIM. Therefore, governmental services are modeled as random variables, one for employers and another for employees. They yield satisfactory services with probability *quality of government services to employers* and *quality of government services to employees*, respectively.

Model Parameters

TAXSIM has a rich set of parameters, presented in Tables 6.2 and 6.3. Some of these are control variables, while others are technical parameters not varied during experiments. Some of the parameters are only applicable to some of the extensions of TAXSIM (see the descriptions).

6.3 Results

In this section, we first provide a snapshot of the kind of *what–if* experiments made possible by TAXSIM. These experiments were reported in detail in Szabó *et al.* (2009a, 2010). This will be followed by an analysis of the rich parameter space of the model, exploring the sensitivities of model behavior as a function of the various parameters. In the end, we turn our attention toward two "virtual policy experiments": one studying the potential consequences of adopting a policy of minimum wages and the other exploring an audit strategy that selects audited companies from the social network of previously discovered evaders.

6.3.1 Scenarios

TAXSIM is capable of modeling very different economic and compliance scenarios and a wide range of employment type distributions, covering the entire spectrum from a totally "hidden economy," via various combinations of illegal and legal employments, to an entirely law-abiding market. Importantly, these are

Table 6.2   Parameters of TAXSIM (part 1)

Parameter	Description
Number of employees	The number of employees in the system
Number of employers	The number of employers in the system
Tax rate (θ)	The percentage of tax to be paid
Fine (π)	The fee to be paid when hidden employment is discovered (Percentage of the tax not paid)
Audit probability (α) (also called *audit frequency*)	The likelihood for an employer to be audited. The expected number of audits per time step is: *audit probability · number of employers*
Audit accuracy	The likelihood for any illicit contract to get discovered when the employer in question is audited
Chance of illness	The probability that an employee needs medical services (per time step)
Employer frequency of requesting government services	The probability that an employer requests services from the government (per time step)
Quality of government services to employees	The probability that the government provides satisfactory services when responding to employee requests
Quality of government services to employers	The probability that the government provides satisfactory services when responding to employer requests
Employee job search probability	The likelihood that an employee shops for a better job when employed
Employee network density	The probability of a link between any pair of employees (in case of Erdős–Rényi networks). The probability of shortcuts (in Watts–Strogatz networks)
Employee network neighborhood size	The size of the abstract neighborhood in which all employees are connected to each other (in Watts–Strogatz networks). A value of 0 means that we work with Erdős–Rényi networks
Employer network density	The probability of a link between any pair of employers (in case of Erdős–Rényi networks). The probability of shortcuts (in Watts–Strogatz networks)
Employer network neighborhood size	The size of the abstract neighborhood in which all employers are connected to each other (in Watts–Strogatz networks). A value of 0 means that we work with Erdős–Rényi networks
Number of employee and employer groups	The number of loosely coupled networks used in the special experiments with extremely large-scale systems (see Sections 6.2.2.6 and 6.2.4.3). The value of 0 (used throughout this chapter) means that this option is not activated

Table 6.3 Parameters of TAXSIM (part 2)

Parameter	Description
Minimum size of companies	The number of employees sought when employers are created
Maximum size of companies	The maximum number of employees allowed for any employer (A value larger than the *number of employees* means no upper limit)
Chance of new employer	Bankrupt employers are removed from the system. They are subsequently replaced by freshly initialized employer agents. However, this replacement occurs in a probabilistic fashion to "smoothen out" the change. This parameter gives the probability (per time step) that a new employer is created (if the current number of employers is less than the value of the *number of employers* parameter)
Salary cost	The nominal wage (for all contracts). A technical scaling parameter only
Medical costs	The amount to be paid for medical services (per time step)
Profit rate	The multiplier generating the selling price of producers (employers) (Larger than 1)
Market demand (#units bought)	The number of product units purchased (from the ensemble of employers) per time step
Market update frequency (time steps)	The number of time steps between two procurement events. The length of the procurement cycle
Social network update frequency (time steps)	The length of the information sharing cycle. The number of time steps between two occasions of agents updating their estimations based on information from their social network
Estimations update frequency (time steps)	The number of time steps between two occasions that the employers update their estimations of the audit probability
Preference updating delta	The number of rows (in the employment table, see Table 6.1) by which agents adjust their *desired tax rate (d)*
Ratio of adaptive audits	The ratio of adaptive audits (over random selections) in experiments with adaptive audit strategies
Minimum tax	The minimum tax (in percentage) to be paid after any employment mix that has a legal component, used in experiments with minimum wage policies. A value of − 1 means there is no minimum wage (and thus, no minimum tax)

emerging properties of the system, not rules coded directly into the model that can be turned on by a control switch. This feature has both advantages and disadvantages. On one hand, the lack of direct control over system-level properties such as tax compliance reflects reality. On the other hand, the emergent nature of the qualitative class of the system (hidden economy, mostly compliant market players, etc.) means that one needs to have some experience with the model to be able to set parameters for experiments with particular types of economies. It also means that exact levels of the response variables (e.g., unemployment rate or tax compliance) cannot be directly set.

In the following experiments, TAXSIM was set up to model a market sector with high levels of noncompliance. With the parameter settings selected (see Table 6.4) the system reaches equilibrium after a transient period of about 300 time steps. While 300 months is a rather long time in real life, we consider this initial period as

Table 6.4   Parameters used in the scenario experiments. Source: Reproduced from Szabó *et al.* (2009a, 2010)

Parameter	Value	Parameter	Value
Number of employees	200	Number of employers	40
Tax rate	45%	Fine	150%
Audit probability	27.5%	Audit accuracy	0.3
Quality of government services to employees	0.1	Quality of government services to employers	0.33
Employer frequency of requesting government services	0.1	Chance of new employer	0.1
Chance of illness	0.2	Medical costs	10
Employee network density	0.1	Employee network neighborhood size	0
Employer network density	0.1	Employer network neighborhood size	0
Number of employee and employer groups	0		
Minimum size of companies	1	Maximum size of companies	3000
Employee job search probability	0.01	Minimum tax	None
Market demand (no. of units bought)	170	Market update frequency (time steps)	6
Salary cost	100	Profit rate	1.1
Social network update frequency (time steps)	12	Estimations update frequency (time steps)	6
Ratio of adaptive audits	0%	Preference updating delta	1

the bootstrapping needed to initialize the simulation. In the (dynamic) equilibrium, the level of unemployment is approximately 15%, while legal employment is at 5%, mixed employment is at around 10%, and illegal employment amounts to 70%. These levels will probably not fit most economies very well. Yet, they suit the kind of "thought experiments" we intend to carry out here. A more thorough analysis of the parameter spectrum and the corresponding model behaviors will be provided in Section 6.3.2.

6.3.1.1   Improving Government Services

We will assume a strong will on the part of policy makers to legalize the dark economy we have set up in TAXSIM. There are some obvious steps lending themselves to decision makers, including stepped up efforts to collect taxes: increased number of audits and improved efficiency of the discovery of noncompliant behavior. Given their obviousness, we will omit these from the following experiments. (We will revisit the issue in Section 6.3.2 when studying the effects of various TAXSIM parameters in detail.) Instead, we imagine a hard-working administration steadily improving the quality of its services. After the bootstrapping period, the quality of governmental services is upgraded by 7.5%, in every 500 time steps. As demonstrated in Figure 6.1 (a) the persistent improvement of governmental services leads to persistent legalization. However, improvement in compliance is not linear. Hidden contracts gradually become mixed and then, following further improvements of service quality, they eventually become legal. This also means that the marginal impact of increasing quality of service differs, especially, if measured only by change in legal or totally hidden employment ratio. (The latter shows decaying rate of improvement, while the former improves progressively.)

Figure 6.2(b) shows the taxes collected over time during the above experiment. The government is clearly rewarded for its efforts. This becomes especially clear when comparing the time series to the baseline scenario without service quality improvements (shown in Figure 6.2(a)). However, the catch is the nonlinearity that is also obvious. For the initial efforts, the return is moderate and slowly increasing. It is only after the third round of improvements (time step 1500) that steep increase in tax income is observable. Not surprisingly, this increase will eventually decay, with the income level getting saturated.

6.3.1.2   Entrance of Companies with Preferential Taxes

Improving governmental services is often desirable, but it might often be a rather costly and uncertain means to legalize an economy. Another desirable goal is to improve the tax and business culture of market players, for example, by attracting players from different, more established economies with higher general tax compliance. One way of achieving this is to offer lower, preferential tax rates, at last for

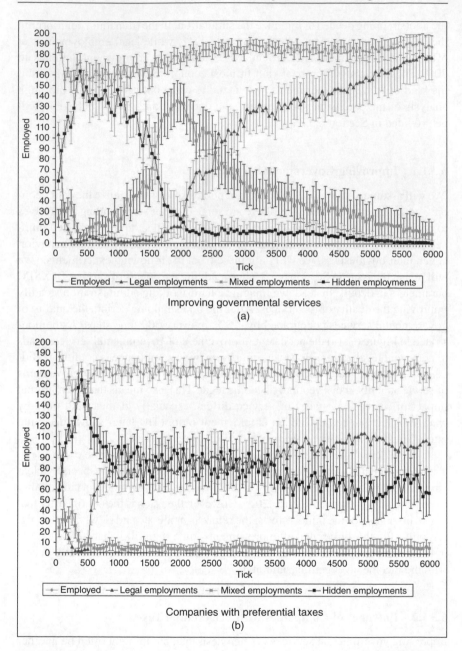

Figure 6.1 (a) Impact of the persistent improvement in the quality of governmental services (improvement at every 500th time steps). (b) Introduction of companies paying preferential taxes (from the 500th time step). In both panels, the horizontal axis shows simulated time, while the vertical one is the number of a) Employed b) Legal employments c) Mixed employments c) Hidden employments. The symbols mark the average and error bars mark the standard deviation of 256 runs. Source: Reproduced from Szabó *et al.* (2009a, 2010)

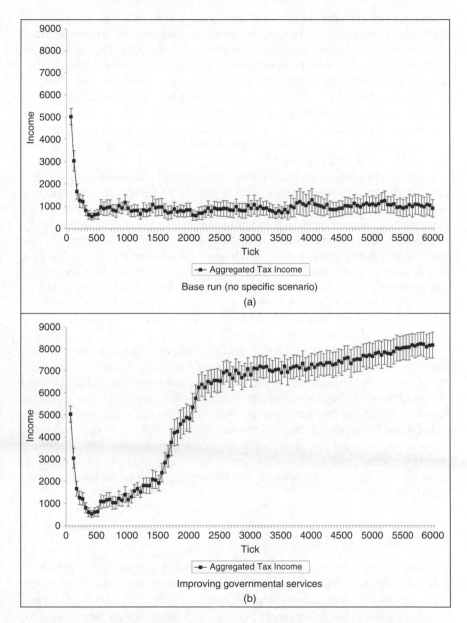

Figure 6.2   Aggregated tax income. (a) Shows the base scenario (reaching equilibrium in about 350 time steps), while (b) displays the impact of the persistent improvement in the quality of governmental services (improvement at every 500th time steps). The symbols mark the average and error bars mark the standard deviation of 256 runs. Source: Szabó *et al.*, (2010). Reproduced with permission of IGI Publication

a certain period, to large international companies in return for their establishing a local branch in the country. This was not an uncommon practice in Eastern Europe after the systems change (Sass, 2003; Semjén and Tóth, 2004; Tóth and Semjén, 2008; Bakos *et al.*, 2008). Unquestionably, attracting foreign investment, creating new jobs, and modernizing market sectors are very valid goals and they can possibly be achieved by attractive tax cuts to multinational companies. Yet, the practice was often accompanied, at least at the level of rhetoric, with the hope that law-abiding foreign companies disseminate their best practices, including their tax practices and improve the standards of their employees and later, via spill-over, those of their competitors as well.[8]

In our computational experiment, three international companies enter the market after the bootstrapping period (at time step 500). Their preferential tax rate is 36% (as opposed to the normal 45%, corresponding to a 20% tax cut). On the other hand, these three employers are fully compliant, rule-abiding operations.[9]

TAXSIM does not offer means to assess the impact of tax-incentivized international enterprises on the general development of an economy. It can, however, help us theorize about the effect of said companies on system level tax compliance and government tax income. The appearance of the three compliant companies do generate a minor increase in governmental income (not shown here), but that is not significant. What is interesting, however, is how they influence the general compliance level of the system.

Surprisingly, the appearance of fully abiding companies increases the number of hidden contracts in comparison to the baseline scenario (see Figure 6.1). The number of legal contracts increases as the new companies gain market share and thus the overall level of illegal employment decreases - there is no question about that. However, mixed employment is almost swept out of the market, as artificial legalization splits the sector into the two extremes. Taking advantage of the lower taxes, the new multinational companies are able to offer competitive salaries: approximately the same as employers who evade most of the taxes. As a consequence, firms offering mixed type employment are driven bankrupt or are forced to make illegal contracts.

6.3.2   Sensitivity Analysis

TAXSIM is a rich model with many parameters and a complex set of behaviors. It is impossible to assess its repertoire without systematically testing its responses to changes in its numerous parameters. A complete analysis is out of the scope of this chapter, but in the following we provide a brief overview of the main points. (More details can be found in Szabó *et al.* (2009b) and Gulyás *et al.* (2015).)

---

[8] The assumption that a newcomer to the market does not change its behavior is arguable. Yet, this was a widespread assumption in Hungary during the first decade after the systems change. The survey-based empirical study of Semjén and Tóth (2009) confirms this assumption.

[9] Technically, these are specially programmed agents that do not even consider tax evasion.

We first perform a two-level factorial analysis of the main input parameters (factors). That is, we select two values (low and high) for each factor and run the simulation ("take samples") for all combinations of these parameter values: 2^n combinations for n factors. (In fact, we run the simulation several times for each combination and average the outcomes for the same combinations.) The purpose of this experiment is to screen the parameter space and identify the factors (parameters) that have the highest effect on the response variables, that is, the ones that are worth focusing our analysis on. This is done by fitting a linear model to the results and sorting the coefficients of the factors and their interactions. While complex systems and thus computational simulations of complex social systems rarely exhibit linear behavior, the simple linear estimation (based on a two-point measurement for each factor) provides a quick and dirty way to initiate the exploration of the model's phase space (Lorscheid et al., 2012).

Table 6.5 summarizes the settings of our two-level factor analysis. Notice that a number of parameters with obvious importance (such as tax rate or those related to system size) were intentionally left out of the factor analysis, by fixing them at a particular value. We did this to focus the exploration on the behavior and interaction of agents: employees and employers, as well as tax authority and the government. Parameters governing the economic environment (e.g., profit rate, salary cost level, market size), as well as technical parameters, such as update frequencies, are left out of the current analysis. Tax rate and fine parameters are exceptions as they are obviously policy parameters. We kept them out of our current scope as their effect is fairly obvious. (See Chapter 5 for a discussion on the interaction of audit probability and fine.)

Also notice that the economic environment defined for this analysis is different from the one in the previous section. Here, we have 300 employees instead of 200 to seek work at the same 40 of employers competing with their products to meet the same demand (170 units per sales cycle). Obviously, these settings create stronger competition in the labor market.

TAXSIM has five system-level response variables: the employment rate, the number of hidden, mixed, and legal employments and the amount of taxes collected. In the present analysis, we focus on the last four of these. The factors with the highest effects are shown on Figure 6.3. Important two-way interactions are also displayed. (The responses are evaluated based on snapshots of the system taken after 6000 time steps and averaged over 10 independent runs with the same parameter setting.)

Obviously, the four response variables studied are interrelated by definition and thus it is expected that the factors driving them would be very similar as well. This is confirmed by our results, although minor variations may and do exist.

The main observation of the analysis is that it is the interaction of the frequency of audits and the accuracy of the tax authority that has the strongest effect on all response variables. These two factors have a strong effect in their interactions with

Table 6.5 Parameters of the two-factorial experiments

Tested factors			
Parameter	Value (low, high)	Parameter	Value (low, high)
Audit probability	0%, 25%	Audit accuracy	0, 0.45
Quality of government services to employees	0.1, 0.9	Quality of government services to employers	0.1, 0.9
Employee job search probability	0.01, 0.3	Minimum tax	None, 20%
Employee network density	0.01, 0.3	Employee network neighborhood size	0, 2
Employer network density	0.01, 0.3	Employer network neighborhood size	0, 2
Number of employee and employer groups	0, 4		

Fixed parameters			
Parameter	Value	Parameter	Value
Tax rate	45%	Fine	50%
Number of employees	300	Number of employers	40
Chance of illness	0.2	Medical costs	10
Minimum size of companies	1	Maximum size of companies	3000
Employer frequency of requesting government services	0.1	Chance of new employer	0.1
Market demand (no. of units bought)	170	Market update frequency (time steps)	6
Salary cost	100	Profit rate	1.1
Social network update frequency (time steps)	12	Estimations update frequency (time steps)	6
Ratio of adaptive audits	0%	Preference updating delta	1

other factors as well. A third parameter with a strong influence is the quality of governmental services provided for employers. Let us now focus our attention on these identified factors of importance.

Figure 6.4(b) shows the percentage of nonlegal contracts (hidden and mixed employments combined) in case of various combinations of the two most relevant parameters. The rest of the parameters are fixed according to Tables 6.5 and 6.6. White shading stands for a fully illegal economy, while the darker shades stand for increasing levels of legality.

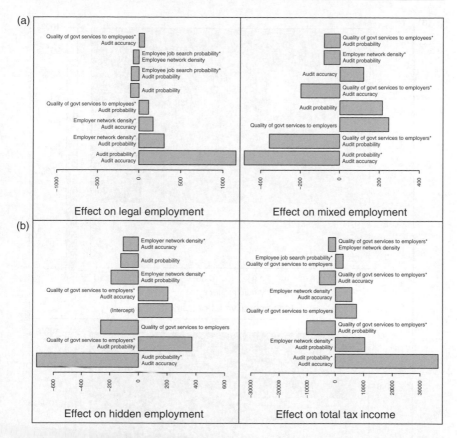

Figure 6.3 Top eight factors with the largest effects in case of the main response variables. Results from a two-level factorial analysis (only two-way interactions shown). The values for the response variables are snapshots after 6000 time steps and averaged over 10 runs. The abbreviation govt stands for government.

It is clear that for most audit probabilities and for most audit accuracies the economy is clean with nearly 100% legal contracts. However, there is a continuous region where illegal contracts exist to various degrees. At first sight it might be surprising that shadow economy may only exist at the "edges" of the parameter space, but this makes sense. While a 2% likelihood of getting audited *every month* only constitutes a tiny fraction of the technically meaningful parameter range, it also means that every company gets audited in every fourth year. Similarly, it is also reasonable that with an above around 40% chance of wrongdoing being discovered during an audit, it does not make sense any more for agents to cheat if there is a realistic chance to get audited. Of course, the interaction of the two parameters is clearly important. The shadow economy can prevail when both of these factors

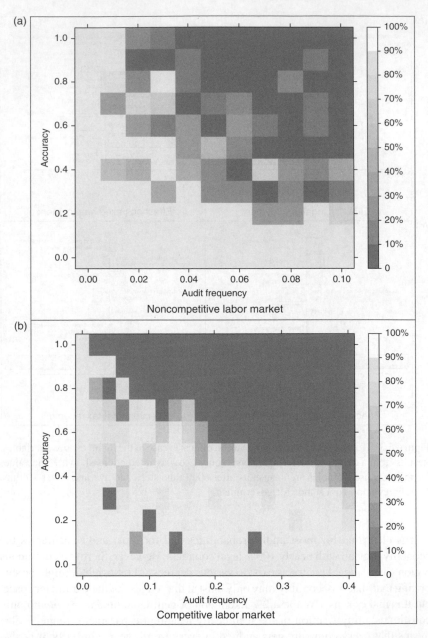

Figure 6.4 Percentage of nonlegal contracts (hidden and mixed employments combined), relative to the total level of employment, in case of various combinations of the audit probability and accuracy of audits. The values shown are snapshots taken after 6000 time steps and averaged over 10 independent runs. The rest of the parameters are fixed according to Tables 6.5 and 6.6. (a) Pictures the original market of Section 6.3.1, while (b) shows the case of the competitive labor market of this section. (Note the different horizontal scales.)

Table 6.6 Parameter values for the analysis of the main parameters

Parameter	Value	Parameter	Value
Minimum tax	None		
Employee network density	0.1	Employee network neighborhood size	0
Employer network density	0.1	Employer network neighborhood size	0
Number of employee and employer groups	0	Employee job search probability	0.01

are relatively low – and it turns out that this means that none of them needs to be extremely low.

It is also worth pointing out that at practical levels of audit probability (and accuracy) the economy is not fully compliant with the current set of parameters. This means that in this region other factors may play an important role. This can also be seen from the "turbulent" nature of the transition regime at the interaction of the two parameters (more about this later).

Figure 6.4(a) shows the percentage of nonlegal contacts as a function of audit probability and accuracy for the market of Section 6.3.1. Here labor competition is more lax as employers can sell enough products to employ basically all workers. (The two-level factor analysis for this market, not reported here, produces slightly different factor effects than the ones reported in Figure 6.3, but the main observations remain.)

Comparing Figure 6.4(a) and (b), we observe a significantly larger region of illicit activities in the competitive market. This is due to the fact that unemployment forces employees to accept job offers even below their compliance preferences. In both (a) and (b), parameter combinations coding low risk of discovery correspond to frequent nonlegal employment contracts. Their numbers can be significantly lowered, even to 0%, by raising the likelihood of discovery (by increasing any of the studied parameters). It is worth noting, however, that the threshold of audit probability below which illegal employment exits in significant proportions is an order of magnitude higher in case of the competitive labor market.

An important further observation from Figure 6.4 is about the nature of the transition from a hidden economy to a completely legalized labor market. As illustrated by Figures 6.5 and 6.6, this is a turbulent region of the model's phase space, where minor changes in parameter settings may result in dramatic changes in responses. The turbulent region is extended at the interaction of the two parameters, but gets narrower as the change in one factor dominates that in the other. In regions where the interaction effect is weak, we even observe sharp phase transitions. This is more pronounced in case of the competitive labor market.

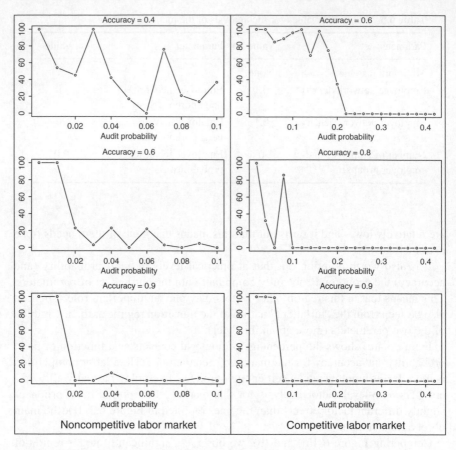

Figure 6.5 Transition from a hidden to a legal economy as a function of *audit probability*. Cross sections of the phase space shown in Figure 6.4

6.3.2.1 The Effect of Governmental Services

We now turn our attention toward the factor with the third largest effect in our factorial experiment: the level of governmental services. In Section 6.3.1.1, we introduced gradual improvements of governmental services when market equilibrium has been reached. In contrast, we now compare the emerging equilibrium outcomes of separate runs with various service quality levels.

In TAXSIM the government provides two types of services: one for employees and another for employers. In the interest of getting a full picture, we vary the quality of both of these services systematically. Figure 6.7 shows the percentage of legal, mixed, and hidden contracts, as well as the amount of taxes collected, as a function of the various levels of the two types of government services. The parameters controlling the quality of the different services are varied between their theoretical extremes (between 0 and 1). The horizontal and vertical axes represent

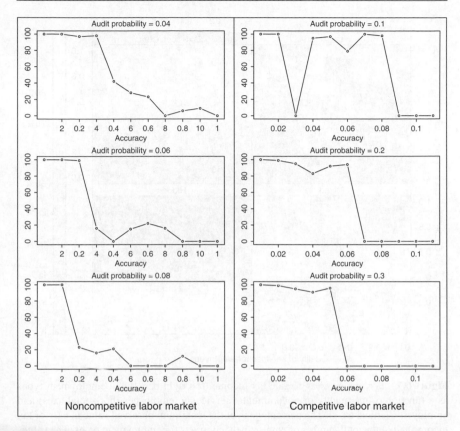

Figure 6.6 Transition from a hidden to a legal economy as a function of *audit accuracy*. Cross sections of the phase space shown in Figure 6.4

services for employees and those for employers, respectively. The grayscale-coded percentage values shown are snapshots taken after 6000 time steps and averaged over 10 independent runs for each parameter combination. The rest of the parameters are fixed according to Tables 6.5 and 6.6 with *audit probability* at 25% and *audit accuracy* at 0.45.

As demonstrated in the figure, the quality of services for employers dominates the quality of those provided for employees. This is likely due to the heavy competition in the labor market: employees are forced to take advantage of any job opportunity and thus are less driven by their own compliance preferences during job negotiations.

It is also clear that the better the quality of services for employers, the more compliant the market sector becomes. In case of low quality of services employment contracts are mostly illegal. In the mid-quality range, contracts become mixed, while with high quality of services legal contracts tend to dominate the market. (Yet, with the given parameter settings totally legal markets are not very frequent.)

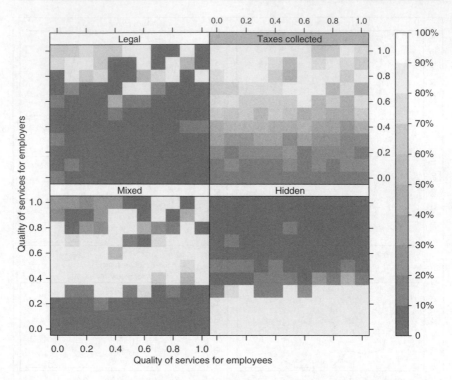

Figure 6.7 Percentage of aggregate tax income (top right panel) and employment types as a function of the quality of governmental services to employees (horizontal axes) and employers (vertical axes). The percentage is of the maximum observed value in case of tax income and of the total number of employment contracts for employment types. The values shown are from snapshots taken after 6000 time steps and averaged over 10 runs. The rest of the parameters are fixed according to Tables 6.5 and 6.6 with *audit probability* at 25% and *audit accuracy* at 0.45

The above general trends are modulated by the quality of services to employees. This becomes visible in the transitional regions, where better services for employees pay off by more legalized emergent economies.

The plot of tax incomes is much smoother, but it shows the same general trends.

6.3.3 *Adaptive Audit Strategy*

The experiments discussed in this section were done with one of TAXSIM's extension models. As discussed in Section 6.2.1.3, an adaptive audit strategy means that the tax authority does not select audit targets entirely at random, but prefers to check employers directly connected (in their social network) to employers previously caught with noncompliant practices. (Initial targets

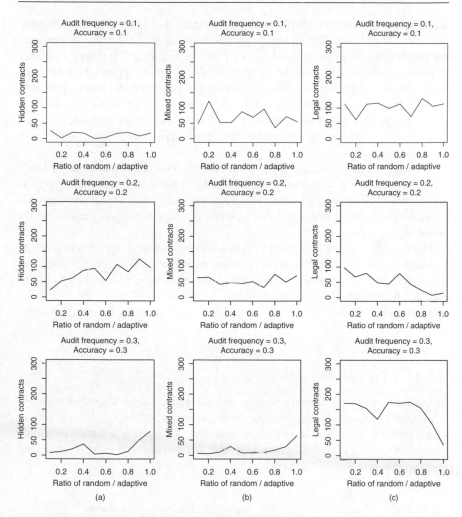

Figure 6.8   The effect of an adaptive audit strategy. The number of hidden (a), mixed (b), and legal (c) employment contracts for various ratios of random and adaptive audit strategies (horizontal axes: 0 stands for all-adaptive selection, and 1 for all-random selection). The panels in the different rows depict results for different levels of *audit frequency* and *audit accuracy* combinations. The values plotted are snapshots after 6000 time steps and are averaged over 10 independent runs

are selected at random and the pick from among the neighbors of previous wrongdoers is still random.) The percentage of random and adaptive audit target selection is governed by a parameter.

Figure 6.8 shows the results of experiments exploring the effect of various levels of adaptive audit target selection. The horizontal axes show the ratio of random and adaptive audit selection. (The value of 1 means that every audit is selected at random, while in case of 0, all targets are selected adaptively.) On the vertical axes the numbers of the given contract types are shown. Figure 6.8(a) plots the number of hidden contracts, (b) plots the number of mixed, while (c) plots the number of legal contracts. Panels in the different rows show results for different combinations of *audit frequency* and *audit accuracy* (with both values set to 0.1, 0.2, and 0.3, respectively). The rest of the parameters are fixed as shown on Tables 6.5 and 6.7. The values plotted are snapshots after 6000 time steps and are averaged over 10 independent runs.

According to the figure, the larger the ratio of adaptive target selection (the lower the percentage of random selection), the less hidden and the more legal contracts are observed. Obviously, this trend is more pronounced when the tax authority has a larger impact on the economy, that is, when both audit frequency and accuracy are of higher values.

These results suggest that it pays off to select employers from the social neighborhood (e.g., business partners) of previous wrongdoers. It is true, however, that the current auditing strategies implemented are rather simplistic and lack empirical backing. Nonetheless, the experiments confirm the findings of Chapter 4 and show that the TAXSIM framework is capable of implementing and testing various target selection policies.

6.3.4 Minimum Wage Policies

In our last set of experiments, we use another extension of TAXSIM, one that implements minimum wage policies. As discussed in Section 6.2.4.3, the impact of a minimum wage policy on the tax system is that there is a forced minimum amount of tax (that of the minimum wage) that is to be paid after every employment contract that contains a legal component.

Table 6.7 Parameter values for the adaptive audit strategy experiments

Parameter	Value	Parameter	Value
Minimum tax	None		
Employee network density	0.02	Employee network neighborhood size	0
Employer network density	0.1	Employer network neighborhood size	0
Number of employee and employer groups	0	Employee job search probability	0.1
Quality of government services to employees	0.1	Quality of government services to employers	0.2

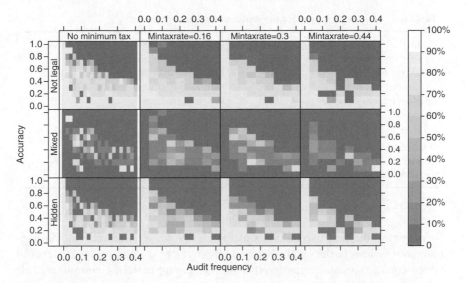

Figure 6.9 The effect of minimum wage policies. The panel columns represent a minimum tax content of 0%, 16%, 30%, and 44% from left to right. The first row of panels plots the percentage of illegal employment (hidden and mixed contracts combined), while middle and bottom rows separate the two types. The horizontal axes show the value of *audit frequency*, while the vertical axes plot *accuracy*. The grayscale-coded values shown are snapshots after 6000 time steps and show averaged values of 10 independent runs. The rest of the parameters are fixed according to their levels in Tables 6.5 and 6.8

Figure 6.9 summarizes our experiments with a minimum tax content of 16%, 30%, and 44% (second, third, and fourth column of panels, respectively). For purposes of comparison, the first column of panels shows the baseline case of no minimum wage (minimum tax rate of 0%). The first row of panels plots the percentage of illegal employment (hidden and mixed contracts combined), while middle and bottom rows of panels display the percentage of mixed and hidden contracts separately. In case of all panels, *audit frequency* is plotted on the horizontal axis, while *accuracy* is plotted on the vertical axis. The grayscale-coded values shown are snapshots after 6000 time steps and show averaged values of 10 independent runs. The values of the other parameters are fixed according to Tables 6.5 and 6.8.

Our results suggest that the presence of minimum wages (and thus, a forced minimum tax rate) slightly legalizes the simulated economy in the given parameter region of TAXSIM. This is most visible from the decaying white region of the top row, in between the first and second columns. This means that we can observe low percentages of illicit activity already in combination with lower values of audit frequency and accuracy. At the same time, the slope of the transition from nonlegal to legal becomes gradually steeper as the minimum tax paid increases.

Table 6.8 Parameter values for the minimum wage experiments

Parameter	Value	Parameter	Value
Employee network density	0.05	Employee network neighborhood size	0
Employer network density	0.05	Employer network neighborhood size	0
Number of employee and employer groups	0	Employee job search probability	0.1
Quality of government services to employees	0.1	Quality of government services to employers	0.2

Comparing the first two columns of Figure 6.9 we see that introducing minimum wages lowers the level of mixed employment in relation to the level of hidden contracts in an economy otherwise becoming more tax compliant. This means that market players unable to pay the tax content of the minimum wage are forced to give up mixed contracts and are driven entirely to the shadow economy. The further columns of the figure reveal that this tendency is reversed with increased minimum tax rates. This is because the higher general level of taxation makes competition stronger. It is worth noting, however, that mixed employment is always rather chaotic in the region of transition, that is, in the curved band corresponding to the transition between legal and nonlegal employment in the top row. This region is full of high peaks and deep valleys, which points to the strong path dependency of the runs in this parameter region.

Our results contribute to a debate in the literature: whether higher minimum wages result in higher unemployment or not (Card and Krueger, 1995). In TAXSIM, we do not separately measure the *reported* employment level, but the increased level of hidden employment is clear. This corresponds to the empirical findings of Khamis (2008) and Maloney and Mendez (2004), while it is in contrast with the results of Lemos (2004) based on Brazilian data. It is important to note, however, that the decreasing number of reported contracts emerges spontaneously in TAXSIM in response to the enforced minimum amount of tax.

6.4 Conclusions

This chapter introduced TAXSIM, an agent-based model of tax evasion in a market sector with constant external demand. In this model the level of taxes paid is subject to an agreement between individual employers and their employees. The agreement is driven by the expected costs of evading (based on the agents' estimation of the likelihood of discovery and on the fines to be paid), by the economic situation (the savings of the employee and the market success of the employer), as well as by the agents' individual preferences (dependent, in part, on their satisfaction with the services offered by the government).

Following the description of TAXSIM using the ODD+D protocol (see Müller *et al.* (2013)) we demonstrated its capabilities by discussing a selection of results from using the model.

We first discussed two scenarios from Szabó *et al.* (2008, 2010), when policy changes are implemented during the course of the simulation. The first showed the effect of the gradual improvement of the services offered by the government, while the second analyzed the entry of a few tax-preferred employers into the market. In the first case, we found that the extended effort of the government to improve the quality of its services pays off with a gradually legalizing economy. However, neither tax income nor the number of legal employment contracts increases linearly, making the immediate observable results of the new policy dependent on the actual, often unobserved, state of the economy.

The effect of improving governmental services was further analyzed in Section 6.3.2.1, confirming our findings in a more general setting and showing that services provided to employers dominate the impact of those provided to employees (at least in the case of the strong labor competition analyzed here).

We also found that the tax preference given to "multinational employers" may lead to a segmentation of employment contracts into legal and hidden (as opposed to mixed) employments, the tax-preferred players driving their competitors toward the shadow economy.

The scenarios were followed by an exploration of the model's complex parameter space. Our factorial analysis showed that the most important factors driving the outcome were *audit probability* and *accuracy*, followed by the *quality of services for employers*. We could also observe, however, that at practical levels of audit probability there is ample room for other factors to influence tax compliance.

In the last two sections, we discussed two extensions of TAXSIM, demonstrating its flexibility to study policy questions at the theoretical level. In the first, we explored the impact of a different, adaptive strategy of the tax authority in selecting its audit targets. We found that even a relatively simple strategy can noticeably improve the level of compliance over a purely random audit target selection. This corresponds to the findings of Chapter 4.

With the other TAXSIM extension, we studied how minimum wage policies might affect tax compliance. In the taxation context, a minimum wage law means that there is a certain level of tax to be paid after any employment relationship containing at least one legal component. (Totally hidden contracts are not affected.) As a consequence, we found that the introduction of minimum wages drives the economy toward noncompliance by transforming mixed employment to hidden ones. However, this effect is reversed as minimum wages (i.e., their tax content) are increased.

There are many opportunities for further studies with and for the improvement of TAXSIM in the future. For example, more realistic (adaptive) audit strategies could be investigated. Another possible extension would focus on the moral costs of tax evasion. To this end more detailed models of "moral obligation" and social norms should be incorporated (Gordon, 1989; Bosco and Mittone, 1997; Lago-Penas and Lago-Penas, 2010; Méder et al., 2012). It would also be interesting to connect the quality of government services to the amount of actual taxes collected.

TAXSIM is a rich theoretical environment to experiment with the complex economics of taxation. Its uniqueness is in its unusually sophisticated structure, so that it does not only include a complex decision-making equation by employees and employers that requires the balancing of economic incentives (the amount of taxes evaded and the expected costs of discovery) with the agents' internal beliefs (based on individual and social experiences with public services) but also an information exchange among agents through their social networks. Moreover, it also makes different auditing strategies possible for the tax authority. At the same time, it is perhaps this complex structure that limits TAXSIM's wider applicability. (A mathematical analysis of the kind of Gaujal et al. (2014) remains an open challenge.) Nonetheless, it is clear that the family of TAXSIM models is able to generate a wide range of interesting and relevant emergent outcomes capable of feeding our theoretical appetite. And certainly, there are still a number of rich courses to follow.

Acknowledgments

This work benefited from grants GVOP-3.2.2-2004.07-005/3.0 and OTKA T62455 (Hungarian Government) and QosCosGrid, EMIL (FP6 STREP #033883, #033841).

References

Alm, J. (1988) Compliance costs and the tax avoidance-tax evasion decision. *Public Finance Quarterly*, **16** (1), 131–166.

Allingham, M.G. and Sandmo, A. (1972) Income tax evasion: a theoretical analysis. *Journal of Public Economics*, **1**, 323–338.

Bakos, P., Bíró, A., Elek, P., and Scharle, Á. (2008) *A magyar adórendszer hatékonysága*, Közpénzügyi füzetek, vol. **21**, ELTE TÁTK, Budapest, pp. 32. (In Hungarian).

Balsa, J., Antunes, L., Respício, A., and Coelho, H. (2006) Tactical Exploration of Tax Compliance Decisions in Multi-Agent Based Simulation. Proceedings of MABS'06, The 7th International Workshop on Multi-Agent-Based Simulation.

Barabási, A.-L. and Albert, R. (1999) Emergence of scaling in random networks. *Science*, **286** (5439), 509–512.

Bloomquist, K.M. (2006) A comparsion of agent-based models of income tax evasion. *Social Science Computer Review*, **24** (4), 411–425.

Bosco, L. and Mittone, L. (1997) Tax evasion and moral constraints: some experimental evidence. *Kyklos*, **50** (3), 297–324.

Card, D. and Krueger, A.B. (1995) *Myth and Measurement – The New Economics of the Minimum Wage*, Princeton University Press, Princeton, NJ.

Cowell, F. (1985) The economic analysis of tax evasion. *Bulletin of Economic Research*, **37** (3), 305–321.

Davis, J.S., Hecht, G., and Perkins, J.D. (2003) Social behaviors, enforcement and tax compliance, dynamics. *Accounting Review*, **78**, 39–69.

Dugundji, E.R. and Gulyás, L. (2008) Sociodynamic discrete choice on networks in space: impacts of agent heterogeneity on emergent outcomes. *Environment and Planning B: Planning and Design*, **35** (6), 1028–1054.

Erdős, P. and Rényi, A. (1959) On random graphs. I. *Publicationes Mathematicae*, **6**, 290–297.

Gaujal, B., Gulyás, L., Mansury, Y., and Thierry, E. (2014) Validating an agent-based model of the Zipf's Law: a discrete Markov-chain approach. *Journal of Economic Dynamics and Control*, **41**, 38–49.

Gordon, J.P.P. (1989) Individual morality and reputation costs as deterrents to tax evasion. *European Economic Review*, **33** (4), 797–805.

Gulyás, L., Máhr, T., and Tóth, I.J. (2015) Factors to curb tax evasion: evidences from the TAXSIM agent-based simulation model. HAS Discussion Papers 1521, Institute of Economics, Centre for Economic and Regional Studies, Hungarian Academy of Sciences. (In Hungarian).

Khamis, M. (2008) Does the minimum wage have a higher impact on the informal than on the formal labor market? Evidence from quasi-experiments. IZA Discussion Paper, No. 3911.

Korobow, A., Johnson, C., and Axtell, R. (2007) An agent-based model of tax compliance with social networks. *National Tax Journal*, **60** (3), 589–610.

Kurowski, K., de Back, W., Dubitzky, W., Gulyás, L., Kampis, G., Mamonski, M., Szemes, G., and Swain, M. (2009) Complex system simulations with QosCosGrid, in *Proceedings of the 9th International Conference on Computational Science: Part I (ICCS '09)* (eds G. Allen, J. Nabrzyski, E. Seidel, G.D. Albada, J. Dongarra, and P.M. Sloot), Springer-Verlag, Berlin, Heidelberg, pp. 387–396.

Kurowski, K., Bosak, B., Grabowski, P., Mamonski, M., Piontek, T., Kampis, G., Gulyás, L., Coti, C., Herault, T., and Cappello, F. (2011) QosCosGrid e-science infrastructure for large-scale complex system simulations, in *Large-Scale Computing Techniques for Complex System Simulations* (eds W. Dubitzky, K. Kurowski, and B. Schott), John Wiley & Sons, pp. 163–183.

Lago-Penas, I. and Lago-Penas, S. (2010) The determinants of tax morale in comparative perspective: evidence from European countries. *European Journal of Political Economy*, **26**, 441–453.

Lemos, S. (2004) The effects of the minimum wage in the formal and informal sectors in Brazil. IZA Discussion Paper, No. 1089.

Lorscheid, I., Heine, B.-O., and Meyer, M. (2012) Opening the 'black box' of simulations: increased transparency and effective communication through the systematic design of experiments. *Computational and Mathematical Organization Theory*, **18**, 22.

Maloney, W. and Mendez, J. (2004) Measuring the impact of minimum wages. Evidence from Latin America, in *Law and Employment: Lessons from the Latin America and the Caribbean* (eds J.J. Heckman and C. Pages), University of Chicago Press, pp. 109–130.

Mansury, Y. and Gulyás, L. (2007) The emergence of Zipf's Law in a system of cities: An agent-based simulation approach. *Journal of Economic Dynamics and Control*, **31** (7), 2438–2460.

Méder, Z.Z., Simonovits, A., and Vincze, J. (2012) *Tax Morale and Tax Evasion: Social Preferences and Bounded Rationality*, MT-DP-2012/3, Institute of Economics, Research Centre for Economic and Regional Studies, Hungarian Academy of Sciences.

Mittone, L. and Patelli, O. (2000) Imitative behaviour in tax evasion, in *Economic Modeling with Swarm* (eds B. Stefansson and F. Luna), Kluwer, Amsterdam.

Müller, B., Bohn, F., Dreßler, G., Groeneveld, J., Klassert, C., Martin, R., Schlüter, M., Schulze, J., Weise, H., and Schwarz, N. (2013) Describing human decisions in agent-based models-ODD+ D, an extension of the ODD protocol. *Environmental Modelling & Software*, **48**, 37–48.

North, M.J., Collier, N.T., and Vos, J.R. (2006) Experiences creating three implementations of the repast agent modeling toolkit. *ACM Transactions on Modeling and Computer Simulation*, **16** (1), 1–25.

Sandmo, A. (1981) Income tax evasion, labor supply and the equity-effiency tradeoff. *Journal of Public Economics*, **16**, 265–288.

Sass, M. (2003) Versenyképesség és a közvetlen külföldi működőtőke-befektetésekkel kapcsolatos gazdaságpolitikák. PM Research Papers, Vol. 3, September, Budapest, p. 22. (In Hungarian).

Semjén, A. and Tóth, I.J. (2004) Rejtett gazdaság és adózási magatartás. Magyar közepes és nagy cégek adózási magatartásának változása 1996-2001. Elemzések a rejtett gazdaság magyarországi szerepéről, MTA KTI, Budapest, Paper #4, p. 70. (In Hungarian).

Semjén, A. and Tóth, I.J. (2009) Institutional environment, contractual discipline and tax behavior (Intézményi környezet, szerződéses fegyelem és adózási magatartás), in *Hidden economy, Undeclared employment and Non-reported income: government policies and the reaction of economic agents (Rejtett gazdaság, be nem jelentett foglalkoztatás és jövedelemeltitkolás – kormányzati lépések és a gazdasági szereplők válaszai)*, Book Series of Institute of Economics (eds. A. Semjén and I.J. Tóth), Hungarian Academy of Sciences. (In Hungarian).

Szabó, A., Gulyás, L., and Tóth, I.J. (2008) TAXSIM Agent Based Tax Evasion Simulator. Proceedings of the ESSA 2008 Conference, Brescia, Italy.

Szabó, A., Gulyás, L., and Tóth, I.J. (2009a) Az adócsalás elterjedtségének változása - becslések a TAXSIM ágensalapú adócsalás-szimulátor segítségével, Rejtett Gazdaság, MTA KTI, Budapest, pp. 65–83. (In Hungarian).

Szabó, A., Gulyás, L., and Tóth, I.J. (2009b) Sensitivity analysis of a tax evasion model applying automated design of experiments, in *Progress in Artificial Intelligence: 14th Portuguese Conference on Artificial Intelligence*, EPIA 2009, LNAI 5816 (eds. L.S. Lopes, N. Lau, P. Mariano, and L.M. Rocha), Springer-Verlag, Berlin Heidelberg pp. 572–583.

Szabó, A., Gulyás, L., and Tóth, I.J. (2010) Simulating tax evasion with utilitarian agents and social feedback. *International Journal of Agent Technologies and Systems*, **2** (1), 16–30.

Szabó, A., Gulyás, L., and Tóth, I.J. (2011) An Agent-Based Tax Evasion Model Calibrated Using Survey Data. Conference on The Shadow Economy, Tax Evasion and Money Laundering (SHADOW 2011).

Tóth, I.J. and Semjén, A. (2008) Tax behaviour and financial discipline of Hungarian enterprises, in *The Hungarian SME Sector Development in Comparative Perspective* (ed. L. Csaba), CIPE/USAID – Kopint-Datorg Foundation, Budapest, pp. 103–134.

Watts, D.J. and Strogatz, S.H. (1998) Collective dynamics of 'small world' networks. *Nature*, **393** (4), 440–442.

Wilensky, U. (1999) *"NetLogo" Center for Connected Learning and Computer-Based Modeling*, Northwestern University, Evanston, IL, http://ccl.northwestern.edu/netlogo/ (accessed 20 July 2017).

7

Development and Calibration of a Large-Scale Agent-Based Model of Individual Tax Reporting Compliance

Kim M. Bloomquist

7.1 Introduction

Since the publication of the groundbreaking theoretical work by Allingham and Sandmo (1972) and Srinivasan (1973), much has been learned about the determinants of taxpayer compliance.[1] Despite these advances, progress has lagged in transforming this knowledge into computational tools that tax officials can use to conduct in silico tests of proposed tax service and enforcement programs prior to implementation on potentially millions of taxpayers. Alm (1999) suggests that the key reason for this lack of progress is the inability of existing analytical (i.e., mathematical) models to incorporate sufficient real-world taxpayer behavior and he goes on to point out that past efforts to introduce greater realism into the standard rational choice model of taxpayer decision-making have tended only to increase the ambiguity of the model's predictions. A similar observation has been made by Janssen and Ostrom (2006) and Axtell (2000) for complex social and ecological systems in general. Increasingly, researchers are concluding that

---

[1] Surveys of the tax compliance literature are found in Andreoni *et al.* (1998), Alm (1999), Sandmo (2005), and Slemrod (2007). The latest survey is available in Chapter 1.

Agent-Based Modeling of Tax Evasion: Theoretical Aspects and Computational Simulations,
First Edition. Edited by Sascha Hokamp, László Gulyás, Matthew Koehler, and Sanith Wijesinghe.
© 2018 John Wiley & Sons Ltd. Published 2018 by John Wiley & Sons Ltd.

agent-based modeling and simulation (ABMS) is a methodology that is well suited for modeling complex social phenomena, of which taxpayer compliance is a prime example (see Alm, 2010).

This chapter describes the development and calibration of a large-scale ABM that simulates the income tax reporting behavior of a community of 85,000 individual taxpayers. The Individual Reporting Compliance Model (IRCM) includes many enforcement mechanisms used by tax agencies, such as audits and information reporting, as well as detailed information on the reporting compliance for major income and offset[2] items. A more detailed description of the IRCM is found in Bloomquist (2012). Other articles featuring the IRCM are Bloomquist (2013) and Bloomquist and Koehler (2015).

The decision to use ABMS (i.e., an object-oriented approach) to model taxpayer reporting behavior over a variable-oriented approach such as system dynamics (Cioffi-Revilla, 2014) was made based on reasoning similar to that outlined in Rand and Rust (2011). In their paper, Rand and Rust cite six characteristics of a social system that make it suitable for analysis using ABMS. The one necessary feature is that the system must be one that is temporally dynamic as it can (or has the potential to) give rise to multiple equilibria over time. Therefore, the system must be one that is temporally dynamic. Adapting to changing environmental conditions (i.e., exhibiting learning behavior) is a characteristic of social systems that the authors see as *sufficient* for analysis using ABMS as adaptive agents. Four other characteristics are seen as *indicative* of social systems that are suited for analysis using ABMS. They are the following:

(1) Heterogeneity. Individuals or groups of individuals that exhibit different responses to identical environmental conditions or that are subject to different constraints. For example, among individual taxpayers the opportunity to evade varies greatly for employees whose earnings are reported to the Internal Revenue Service (IRS) by employers, and sole proprietors that are subject to little or no third-party information reporting.

(2) Local and Potentially Complex Interactions. When individuals exchange information through networked interaction with other individuals or businesses, this can substantially alter behavior over time and space. Stolen identity tax refund schemes is an example of such behavior.

(3) Rich Environment. An environment that is "rich" in detail is one that captures a broad range of situations that potentially influence behavior. Among taxpayers this can be interactions with the tax agency itself, tax preparers, family members, acquaintances, coworkers, and professional organizations.

(4) Medium Numbers. Rand and Rust correctly point out that ABM is not an appropriate tool when only one or two agents are involved since such small-scale interactions are better analyzed using game theory. The IRCM

---

[2] Examples of offset items include deductions, exemptions, and tax credits.

itself may be classified as a medium- to large-scale model since it investigates the tax reporting behavior of approximately 10^5 agents. However, ABMs with millions of agents are now being built,[3] and limiting the scope of ABMs to "medium-size" systems seems an artificial limitation.

The goal for the design of the IRCM was to have a model that would represent the major real-world features and institutions of modern tax administration. Modeling an entire community of taxpayers makes it possible to represent social networks that empirical research has shown to significantly influence tax compliance behavior (Bernard *et al.*, 2007). The IRCM does this by including formal (and observable) relationships between taxpayers and commercial tax preparers and employers as well as informal (and less directly observable) social networks among taxpayers in both workplace and residential settings. Links between taxpayers and paid preparers and taxpayers (employees) and employers were based directly on tax return data but all unique identifying information (including the identity of the study area itself) has been removed. Tax return information for all 85,000 taxpayers is amply detailed with each tax return containing 180 distinct elements. Misreporting behavior (both over and underreporting) is based on results from random taxpayer audits conducted by the IRS for tax year (TY) 2001 (Internal Revenue Service, 2007). Last, but not least importantly, the IRCM provides the main tax enforcement tools including taxpayer audits, third-party information reporting, and tax withholding.

7.1.1   Taxpayer Dataset

To preserve taxpayer anonymity and yet facilitate model verification and validation, the IRCM uses a dataset of artificial taxpayers. The dataset of artificial taxpayers was created by substituting cases from the US IRS Statistics of Income (SOI) Public Use File (PUF) for actual tax returns of the 85,000 taxpayers featuring in the study area. Although most fields in the PUF are derived from tax forms, SOI modifies the data in order to protect the identity of individuals. Substitution was performed by first partitioning tax return and PUF records and selecting (with replacement) the PUF record that most closely matches each taxpayer record in the study area. Further details on the statistical matching algorithm are provided in Bloomquist (2012). Table 7.1 compares the resulting dataset of artificial taxpayers to the actual tax return data by major income and offset item.

From Table 7.1 it can be seen that for the largest line items (e.g., wages, interest, Schedule C income, capital gains, pension income, Schedule E income, deductions, exemptions, and total adjusted gross income (AGI)) there is close agreement in the number of returns (with non-zero values) and total dollar

---

[3] See Axtell (2013).

Table 7.1 Comparison of actual versus artificial taxpayer data for study region

Income item	Actual data N (non-zero)	Actual data Sum ($1000)	Artificial data N (non-zero)	Artificial data Sum ($1000)	Percent difference in sums (%)
Wages	72,058	$2,744,170	71,773	$2,738,049	−0.2
Interest	47,768	$138,156	42,582	$125,803	−8.9
Dividends	22,951	$77,716	19,590	$65,905	−15.2
Tax refunds	14,955	$6,098	10,764	$7,287	19.5
Alimony	238	$2,748	155	$2,071	−24.6
Schedule C	8,728	$92,480	7,610	$90,104	−2.6
Capital gains	17,636	$95,117	14,520	$89,043	−6.4
Other gains	930	$81	690	$802	887.6
IRA	6,820	$68,681	5,315	$59,328	−13.6
Pensions	18,604	$277,083	16,597	$269,574	−2.7
Schedule E	8,769	$116,042	7,185	$120,370	3.7
Schedule F	1,143	$1,154	841	$2,252	95.2
Unemp. comp.	6,203	$19,783	4,774	$15,311	−22.6
Social security	8,461	$73,374	7,821	$68,003	−7.3
Other income	4,576	$9,194	4,573	$222	−97.6
Total AGI	84,842	$3,695,035	84,846	$3,635,509	−1.6
Deductions	84,851	$731,363	84,907	$743,302	1.6
Exemptions	75,870	$455,524	75,905	$453,310	−0.5

amount.[4] In addition, in 20 of 21 postal code zones that make up the study region (not shown in Table 7.1), the percentage difference in Total AGI between the actual and artificial data is in low single digits (Bloomquist, 2012).

7.1.2   Agents

Figure 7.1 graphically displays the IRCM agent architecture.[5] A single *Region* is composed of multiple non-overlapping zones (e.g., a postal code zone). Each *Zone* is the place of residence for a group of filers. Each *Zone* also includes any tax preparers and employers operating within its borders. A *Preparer* agent prepares tax returns for its *Filer* clients. *Employer* agents represent firms having one or more employee tax filers. The *TaxReturn* class defines the characteristics of all tax returns, which are reviewed by a tax agency (an instance of the *TaxAgency* class) and may be selected for an audit.

---

[4] Schedule C income is income earned by sole proprietors. Schedule E income is income derived from partnerships and small corporations. Schedule F income is farm income. IRA is taxable income from individual retirement accounts (IRAs). Unemp. Comp. is unemployment compensation. AGI is adjusted gross income or total income net of statutory adjustments.

[5] In Figure 7.1 the symbols 1, ◊, and * refer to 1 (agent), *aggregation*, and *many* agents, respectively.

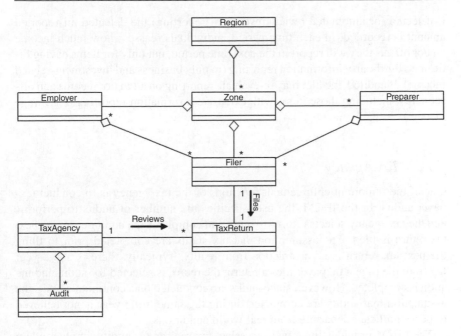

Figure 7.1   IRCM agent hierarchy

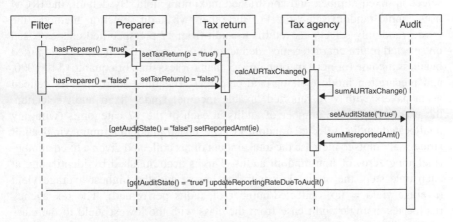

Figure 7.2   Interaction between filer and tax agency

The interaction between filers and the tax agency is illustrated in Figure 7.2.[6] The filer either self-prepares or uses a paid preparer. The tax agency reviews all filed tax returns to determine if any discrepancies are present on items having third-party information reporting. If a return is audited and underreporting

---

[6] In Figure 7.2 the connector label *setReportedAmt(ie)* means "execute the method that sets the amount reported for an imputable income or offset line item *element* on an individual's tax return."

is detected (or automated review reveals misreporting) the detected misreported amount is recorded. In each time period, individuals reassess how much income and/or offsets they will report in the next time period, but only for items having little or no third-party information reporting (mainly business and investment-related income). If audited, the filer may re-evaluate reporting on all major income and offset items, including those items with extensive information reporting (e.g., wage income).

7.1.3   Tax Agency

One of the important enforcement functions of the tax agency is to conduct taxpayer audits. In the IRCM, the user specifies the number of audits to perform[7] and the tax agency selects returns to audit and completes the audit before the next tax return is filed. The assumption that tax audits are completed prior to filing the next tax return is an abstraction from reality. Typically, there is a two-year lag from the time a taxpayer files a return, the return is selected for audit, and the audit is completed. However, since audits are conducted on a continuous basis, the assumption that audits are completed during the same filing year is not believed to be a significant departure from real-world conditions.

The IRCM provides three ways to select tax returns for audit: simple random selection, fixed number, and constrained maximum yield. By default, the IRCM uses simple random selection. For the other two audit selection methods, the model randomly selects tax returns to audit from 17 pre-specified classes based on selected return characteristics: deduction type (standard or itemized), reported business income (or not), income greater than (or less than or equal to) $100,000, and preparation mode (self or paid-preparer). The final audit class is defined as taxpayers with zero-reported taxable income. Under fixed audit selection, the user specifies the number of audits in each of the 17 categories (with any unallocated audits selected randomly). The constrained maximum yield audit strategy attempts to increase the total amount of tax collected given a fixed number (including zero) of non-random audits. This is accomplished by identifying, at each time step, the audit class with the lowest and the highest average yield (average yield = tax collected/number of audits performed). The tax agency reallocates a single audit case from the class with the lowest yield to the class with the highest yield. The process of reallocating a single audit case for the next time step continues until the minimum coverage for the lowest yielding audit class is reached. The tax agency then repeats this same process with the second lowest yielding audit class and so on. Similarly, if the user-specified maximum coverage rate is reached the tax agency reallocates audits to the second highest yielding class, and so on.

---

[7] The number of audits to perform can be entered either as a whole number or as a percentage of the number of filers.

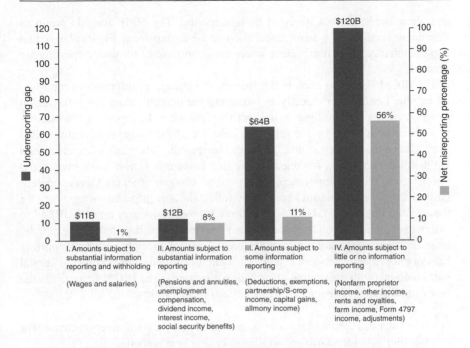

Figure 7.3   Tax year 2006 individual income tax underreporting gap

In addition to performing audits, the tax agency in the IRCM also performs automatic checks of taxpayer reported income using available third-party information documents. In fact, although infrequently discussed in the academic literature on tax evasion, the most effective means for promoting high levels of voluntary compliance is, in fact, the existence of third-party information reporting. Figure 7.3 displays the relationship between reporting non-compliance and amount of information reporting for a weighted representative sample of TY 2006 US individual taxpayers (Black *et al.*, 2012). In Figure 7.3, the "Underreporting Gap" refers to the dollar amount of underreported tax and "net misreporting percentage" (NMP) is a measure that allows for the comparison of relative non-compliance among the various line items.[8] Wage and salary income is least likely to be misreported (NMP equal to 1%) because it is subject to both third-party information reporting and withholding. Items subject to third-party information reporting, but not to withholding (e.g., pension income, social security income, and interest and dividend income), have an NMP of 8%. Items subject to partial reporting by third parties (e.g., capital gains) have a still higher NMP of 11%. Lastly, items not subject to withholding or third-party information reporting (e.g., sole proprietor income and "other" income) are the least visible

---

[8] The NMP is defined as the net amount of income misreported in the taxpayer's favor divided by the sum of the absolute values of the amounts that should have been reported.

and, therefore, are most likely to be misreported. The NMP for this group of line items is 56%. The main conclusion to be drawn from Figure 7.3 is that non-compliance is most prevalent where the opportunities for underreporting are greatest.[9]

The IRCM is able to explore the impact of a change in information reporting on income items either directly by changing the default values for information coverage and/or withholding or indirectly by mapping the reporting characteristics of one item based on the reporting behavior of the taxpayer on another line item. Under the direct method, the user changes the information coverage and withholding parameters for one or more line items using drop-down menus and check boxes. With this approach, the reporting behavior of all taxpayers is potentially changed (with random variation) to reflect the new information reporting for the selected line item(s). Under the indirect approach, the user tells the IRCM to assign user reporting behavior for one line item to another line item. For example, let us say the analyst wants to use the indirect approach to apply observed reporting behavior for dividend income (with substantial information reporting) to capital gains income (with only some information reporting). The IRCM determines the new capital gains reporting behavior for each filer as follows:

(1) If the filer has dividend income, then assume that the current reporting rate for this filer's dividend income applies to capital gains income.
(2) If the filer does not have dividend income, then query the members of the filer's neighbor reference group to see if someone has dividend income. Use the reporting characteristics of the first neighbor reference group member with dividend income.
(3) If no one in the filer's neighbor reference group has dividend income, then query the members of the filer's coworker reference group. Use the first coworker's dividend income reporting behavior, if one is found.
(4) If no coworker has dividend income (or the filer has no coworkers, i.e., self-employed), then query the clients of the filer's tax preparer (if not self-prepared). Use the first client's dividend income reporting behavior, if one is found.
(5) If no neighbor, coworker, or preparer client has dividend income, randomly query other filers in the region until someone with dividend income is found. Use the dividend income reporting characteristics of the randomly selected filer as this filer's capital gains reporting characteristics.

Once the user finishes specifying the level of information reporting for selected line items, IRCM uses these up-dated parameters to recalculate reported amounts

---

[9] The data in Figure 7.3 illustrate the relationship between taxpayer reporting behavior and third-party information reporting for TY 2006. Table 7.2 (column 2) in this chapter provides similar results (although at the line item level) for TY 2001. Information on prior IRS tax gap studies are available on the web at https://www.irs.gov/uac/irs-the-tax-gap.

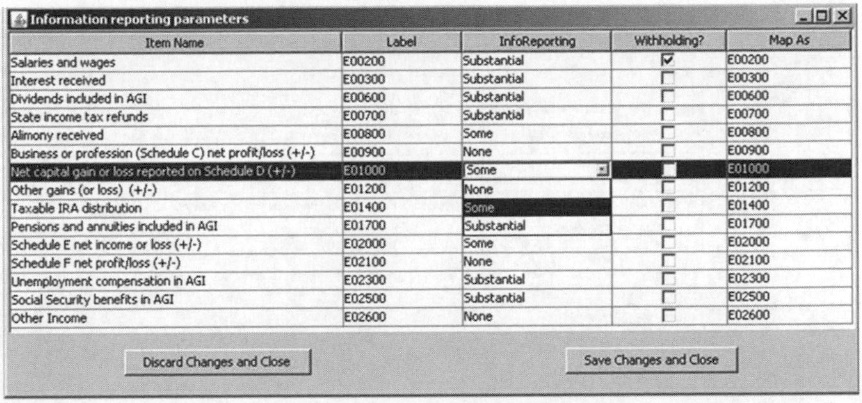

Information reporting parameters

Item Name	Label	InfoReporting	Withholding?	Map As
Salaries and wages	E00200	Substantial	☑	E00200
Interest received	E00300	Substantial	☐	E00300
Dividends included in AGI	E00600	Substantial	☐	E00600
State income tax refunds	E00700	Substantial	☐	E00700
Alimony received	E00800	Some	☐	E00800
Business or profession (Schedule C) net profit/loss (+/-)	E00900	None	☐	E00900
Net capital gain or loss reported on Schedule D (+/-)	E01000	Some	☐	E01000
Other gains (or loss) (+/-)	E01200	None	☐	E01200
Taxable IRA distribution	E01400	Some	☐	E01400
Pensions and annuities included in AGI	E01700	Substantial	☐	E01700
Schedule E net income or loss (+/-)	E02000	Some	☐	E02000
Schedule F net profit/loss (+/-)	E02100	None	☐	E02100
Unemployment compensation in AGI	E02300	Substantial	☐	E02300
Social Security benefits in AGI	E02500	Substantial	☐	E02500
Other Income	E02600	None	☐	E02600

Discard Changes and Close   Save Changes and Close

Figure 7.4 IRCM information reporting parameters screen

prior to running a simulation. Figure 7.4 shows the IRCM's Information Reporting Parameters Screen that allows a user to make changes to information reporting and withholding for the major income line items.

7.1.4 Taxpayer Reporting Behavior

In the IRCM, taxpayer reporting behavior is modeled using either the SOI reporting regime (default) or the rule-based reporting regime. When the SOI reporting regime is selected, the IRCM uses values from the PUF data to instantiate filer reported income and offset amounts. The SOI reporting regime option is useful for performing model validation, for example, by comparing the line item NMPs calculated by the model to NMPs in published IRS studies on the tax gap (Internal Revenue Service, 2007). The SOI reporting regime also provides a benchmark for model calibration (discussed later in this chapter). The baseline reporting rate calculated using the SOI reporting regime is assumed to be the individual filer's preferred reporting behavior given the enforcement environment in effect at time $t = 0$. If never audited or if the filer's coworkers or neighbors are never audited, then the baseline reporting rate on each line item remains unchanged throughout the filing "lifetime" of the individual.

By selecting the rule-based reporting regime the user tells the IRCM to determine filer reported amounts and baseline reporting behavior using six user-specified parameters. The user sets the values of these parameters using sliders on the Filer Parameters screen (Figure 7.5). The top three sliders set the probability of misreporting success for income and offset items characterized by the extent of third-party information reporting (No Information Reporting, Some Information Reporting, or Substantial Information Reporting). For example, if the Substantial Information Reporting slider is set to a value of 10, then the

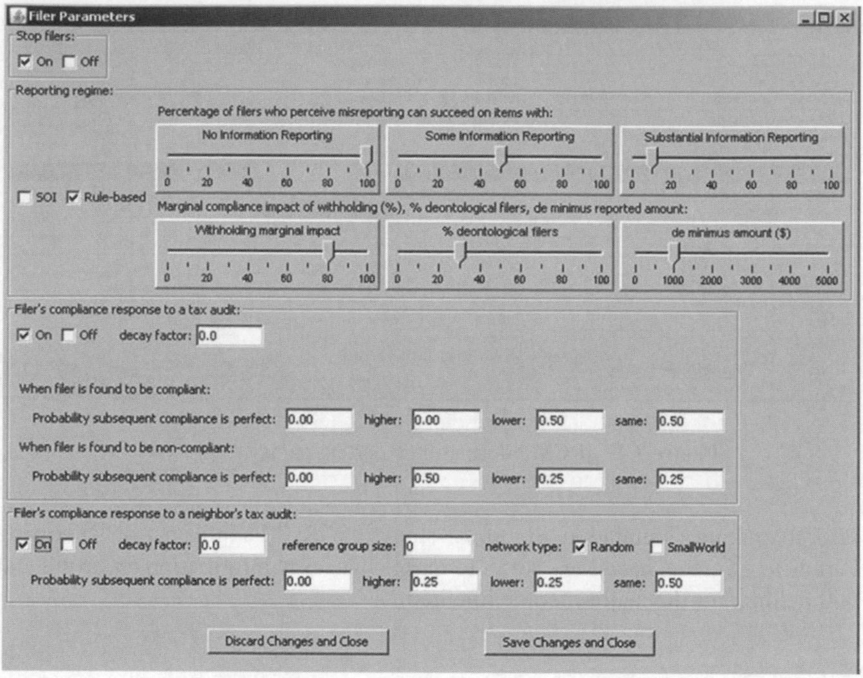

Figure 7.5   IRCM filer parameters screen

model assumes 10% of filers believe that misreporting on items with substantial
information reporting will be successful and 90% of filers hold the opinion that
misreporting will not succeed. This difference in perception among filers may
stem from different levels of knowledge and experience or due to qualitative
differences in information reporting within a given income or offset line item.

The second row of sliders in the Reporting Regime section of the Filer
Parameters screen defines additional influences on filers' reporting behavior. The
"Withholding marginal impact" slider sets the marginal impact of withholding on
reporting compliance. For example, Figure 7.3 above suggests that the marginal
impact of withholding is between 80 and 90% based on the reduction in NMP
from 8% for items subject to substantial misreporting (e.g., unemployment
compensation, dividends and interest income) to an NMP of 1% for wage and
salary income.[10] The "% deontological filers" slider sets the percentage of filers
whose reporting compliance is motivated by non-economic factors. While the
term "deontological" suggests that the primary motivating principle is duty-based
influences, equity or personal integrity can be included here as well. If this slider
is set to a value of 30, then IRCM randomly selects 30% of filers to become

[10] That is, the impact of withholding on reporting compliance is in addition to the impact of third-party
information reporting.

deontological filers. Such filers are assumed to fully and accurately report all income and offset items.[11] Finally, the slider de minimis amount is used to set a minimum threshold amount for reporting for items with no information reporting. If the calculated reported amount for a given live item falls below the de minimis threshold, the filer is assumed to report zero for that item. More detail on the procedure the IRCM uses to derive line item-specific reporting rates is available in Bloomquist (2012).

7.1.5 Filer Behavioral Response to Tax Audit

Since taxpayers cannot know for certain that actions they take (or not take) will cause the tax agency to select their tax return for an audit, the reporting behavior of these taxpayers is modeled as a partially observable Markov decision process (POMDP) (Ghallab *et al.*, 2004).

A POMDP is a five-tuple $\sum = (S, A, P, C, O)$ where:

- S is a finite set of states (audited or not audited)
- R is a finite set of compliance responses (perfect, increase, decrease, no change)
- P is a probability distribution where for each $s \in S$, if there exists $r \in R$ and $\hat{s} \in S$ such that $P_r(\hat{s} \mid s) \neq 0$, we have $\sum_{\hat{s} \in S} P(s, r, \hat{s}) = 1$
- $C_r(s, \hat{s})$ is the cost/reward (or expected cost/reward) experience from transition to state $\hat{s}$ from state s with transition probability $P_r(\hat{s} \mid s)$. The quantity $P_r(\hat{s} \mid s)$ is the probability that if response r is taken in state s, then state $\hat{s}$ will result. For example, if a taxpayer decides to increase compliance following a tax audit, one can infer that the action is being taken in order to reduce the probability of being selected for an audit (and the associated costs) in future time periods.
- O is a set of observations with probabilities $P_r(o \mid s)$, for any $r \in R$, $s \in S$, and $o \in O$. $P_r(o \mid s)$ represents the probability of observing o in state s after taking response r. Finally, it is required that the sum of probabilities over the set of observations is 1, that is, $\sum_{o \in O} P_r(o \in s) = 1$.

Since the observations in a POMDP represent probability distributions, rather than exact states of the system, the probability distributions are called *belief states* and are updated using Bayes rule. The use of Bayes rule implies that the probabilities represented by $P_r(o \in s)$ are not static but change as knowledge of the enforcement environment changes.

In the IRCM, neither the belief states (O) nor the cost functions (C) of individual filers are modeled explicitly but are implied by filers' stochastically modeled "choices." Stochastic choice modeling is used since relatively little is known about how taxpayers perceive the tax enforcement environment and what factors motivate changes in observed behavior.

---

[11] Empirical research (Alm *et al.*, 2015) has shown that even when evasion is the "rational" choice, some taxpayers report 100% of their income.

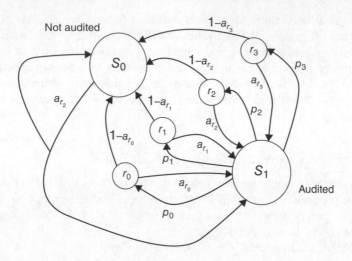

Figure 7.6   POMDP of the filer's response to the tax audit environment

Figure 7.6 graphically illustrates the POMDP for the filer's response to a tax audit. The two states are not audited (S_0) and audited (S_1). A taxpayer is in one of these two states in each time period. The filer's belief about the probability of audit is defined as $a_{r_k} = b(a \in r_k)$, implying that a filer's perceived probability of being selected for a tax audit depends on her belief about how the baseline audit probability (a) changes with a change in reporting behavior (response) r_k.

In Figure 7.6 it is assumed $r_0 \equiv$ no change in reporting compliance, $r_1 \equiv$ an increase in reporting compliance, $r_2 \equiv$ decrease in reporting compliance and $r_3 \equiv$ perfect reporting compliance.[12] If not audited in time t, the filer may start or increase underreporting in time $t + 1$ on income subject to little or no information reporting, assuming the filer has such income from one or more sources. If the filer is audited in time t, the decision to select a response $r_0 \dots r_3$ is determined in the IRCM by a random draw and the user-specified probabilities $p_0 \dots p_3$. Although the IRCM models the filer's response as a stochastic process, actual filers are presumed to select an action r_k based on their (heterogeneous and non-stationary) beliefs about the expected cost associated with that action.

7.1.6   Model Execution

The steps followed in executing a simulation using the IRCM are shown in Figure 7.7. The model reads tax return data for the population of artificial taxpayers and instantiates agents. During instantiation, the IRCM estimates a true amount for the largest income and offset items. The true amount is the

---

[12] These response categories are based on Gemmell and Ratto (2012).

amount reported plus imputed misreporting.[13] Imputed amounts are based on audit results from the TY 2001 National Research Program (NRP) study. Details of the imputation methodology are described in Bloomquist (2012).

Each time step represents one filing cycle (year). Tax calculations are performed twice for all taxpayers, first using reported amounts and again using estimated true amounts. The difference in calculated tax using true and reported amounts is the tax gap for each filer. By default, the IRCM assumes that the difference between the true and reported tax amounts is the amount identified by the tax auditor. An option is provided to account for underreporting not detected by examiners.[14]

Tax audits are performed at the penultimate step in each time loop. During wrap up, the tax agency issues notices to taxpayers who are not audited but where computer checking of tax returns against information documents detects some underreporting.[15] In addition, filers who stop filing, either because they leave the region or because they no longer have an obligation to file, are replaced by a new filer having identical income and network relationships as the "stop filer" being replaced, but with reporting behavior and memory reset to baseline levels (i.e., no memory of a prior audit experience or audits of reference group members, if that option is selected). The reporting behavior of filers who are not "stop filers" is also updated at each time step, as is the audit selection strategy of the tax agency.[16] Finally, data collection occurs during the wrap-up phase. When the user-specified number of time steps has completed the model generates output in the form of tables and charts that can be reviewed and saved for further analysis.[17]

7.2 Model Validation and Calibration

A two-stage approach is used to validate and calibrate the IRCM. In stage 1 (validation), the model is executed using values from the PUF (the "SOI reporting regime" option) and the output is compared to IRS estimates of reporting non-compliance published tax gap studies. The method of comparison follows Axtell and Epstein's (1994) hierarchical approach consisting of four increasingly

---

[13] Imputed income items are: wages, interest, dividends, state tax refunds, alimony, sole proprietor (Schedule C) income, capital gains (Schedule D) income, other gains (Form 4797), individual retirement account (IRA) income, pension income, partnership/small corporation/rents and royalty (Schedule E) income, farm (Schedule F) income, unemployment compensation, social security and other income. Imputed offset items are adjustments, deductions, exemptions, and statutory credits (net the Child Tax Credit). Adjustments to certain credits (e.g., Child Tax Credit, Earned Income Credit (EIC), and Additional Child Tax Credit) that reflect a change in income are performed by a tax calculator. Each tax return record consists of 180 elements based on PUF data.

[14] This is done by applying multipliers to positive detected misreported amounts; that is, misreporting in the taxpayer's favor. See Erard and Feinstein (2011) for an overview of the detection controlled estimation (DCE) methodology used to derive the multipliers used by IRCM.

[15] The threshold amount for issuing a notice is set by the user. The default value is $1.

[16] See Bloomquist (2012, 2013) for further discussion of IRCM audit selection strategies.

[17] See Appendix 7A for a more detailed discussion of IRCM's design and implementation according to the ODD protocol (Grimm et al. 2006, 2010).

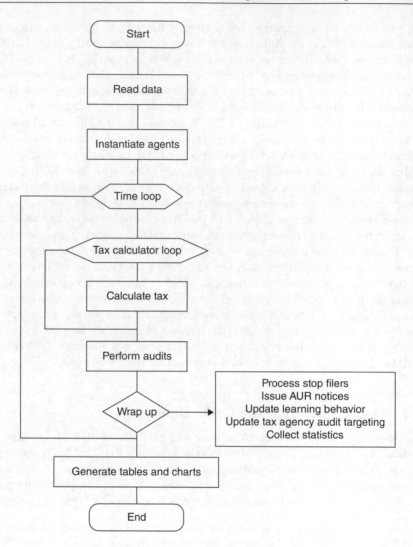

Figure 7.7   IRCM execution sequence: top-level view

detailed levels of validation. A model with Level 0 validity is considered to be a caricature of reality. At this level the model needs to show only that the system as a whole exhibits behavior that is consistent with the available data (e.g., the aggregate response of agents to changing environmental conditions is in the appropriate direction). At Level 1, the model is expected to be in qualitative agreement with empirical macro-structures. This is demonstrated by comparing the distributional characteristics of the actual population to the modeled population. To be valid at Level 2 the model must show quantitative agreement with

Table 7.2 Line item net misreporting percentages: IRS versus IRCM

| | Net misreporting percentage | | |
| | | IRCM reporting regime | |
Income item	IRS	SOI	Rule-based
Wages	1	1	1
Interest	4	3	5
Dividends	4	4	5
Tax refunds	12	14	7
Schedule C	57	63	63
Capital gains	12	13	24
IRA	4	7	4
Pensions	4	3	5
Schedule E	35	28	28
Schedule F	72	63	62
Unemp. comp.	11	15	6
Social security	6	10	5
Other income	64	82	63
Taxable income	11	13	12
Tax	18	17	16
Adjustments	−21	−24	−41
Deductions	5	3	5
Exemptions	5	4	5

empirical macro-structures. Finally, at Level 3, the model exhibits quantitative agreement with empirical micro-structures, as determined from cross-sectional and longitudinal analysis of the agent population.

IRCM's on-board graphical and statistical routines are used to demonstrate model validity through Level 2. Validation at Level 3 requires panel data on an individual's tax reporting behavior, which is a standard not yet available to researchers. Table 7.2 summarizes results for a Level 2 validation that compares line item NMPs produced by the model to NMPs calculated by IRS in the TY 2001 tax gap study (Internal Revenue Service, 2007).[18] Focusing on the column in Table 7.2 labeled "SOI" we see that the model overestimates the NMPs on some items (e.g., Schedule C (sole proprietor) income, taxable IRA income, unemployment compensation, taxable social security benefits, and other income) and underestimates on others (e.g., Schedule E (partnership/small corporation) income, Schedule F (farm) income, deductions, and exemptions). The model-generated NMP for total tax is within one percentage

---

[18] IRCM-based values in Table 7.2 represent an average of five simulations using different random number seeds.

point of the IRS estimate for the SOI regime and two percentage points for the Rule-Based regime. One reason for this difference is a lack of data on the PUF specific to children eligible for the EIC. Another reason is an overall average effective tax rate for the study area, which is slightly lower than the national average.[19]

The goal in stage 2 (calibration) is to find a combination of values for the six "rule-based reporting regime" parameters[20] that can closely replicate IRCM output using the "SOI reporting regime" option. Formally, it is preferable to minimize the sum of differences in reported incomes between the SOI and rule-based reporting regimes:

$$\min \sum_i (\Lambda_i^{RB} - \Lambda_i^{SOI}) \tag{7.1}$$

In Eq. (7.1), Λ_i^{RB} is the calculated reported amount using the rule-based reporting regime in IRCM for income type i and Λ_i^{SOI} is the calculated reported amount for income type i using the SOI reporting regime. A solution for Eq. (7.1) is found by inspection using multi-stage Monte Carlo simulation, the details of which are described in Bloomquist (2012). The column of Table 7.2 labeled "rule-based" shows the resulting line item NMPs for model calibration.

7.3   Hypothetical Simulation: Size of the "Gig" Economy and Taxpayer Compliance

Recently, much attention has been focused on the growing number of people employed in the so-called "Gig" Economy. Gig workers typically are self-employed as independent contractors or freelancers and are often associated with services marketed online. Some better-known examples of Gig Economy firms include: Uber, Lyft, and Airbnb. Other categories of jobs associated with the Gig Economy include "contingent" workers such as agency temps, on-call workers, contract company workers, and part-time laborers. Although little official data exists on the number of gig workers, a recent estimate using the broadest definition of the US gig workforce finds that contingent workers accounted for over 40% of the total labor force in 2010 (U.S. Government Accountability Office, 2015) and grew twice as fast as overall employment (14.4% vs 7.2%) from 2002 to 2014 (Rinehart and Gitis, 2015).

From a tax compliance perspective, why does it matter that gig workers seem to be accounting for an increasing share of the labor force? In short, it matters if jobs performed by gig workers as independent contractors are substitutes for jobs previously performed by full-time employees. By law, employers are responsible for their employees' tax withholding and information reporting whereas independent contractors are responsible for their own withholding and are not subject to

---

[19] The nation's estimated mean effective tax rate in TY01 is 14.95% versus 14.69% for the study region.
[20] See discussion in Section 7.1.4.

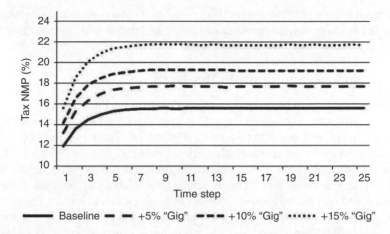

Figure 7.8 Model time series of tax NMPs for alternative increases in share of gig economy workers

third-party information reporting.[21] IRS random audit studies have conclusively established that incomes subject to information reporting are reported by taxpayers at much higher levels than incomes not subject to information reporting. For example, Figure 7.3 shows the NMP for income subject to substantial information (e.g., interest income, pension income) was 8% compared to 56% for sole proprietor income (no information reporting). The NMP for wage income, subject to both information reporting and withholding, is only 1%.

The IRCM was used to simulate the impact on tax reporting compliance of a shift in the composition of the labor force to include progressively larger shares of gig economy workers. Figure 7.8 shows the model output for tax NMP for the baseline and three alternative scenarios that represent a shift of 5%, 10%, and 15% of current full-time employees to gig workers. In all scenarios simulations are performed using default values for the rule-based reporting regime. The values displayed in Figure 7.8 are averages for five independent simulations using different random number seeds.

Figure 7.8 shows that from a baseline tax NMP of about 16% (84% reporting compliance rate), each five percentage point increase in the share of gig workers causes the tax NMP to increase by about two percentage points. At the national level in the United States each additional percentage point of voluntary compliance brings in approximately $30 billion in tax receipts (Koskinen, 2015). Thus, when

---

[21] However, recent legislative changes in the United States have extended third-party information to certain payment card (e.g., credit, debit) and other electronic transactions used by independent contractors and other businesses. These new requirements began to be phased beginning 2011 (Internal Revenue Service, 2015).

considered from a national perspective, even small reductions in the relative size of the full-time employee labor force can result in significant losses in tax revenue.

7.4   Conclusion and Future Research

The purpose of this chapter has been to demonstrate the feasibility of using ABMs to simulate a taxpayer reporting compliance while incorporating the complexities of real-world tax systems. The major features of the IRCM are: (i) a community-based approach that allows network relations to be modeled explicitly, (ii) imputation of misreported income and offsets using results from random taxpayer audits thus enabling micro-level analysis of taxpayer reporting behavior, and (iii) the creation of a dataset of artificial taxpayers that can be used to independent model verification and validation. In fact, the IRCM has undergone independent verification and validation testing by analysts at The MITRE Corporation. The model, originally written in Java using Repast Simphony 1.0 (North *et al.*, 2007), has been ported successfully to both Windows and Mac-OS platforms running Repast 2.0 and MASON (Cline *et al.*, 2014).

The IRCM is capable of performing a wide range of "what–if" analyses involving various aspects of taxpayer reporting compliance. This capability was demonstrated in a simulation experiment that estimated the impact on voluntary reporting compliance of progressively larger relative shares of so-called "gig economy" workers. The simulation found that switching 5% of the labor force from full-time employees to gig workers lowers the voluntary compliance rate by roughly two percentage points. In the United States, each percentage point of voluntary compliance translates to approximately $30 billion in tax receipts.

The value of a model such as the IRCM increases as our knowledge of taxpayer behavior improves. Specific topics that would improve the predictive capability of ABMs for taxpayer compliance use include (i) research on the indirect effect of taxpayer audits, (ii) research on attitudinal and social factors associated with tax morale, and (iii) how the provision of taxpayer services interacts with tax enforcement to achieve observed levels of voluntary compliance. Finally, building a massive-scale ABM ($\sim 10^8$ taxpayers) is now feasible due to the availability of multi-processor computing environments and software, such as Repast HPC, designed for such platforms (see also Chapter 8). The development of the IRCM shows that building such national-scale models is a goal that is now within reach.

Acknowledgments

The author is grateful to Professors Robert L. Axtell and Claudio Cioffi-Revilla for their guidance and encouragement on this project, which was done in partial fulfillment of his Ph.D. in Computational Social Science at George Mason University, Fairfax, Virginia.

References

Allingham, M.G. and Sandmo, A. (1972) Income tax evasion: a theoretical analysis. *Journal of Public Economics*, **1**, 323–338.

Alm, J. (1999) Tax compliance and administration, in *Handbook on Taxation* (eds W.B. Hildreth and J.A. Richardson), Mercel Dekker, New York, pp. 741–768.

Alm, J. (2010) Testing behavioral public economics theories in the laboratory. *National Tax Journal*, **63**, 635–658.

Alm, J., Bloomquist, K.M., and McKee, M. (2015) On the external validity of laboratory tax compliance experiments. *Economic Inquiry*, **53**, 1170–1186.

Andreoni, J., Erard, B., and Feinstein, J. (1998) Tax compliance. *Journal of Economic Literature*, **36**, 818–860.

Axtell, R.L. (2000) *Why Agents? On the Varied Motivations for Agent Computing in the Social Sciences*, Center on Social and Economic Dynamics Working Paper No. 27, Brookings Institution, Washington, DC.

Axtell, R.L. (2013) Team dynamics and the empirical structure of U.S. firms, Department of Computational Social Science Working Paper, George Mason University, http://tinyurl.com/jee8anf (accessed 20 June 2016).

Axtell, R.L. and Epstein, J.M. (1994) Agent-based modeling: understanding our creations. *Bulletin of the Santa Fe Institute*, **9** (2), 28–32.

Bernard, F., Lacroix, G., and Villeval, M.C. (2007) Tax evasion and social interactions. *Journal of Public Economics*, **91**, 2089–2112.

Black, T., Bloomquist, K.M., Emblom, E. *et al.* (2012) Federal tax compliance research: tax year 2006 tax gap estimation. IRS Research, Analysis & Statistics working paper, https://www.irs.gov/pub/irs-soi/06rastg12workppr.pdf (accessed 17 June 2016).

Bloomquist, K.M. (2012) Agent-based simulation of tax reporting compliance. Doctoral dissertation. George Mason University, http://digilib.gmu.edu/xmlui/handle/1920/7927 (accessed 17 June 2016).

Bloomquist, K.M. (2013) *Incorporating Indirect Effects in Audit Case Selection: An Agent-Based Approach*, The IRS Research Bulletin, Publication 1500, pp. 103–116.

Bloomquist, K.M. and Koehler, M. (2015) A large-scale agent-based model of taxpayer reporting compliance. *Journal of Artificial Societies and Social Simulation*, **18**, 20, http://jasss.soc.surrey.ac.uk/18/2/20.html (accessed 17 June 2016).

Cioffi-Revilla, C. (2014) *Introduction to Computational Social Science: Principles and Applications*, Springer-Verlag, London.

Cline, J.S., Bloomquist, K.M., Gentile, J.E. *et al.* (2014) From Thought to Action: Creating Tax Compliance Models at National Scales. *Paper presented at the 11th International Conference on Tax Administration*, Sydney, Australia.

Erard, B. and Feinstein, J.S. (2011) *The Individual Income Reporting Gap: What We See and What We Don't*, The IRS Research Bulletin, Publication 1500, pp. 129–142.

Gemmell, N. and Ratto, M. (2012) Behavioral responses to taxpayer audits: evidence from random taxpayer inquiries. *National Tax Journal*, **65**, 33–58.

Ghallab, M., Nau, D., and Traverso, P. (2004) *Automated Planning: Theory and Practice*, Morgan Kaufmann Publishers Inc., San Francisco, CA.

Grimm, V., Berger, U., Bastiansen, F., Eliassen, S., Ginot, V., Giske, J., Goss-Custard, J., Grand, T., Heinz, S., Huse, G., Huth, A., Jepsen, J.U., Jørgensen, C., Mooij, W.M., Müller, B., Pe'er, G., Piou, C., Railsback, S.F., Robbins, A.M., Robbins, M.M., Rossmanith, E., Rüger, N., Strand, E., Souissi, S., Stillman, R.A., Vabø, R., Visser, U., and DeAngelis, D.L. (2006) A standard protocol for describing individual-based and agent-based models. *Ecological Modelling*, **198**, 115–126.

Grimm, V., Berger, U., DeAngelis, D.L., Polhill, G., Giske, J., and Railsback, S.F. (2010) The ODD protocol: a review and first update. *Ecological Modelling*, **221**, 2760–2768.

Internal Revenue Service (2007) Reducing the Federal Tax Gap: A Report on Improving Voluntary Compliance, https://www.irs.gov/pub/irs-news/tax_gap_report_final_080207_linked.pdf (accessed 17 June 2016).

Internal Revenue Service (2015) Third Party Reporting Information Center - Information Documents, http://tinyurl.com/zudmhob (accessed 17 June 2016).

Janssen, M.A. and Ostrom, E. (2006) Empirically based, agent-based models. *Ecology and Society*, **11** (2), 37.

Koskinen, J.A. Written testimony of Koskinen, J.A. (2015) Commisioner, Internal Revenue Service, before the House Oversight and Government Reform Committee on the Government Accountability Office's High-Risk List, http://tinyurl.com/judgku7 (accessed 17 June 2016).

North, M.J., Tatara, E., Collier, N.T., and Ozik, J. (2007) Visual Agent-Based Model Development with Repast Simphony. *Proceedings of the 2007 Conference on Complex Interaction and Social Emergence*, Argonne National Laboratory, Argonne, IL.

Rand, W. and Rust, R.T. (2011) Agent-based modeling in marketing: guidelines for rigor. *International Journal of Research in Marketing*, **28**, 181–193.

Rinehart, W. and Gitis, B. (2015) *Independent Contractors and the Emerging Gig Economy*, American Action Forum, http://tinyurl.com/zevgo4s (accessed 17 June 2016).

Sandmo, A. (2005) The theory of tax evasion: a retrospective view. *National Tax Journal*, **58**, 643–663.

Slemrod, J. (2007) Cheating ourselves: the economics of tax evasion. *Journal of Economic Perspectives*, **21**, 25–48.

Srinivasan, T.N. (1973) Tax evasion: a model. *Journal of Public Economics*, **2**, 339–346.

U.S. Government Accountability Office (2015) Contingent Workforce: Size, Characteristics, Earnings, and Benefits, GAO-15-168R, http://www.gao.gov/assets/670/669899.pdf (accessed at 17 June 2016).

Appendix 7A

Overview, Design Concepts, and Details (ODD)

7A.1   Purpose

The Individual Reporting Compliance Model (IRCM) is designed to enable tax administrators to explore alternative enforcement strategies (e.g., audit case selection, computerized validation through use of third-party information reporting) for improving the compliance of individual taxpayers.

7A.2   Entities, State Variables, and Scales

The IRCM has five major types of entities: *Region*, *Filer*, *Tax Agency*, *Preparer*, and *Employer*. A *Region* is an integral unit of geography (e.g., state, county, or city) that is composed of one or more *Zone* entities. *Zones* are nonoverlapping areal subunits located entirely within the *Region* (e.g., postal zip code zones). A *Filer* in an IRCM represents an individual tax filer. In the current version of the IRCM, there are 84,912 filers who reside in the test-bed region. Each filer files a tax return (an instance of the *TaxReturn* class) which, in turn, contains 180 items (elements). A *Preparer* prepares a client's tax returns unless the filer self-prepares. An *Employer* employs filers, except for the self-employed. Filers, preparers, and employers are allocated to zones based on identifiers contained in actual tax return data. A single tax agency (an instance of the *TaxAgency* class) reviews and validates filed tax returns for accuracy against available third-party information documents and audits tax returns. Each simulation time step represents a tax filing year. The number of time steps is a user input.

7A.3   Process Overview and Scheduling

The main process is tax return filing, which is performed once per time step. A second set of processes involves the actions of the tax agency, which reviews all filed tax returns and selects returns for audit. The tax agency's review of tax returns involves comparing the amount reported on each major line item with the amount reported on third-party information documents, if they exist for a given item. Discrepancies are flagged and a notice is issued if the discrepancy exceeds a user-specified threshold. There are three types of audit selection strategies: Random, Fixed, and Constrained Maximum Yield (CMY). The number of audits to perform (N) is a user input. Under Random selection the tax agency selects N returns at random. Under Fixed selection the tax agency selects a user-specific fixed number of returns in each of 17 nonoverlapping audit classes. The CMY selection strategy uses a simple greedy-type algorithm that targets taxpayers in audit classes having the highest average yield (tax). The order in which the returns are filed or processed is unimportant; therefore, scheduling is not a consideration in the IRCM.

7A.4   Design Concepts

7A.4.1   Basic Principles

People exhibit heterogeneous reporting behaviors when filing their tax returns. Some appear to behave as rational decision makers, others comply out of a sense of duty or fear, and some pattern their reporting behavior by taking cues from family and friends. In addition to varying motivational factors, taxpayers have different opportunities for evasion based largely on the source of their income. Finally, taxpayers learn through repeated interactions with other taxpayers and with paid preparers what types of behaviors are more likely to draw the tax agency's attention. Analytical models in the tradition of Allingham and Sandmo (1972) and Srinivasan (1973) assume that taxpayers are independent, rational, and self-interested actors motivated to comply solely due to probability of detection and associated fines. However, empirical evidence from laboratory experiments, field studies, and random taxpayer audits suggests that a variety of noneconomic considerations also influence taxpayer reporting decisions. Agent-based models, such as the IRCM, are capable of incorporating both rational and behavioral motivations in a heterogeneous population of taxpayers.

7A.4.2   Emergence

The main emergent feature is a stochastically stable level of compliance (for major line items and total tax) that reflects user-specified assumptions for the level, quality, and effectiveness of tax agency enforcement activities and individuals' behavioral and filing characteristics.

7A.4.3   Adaptation

Filers adapt their reporting behavior to the perceived enforcement environment as determined from repeated interactions with the tax agency and (optionally) with their neighbors and coworkers.

7A.4.4   Objectives

The overall objective for each filer is to achieve a level of tax compliance consistent with their perception of the tax enforcement environment as well as their individual behavioral and filing characteristics.

7A.4.5   Learning

Filers may adjust their reporting behavior if they are audited or someone they know (e.g., a neighbor or coworker) is audited. This learning behavior is modeled as a partially observable Markov decision process (POMDP) $\sum = (S, R, P, C, O)$, where

- S is a finite set of states (audited or not audited)
- R is a finite set of compliance responses (perfect, increase, decrease, no change)
- P is a probability distribution where for each $s \in S$, if there exists $r \in R$ and $\acute{s} \in S$ such that $P_r(\acute{s} \mid s) \neq 0$, we have $\sum_{\acute{s} \in S} P(s, r, \acute{s}) = 1$
- $C_r(s, \acute{s})$ is the cost/reward (or expected cost/reward) experience from transition to state $\acute{s}$ from state s with transition probability $P_r(\acute{s} \mid s)$. The quantity $P_r(\acute{s} \mid s)$ is the probability that if response r is taken in state s, then state $\acute{s}$ will result. For example, if a taxpayer decides to increase compliance following a tax audit, one can infer that the action is being taken in order to reduce the probability of being selected for an audit (and the associated costs) in future time periods.
- O is a set of observations with probabilities $P_r(o \mid s)$, for any $r \in R$, $s \in S$, and $o \in O$. $P_r(o \mid s)$ represents the probability of observing o in state s after taking response r.

The IRCM does not explicitly model costs (C) and belief states (O) but assumes that these are implicit in the stochastically determined "choices" made by filers. These elements could be added to the model when better data on taxpayer decision-making becomes available. The IRCM allows users to provide independent sets of choice probabilities (P) to reflect different degrees of responsiveness by filers to a tax audit of themselves or someone in a reference group (see Section B.4.10).

7A.4.6 Prediction

The IRCM makes no predictions about future taxpayer behavior but simply models the presumed behavior of taxpayers given certain enforcement conditions.

7A.4.7 Sensing

Sensing occurs when filers become aware that someone in either their coworker or neighbor reference groups has been audited. This "sensing" is achieved by a filer polling her reference group members. If a reference group member has been audited, it is assumed that this information is openly communicated to all other reference group members. Lastly, the tax agency can use audits as a sensing mechanism if the CMY selection strategy is used.

7A.4.8 Interaction

The main types of interactions in the model that can potentially influence the behavior of individuals include (i) tax agency audits of filers and (ii) filers polling members of their reference groups to determine if someone was audited in the previous time period. Implied interactions occur between tax preparers and their clients. However, these preparer–client "interactions" are implied only because they appear as differences in estimated coefficients used to impute misreported amounts for paid prepared and self-prepared taxpayers.

7A.4.9 Stochasticity

Stochasticity is an integral feature of an IRCM. One way the model uses stochasticity is to determine which filers become "stop filers" at each time step. If the stop filer option is activated (the default setting) a uniform random number is drawn and compared to a fixed probability of becoming a stop filer as determined from analyzing filing behavior in the study area. Stop filer probabilities are specific to filing status. Another use of stochasticity is determining which filers are audited at each time step. Audit cases may be selected completely at random or by using one of two targeted strategies. The IRCM has 17 pre-determined audit classes used for targeted audits. These audit classes are groups of filers that share certain characteristics. These include filing status (single, married filing joint/qualified widow(er), head of household, married filing separate, dependent filer), children at home (yes/no), itemized or standard deduction, adjusted gross income (AGI) greater than the median (by filing status), and wage income more than one-half of AGI. Targeted audits may either be fixed in number or use a search algorithm (i.e., CMY) that assigns cases to audit classes with the highest average tax yield. A third use of stochasticity involves modeling filers' response to being audited. The user defines a vector of response probabilities (e.g., perfect compliance, increase compliance, decrease compliance, no change) and the model generates a uniform random number to determine which category of response the filer "selects." When the rule-based reporting regime is selected, the IRCM uses a stochastic process to assign line item reporting behavior to each taxpayer. The model first determines if a filer is a "deontological" filer meaning that the filer has perfect compliance. If a line item is subject to information reporting and/or withholding the IRCM determines how much the filer will report using separate random draws for information reporting and withholding, depending on which conditions apply. Stochasticity is also involved in the process of imputing misreported income and offset amounts. These values are imputed from estimated equations that are fit to empirical cumulative distribution functions (ECDFs). Uniform [0, 1] random numbers are generated and used to select imputed amounts from these equations. Finally, creating reference groups involves stochasticity. Members of a filer's coworker and neighbor reference groups may be structured as either random or "small world" networks. In the former, reference group members are assigned using random selection from a filer's coworkers and neighbors. The process of creating "small world" networks is the same as random except one individual (the "hub") is known to all of a firm's employees or residents of a given zone. The "hub" is determined by random selection.

7A.4.10   Collectives

There are two types of filer reference groups: neighbor and coworker. These are determined at the time of instantiation. Both groups assume the same (user specified) fixed size. If the "stop filer" option is activated (the default setting), then reference group stop filers are replaced over time; however, this does not affect group size or member relationships. Preparer networks are a third type of collective that may be optionally specified. At present, preparer networks only become relevant for scenarios that simulate a preparer-based tax scheme.

7A.4.11   Observation

The IRCM generates an output in the form of tables and figures. These can be copied and pasted into other applications for further analysis. The main interface screen also has a "map" of the study region and component zones. Options are provided that allow a user to drill down to view model output for individual preparers and employers by zone. This capability is especially useful for model verification and validation.

7A.5 Initialization

All agents are instantiated when the user selects a data file to read. The order in which agents are instantiated is as follows:

(1) Regions and zones
(2) Employers
(3) Preparers
(4) Filers
(5) Tax agency.

Once these entities have been created and default values assigned the following relationships are added:

(1) Filer + Zone
(2) Filer (client) + Preparer
(3) Filer (employee) + Employer
(4) Preparer + Zone
(5) Employer + Zone

Last, preparer networks and filer reference groups are created (see Section B.4.10).

7A.6 Input Data

The IRCM uses tax return information from the Statistics of Income (SOI) to describe the filing characteristics of taxpayers in the study region. Public Use File (PUF) records are substituted for the tax returns of filers in the study region using statistical matching (performed outside of the model). In addition to the PUF data, filer data includes pseudo-values for the paid preparer taxpayer identification number (TIN), employer identification number (EIN) and zone id as well as a calculated ratio of primary to secondary earnings and an estimate of the number of children living at home under the age of 17. These non-PUF values are derived from filers' tax returns and are used to preserve key filer relationships that influence reporting behavior and tax calculation. Once the data set is constructed, the name of the data file becomes an input parameter to the model. The IRCM allows the user to create and save all model parameters used to define a scenario in an xml (.*xml*) file. This facilitates the re-creation of scenarios for sensitivity testing and model verification and validation.

7A.7   Submodels

Submodels are provided to analyze alternative behavioral assumptions for paid preparers and employers. The paid preparer submodel enables the user to change the reporting compliance of filers using a paid preparer up or down relative to default levels for all preparers (region) or only for preparers in a specific zone. Networks of preparers (conceptually similar to filer reference groups) can also optionally be created by specifying the network size and the proportion of network members located in the same "home" zone for a given preparer. A fraction of preparers also may be resistant to network influences and an option is available to indicate this as well. The employer submodel permits the user to explore the impact on compliance if some fraction of firms converts their workers from employees to independent contractors (ICs). Conversion of employees to ICs has several advantages for firms; for instane, employers are no longer responsible for making payments of state unemployment tax or withholding of employees' income tax. In addition, ICs, not firms, become responsible for paying the employers' share of Social Security and Medicare taxes. The model represents the conversion of employees to ICs by converting wage income to Schedule C income, determining the baseline reporting rate on this income (based on National Research Program (NRP) random audit data), and using the tax calculator to determine income tax and employment tax liabilities.

8

Investigating the Effects of Network Structures in Massive Agent-Based Models of Tax Evasion

Matthew Koehler, Shaun Michel, David Slater, Christine Harvey, Amanda Andrei and Kevin Comer

8.1 Introduction

With the ongoing increase in computing power, agent-based models have become a preferred tool of choice for the study of complex adaptive systems, especially those systems in which humans are a nontrivial part. Agent-based models are an appropriate choice for these types of systems as they allow the modeler to express the system more naturally, using a logical rule-based approach, rather than with closed form equations that require strong assumptions to be made about the said system (Axtell, 2000b; Epstein, 2006). This is particularly the case when the system is made up of a large (but not infinite) number of discrete, adaptive agents (i.e., humans) that may change, adapt, or coordinate their behaviors over time, recognized as *organized complexity* (Weaver, 1948). These types of systems currently stymie closed form analysis as well as statistical approximation (Weaver, 1948). Under these circumstances the most efficient way to understand the temporal dynamics of the system is to simulate it (Buss *et al.*, 1990).

Agent-Based Modeling of Tax Evasion: Theoretical Aspects and Computational Simulations,
First Edition. Edited by Sascha Hokamp, László Gulyás, Matthew Koehler, and Sanith Wijesinghe.
© 2018 John Wiley & Sons Ltd. Published 2018 by John Wiley & Sons Ltd.

With respect to the analysis of a tax system, agent-based modeling (ABM) may be particularly useful. By representing the system as a collection of interacting individuals, displaying bounded rationality, each embedded within a space and social network, one can get around many of the strong assumptions necessary for classical microeconomic analysis such as infinite computing power, a fixed preference ranking, and perfect rationality (Simon, 1972; Axtell, 2000b; Epstein, 2006). There have been many agent-based models of tax compliance that have been created, perhaps most notably by Korobow *et al.* (2007), Hokamp and Pickhardt (2010), and Bloomquist (2011). Of particular interest here is Korobow *et al.* (2007), who explicitly introduced the use of network dynamics. This work was extended by Andrei *et al.* (2014) and combined with the more parsimonious underlying dynamics of the Hokamp–Pickhardt model (Hokamp and Pickhardt, 2010). We use that model as our base here. The model we created here, based upon previous work that was carried out in NetLogo (Wilensky, 1999), was re-implemented in a modeling framework specifically designed to allow researchers to create very large-scale agent-based models.

8.2   Networks and Scale

A great deal of work has been done demonstrating the impact of network structure on system dynamics. We will not attempt to summarize this vast literature here but rather refer the interested reader to Newman (2010) for a general discussion of network dynamics, or Jackson (2008), and Easley and Kleinberg (2010) for more social and economic treatments of network dynamics. However, relatively little work has been specifically applied to the dynamics of tax compliance (see Korobow *et al.*, 2007, and Andrei *et al.*, 2014, for notable exceptions). Andrei *et al.* (2014) demonstrated that network structure can have a significant effect on tax paying assuming, of course, that there is a social component to tax compliance. This impact was due to how information flowed around networks of different structures. The social networks used by Andrei *et al.* (2014) were, however, quite small comprising approximately 450 agents. While very few tax systems are populated by only 450 tax paying individuals, or even 1800, insights can be made by a study of such small-scale systems. This is due, first, because no part of our universe is so simple that we can understand it in its entirety. As such, careful study necessitates the creation of models that are simplifications of the real system. Secondly, the value of studying such smaller-scale systems is also a function of whether the system is one that is social or asocial. If the system is asocial, meaning that the humans in a said system make decisions individually without consulting with other individuals, then studying smaller-scale systems can be quite insightful as one is studying the individual decisions of humans. The only need for the creation of a system of more than one human is when humans interact to create the environment in which they make decisions, meaning they directly or indirectly influence each other's behavior and decisions. Moreover, if the system is populated with adaptive agents that interact, the system is now complex, and system behavior may change with scale (Anderson, 1972). Therefore, before one can claim to represent the dynamics of the system in question one must be able to test the dynamics at the appropriate scale. With modern computing hardware, it is now possible to represent taxation systems at a 1:1 scale.

To stress the importance, if the system is social, then the humans within it actively engage with one another to make decisions and have direct influence on decision-making. In this case, it may be necessary to more closely replicate the scale and structure of the system in question. This is due to the fact that the social structure may have a significant impact on how the humans interact with each other and, therefore, how information relevant to their respective decision-making flows within the society. Here, following Korobow et al. (2007) and Andrei et al. (2014), we assume that tax paying is a social process and, therefore, the impact of network structure and the scale of the system should be investigated.

The principle contribution of this chapter is the demonstration that, with modern computing hardware and software, researchers in the tax evasion field (and other economic sectors) arc no longer forced to study small-scale systems but may now study systems that approach a 1:1 ratio of artificial agents to "real" taxpayers. This is important as scale may impact one's results. As shown by Gotts and Polhill (2010), system scale in land use with imitative agents can show different results as the scale increases. However, as reported by Nardin et al. (2014) scale may not impact system level results. While more systematic investigation of the importance of scale are not yet available, given the theoretic argument by Anderson (1972) and recent conflicting results by Gotts and Polhill (2010) and Nardin et al. (2014), researchers should be cautious in drawing conclusions about system level results if they have not investigated the impact of scale. Having said that, one should also keep in mind that these sorts of analytic systems should be used for the purpose of gaining insight and for understanding the potential dynamics of a system rather than point prediction. Therefore, how much effort should be spent on investigating the impact of scale needs to be determined on a case-by-case basis at the discretion of the researchers involved. For example, one can gain insight into the dynamics of a Schelling (2006) segregation model at relatively small scales, while one may need much larger scales to understand epidemics in a major metropolitan area (Parker and Epstein, 2011).

While scale is an important issue, there are surprisingly few systematic analyses of how system dynamics may change with scale (notable exceptions, mentioned include Gotts and Polhill, 2010, and Nardin et al., 2014). Since Anderson (1972) published his seminal article, the potential issue of scale has been highlighted as a feature of complex systems that should not be ignored. Unfortunately, we have been unable to locate published work that addresses the problem of determining what the appropriate scale is for modeling the dynamics of a particular system. For example, a single water molecule does not possess the property of wetness. It takes a number of water molecules to create the emergent phenomenon of wetness.

But how many water molecules must be included in a model to create the characteristic of wetness to claim we have captured the relevant features? What if we decide to model wave action instead of wetness? Similarly, a single ant will have a great deal of trouble surviving on its own. An ant colony, however, can be quite adaptive for farming, going to war, creating structures, and calculating optimal foraging paths (see generally, Hölldobler and Wilson, 2009). How many ants must be modeled to capture these emergent properties?

Just like water molecules and ants, there is growing literature suggesting that as human systems change in scale the dynamics associated with them may also change. This work grew out of findings in biology that showed very specific allometric scaling relationships among various organisms. For example, organisms enjoy economies of scale, for example, an elephant is more metabolically efficient than a mouse (West et al., 1997; West and Brown, 2005). Human social systems appear to function in a similar way. Human societies become more resource efficient as they grow in size (Bettencourt et al., 2007). The efficiency is for such things as miles of road, miles of electric lines, and so on. Unlike nonhuman organisms, the "social production" of human societies may be subject to scaling dynamics. Starting with West and continuing with Bettencourt and others, many have reported that signals have been found that indicate that the dynamics of human populations may change in systematic ways as the scale of human society changes. Specifically, for our purposes here, there appears to be specific changes to income, gross adjusted income, and tax revenue as human society changes in scale (this study looked specifically at cities within the United States). As described in Gulden et al. (2015), while incomes tend to grow superlinearly as cities increase in size because the United States has a progressive tax, tax revenue increases at an even faster rate than does income as the size of a city increases. However, there is more to this story, if one categorizes income into a number of income bands then one can see that lower incomes increase linearly while the highest incomes grow superlinearly. It is this growth in the highest income categories that causes overall income to appear to grow superlinearly. This, in turn, will obviously have impacts on the overall growth in tax revenue as cities increase in size. This being the case, it is our contention that researchers should be cognizant of the effects of scale on their analyses and their conclusions and test their models at various scales.

Given the bounded rationality of human actors, scaling up a social system may have other effects, such as changes to network diameter (a measure of how far apart components of the network are). This is one way to think about how long it may take information from one node to move to another node. This is a function of network structure. In a fully connected network, one in which every node is connected directly to every other node, moving information from one node to another is trivial. However, if we make the rather mild assumption that humans have finite ability to process social information, which in this context is the number of other individuals they can know (Dunbar, 1992), then as the size of the whole network grows beyond the number of individuals a human can know that we can no longer have a fully connected network and that information must flow around the network via intermediaries. Clearly, this will impact how information will move from one individual to another. How much this impacts the flow will be a function of the structure of the network. For example, the Small Worlds (Watts and Strogatz, 1998) network structure can be particularly efficient in this case; however, it cannot fully mitigate the impact of network size growing much larger than the cognitive ability of the humans. This reinforces the importance of knowing the network structure underlying the system of interest. For example, if we assume the number of relevant social connections is held fixed and constant at 50 (again this can be any finite number much less than the overall population), as can be seen in Figure 8.1, when network size is less than or equal to the social connections number the graph may be fully connected and information can flow directly from any individual to any other individual. However, as the network increases in size beyond that of the cognitive capabilities of the humans that make up the network the closeness of the network decreases rather abruptly, which is another type of phase transition. As the network size overtakes the social connection number, it is more difficult for information to flow from one individual to all others. These dynamics may be exacerbated by other network formation characteristics such as homopholitic or assortative dynamics (Newman, 2010). This being the case it is important to explore the dynamics of the system of interest at the appropriate scale, or as close as is possible. This is simply a thought experiment that motivated the work reported within this chapter. Here, we explore these dynamics with a simple income tax system and explore how its dynamics change with the structure of the social network and the size of the society.

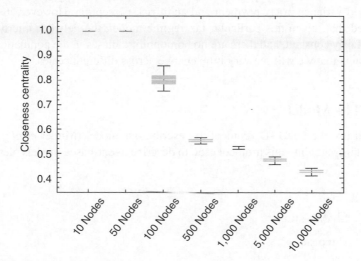

Figure 8.1 Closeness centrality as a function of network size

While this chapter focuses on the analysis of large-scale social systems via distributed ABM, another approach used to analyze very large-scale economic systems is that of econophysics (Mantegna and Stanley, 1999). Econophysics uses the tools of statistical physics to study economic systems. While many insights can be generated by this approach, as discussed in Chapter 11 of this book, the use of these techniques requires strong assumptions about the human component, such as the exclusion of learning and adaptation, heterogeneity of strategies and behavior, and systematic biases that may limit their ability to generalize. This being the case, we feel that this approach should not be the only way to study large-scale human complex systems. Therefore, we focus our efforts here on the creation of large-scale agent-based models in an attempt to show others in the field that such simulations can be created, run, and analyzed efficiently and thereby be an effective analytical tool. Moreover, keeping track of the dynamics of a social system as a temporal trajectory may be very important and likely cannot be done in a way other than via simulation (Buss *et al.*, 1990), therefore, in this case, we turn to agent-based models. Finally, in addition to tax system dynamics, agent-based models have been used to study other puzzling human economic system dynamics. For example, in the United States when the retirement age was lowered from 65 to 62.5, that did not result in an immediate change in the median retirement age. Rather, there was a multi-decade lag before the median age lowered. Axtell and Epstein (1999) was able to demonstrate conditions sufficient to cause such a lag using a parsimonious agent-based model.

In the sections that follow, the model and our analysis of the impact of scale on model dynamics are discussed in more detail. If scale has an impact here, we expect it to manifest in the population variance. More specifically, we expect to see increased variance in tax payment and audit risk perception. However, it should be noted that, given this particular tax regime model, the network micro structure is fairly constant, and there are no homophilitic biases in the population. Our intuition is that we will see very little change across different scales.

8.3   The Model

We will use the ODD+D Protocol to describe our model (Müller *et al.*, 2013). For details regarding this protocol used to describe agent-based models please see Chapter 1.

8.3.1   Overview

8.3.1.1   Purpose

There are two main purposes to the model described here: the first is to explore the impact of network structure on the performance of a taxing regime as assessed by the difference between the actual tax rate and the mean voluntary tax rate. The second is to test the impact of a scale on the performance of the tax system. The final purpose is to introduce a new Python-based ABM framework for more easy creation of ABMs designed to be run on distributed memory computational hardware or cluster computers. This is an important element as scale can have important implications on the performance of a complex system and, as such, researchers and policy analysts need tools to allow them to simulate these types of systems at an appropriate scale. This model was designed for both policy analysts/researchers (the findings that relate to network structure and scale impact on tax dynamics) as well as computer scientists (the impact of system performance as scale changes and the agents are distributed across a number of computational nodes).

8.3.1.2   Entities, State Variables, and Scales

There are two entities in this model. The first is the taxing authority. This entity collects taxes and performs random audits. It should be noted that this entity is virtual. The impact of its existence is felt in the model (taxes are collected and audits occur; however, for efficiency these elements were embedded within the taxpayer rather than having a separate taxing authority). The other type of entity in the model is a taxpayer. The taxpayer agents in the model represent the tax paying citizens in a population. Therefore, they are made up of entities embedded within a social network who communicate with each other. Tax paying being the focus of this model, communication events center around tax paying, specifically, how much of one's income should be disclosed and whether one has been subjected to an audit. There are three types of taxpayers in the model: taxpayers that always declare all of their income, taxpayers that try to hide some or all of their income, and taxpayers that mimic what those in their social network do.

8.3.1.3   Process Overview and Scheduling

As this model was designed to run on a cluster computer, we utilized a mixed agent activation regime. For each processor being used to run the model we utilized a random sequential activation regime. Meaning, that at the start of each time step we created a list of all agents resident on that processor in randomized order and iterated over that list activating each agent once and only once. Therefore, each agent was able to update their state and make decisions regarding tax paying onc time per tax paying year. The agent states within a time step were not buffered. Each agent updated the variables in real time. This meant that when agents queried their social network neighbors they would receive information about the current tax period from agents that were already activated during the time step before the current agent was activated, and information about the previous tax period from agents that have not yet been activated. As agent states are not buffered within a processor, agents were activated in a new random order at each time step on each processor. Due to the expense of inter-processor communications, inter-processor agent states are buffered, which is akin to saying these agents only "discuss" the previous tax year.

Communicating information between processors is time-consuming and limiting this interaction until the end of a time step allows for a much higher level of efficiency in computing. Therefore, if agents have social network neighbors that are resident on a different processor they will only receive information about the previous tax year. This design choice was made for simulation execution efficiency. Given the simplicity of this model, and the fact that agent state is not buffered on the processor either, we felt this would create little disruption in the overall model dynamics but this choice may not work in all situations as this does introduce a potential bias toward agent decision-making with prior tax year information.

Upon activation, agents enter an "update phase." During this phase agents update their declared income, then do an audit check, put themselves back on the scheduler, and finally, if declared income has changed, this change is added to the list of information that should be shared.

8.3.2   Design Concepts

8.3.2.1   Theoretical and Empirical Background

The model was derived from the prior work by Hokamp and Pickhardt (2010) (additional details about this foundational model can be found in Chapter 9), (Korobow *et al.*, 2007; Andrei *et al.*, 2014). Of note, while this study implements the Andrei *et al.* model, it is not, strictly speaking, a replication of that previous work. As stated above, the contribution from this piece is the demonstration that massive-scale agent-based models can be used productively by researchers engaged in tax evasion analyses, and not that the prior results from the work of Andrei *et al.* (2014) can just be replicated. As is clear in the following discussions, we, in fact, do not replicate the results of Andrei *et al.* (2014). While that is interesting, one should not place too much stock in this deviation as we have not attempted a formal replication here. For example, some of our parameter settings and our activation regime differs from the previous work of Andrei *et al.*.

Agents are embedded within a social network. The network is one of seven potential types: Erdös–Rényi (Boccaletti *et al.*, 2006), small worlds (Newman, 2010), power law (Clauset *et al.*, 2009), ring world (Boccaletti *et al.*, 2006), von Neumann (Weisstein, 2002), Moore (Weisstein, 2002), and finally no network may be present (all agents are disconnected from each other). Agents are of three basic types (specific proportions are discussed later): honest (these agents always declare all of their income), dishonest (these agents calculate the lowest amount of income to declare based upon their risk calculations discussed further), and imitating (these agents declare income based upon the behavior they observe among their social network neighbors). Examples of the networks used in the simulation are shown in Figure 8.2. We chose these networks for two reasons: first, these were the networks used by Andrei *et al.* (2014), and second, they represent a nice spread of network structure and characteristics as shown in Figures 8.3 and 8.4. For example, as can be seen in Figures 8.3 and 8.4, these networks have statistically distinct distributions of degree centrality and closeness centrality. Differences in these measures will impact how information can spread across the network, here meaning information about tax evasion and audits. This, in turn, may have an impact on agent learning and decision-making (Barkoczi and Galesic, 2016).

8.3.2.2   Individual Decision-Making

There is one major decision made by agents in this model: that is, how much of their income should be declared and, thereby, taxed. As discussed earlier, there are three types of tax paying agents in this model and how much income an agent decides to declare will depend on what type a given agent is. An honest agent will simply declare all of its income. Dishonest agents will declare the least income they "feel" is necessary. Here, necessity is a function of their risk-aversion and how likely they feel they are to be audited. Conforming agents look to the ratio of declared to actual income of their network neighbors to decide how much income to declare. Details of the decision-making can be found further in the Section 8.3.3.5.

8.3.2.3   Learning

This model includes mild learning. Over model time taxpayers will adjust how much of their income to declare based upon an increase in information about how likely they are to be audited. It should be noted that this learning is heavily biased to overestimate the likelihood of being subjected to an audit.

8.3.2.4   Individual Sensing

Agent sensing takes place via the social network, if present. Agents only attempt to collect information about the tax paying environment and ignore all other features. Specifically, agents attempt to collect information on their social network neighbors regarding audit experiences and tax paying strategy.

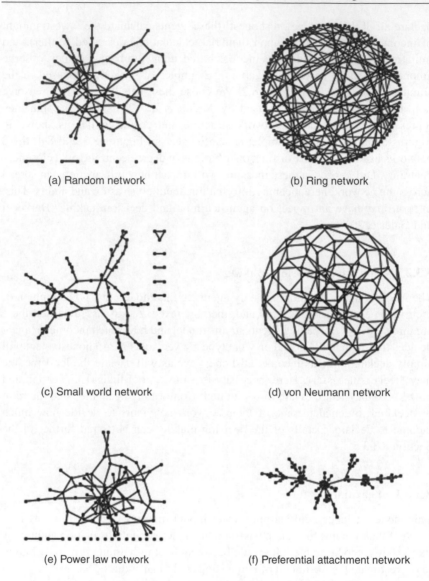

(a) Random network

(b) Ring network

(c) Small world network

(d) von Neumann network

(e) Power law network

(f) Preferential attachment network

Figure 8.2   Sample network structures used in the simulation

8.3.2.5   Individual Prediction

The only prediction agents undertake in the model is determining how likely they think it is that they will be subjected to an audit. Given agent decision-making it is very likely that this value will be much higher than the actual probability of being subjected to an audit.

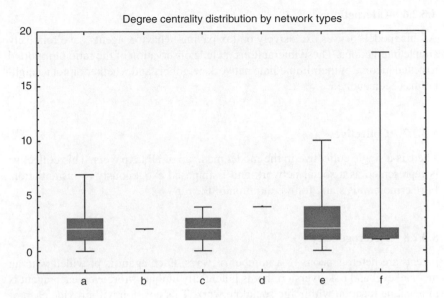

Figure 8.3 Degree centrality distributions for the networks, labels are: (a) random network, (b) ring network, (c) small world network, (d) von Neumann network, (e) power law network, (f) preferential attachment network

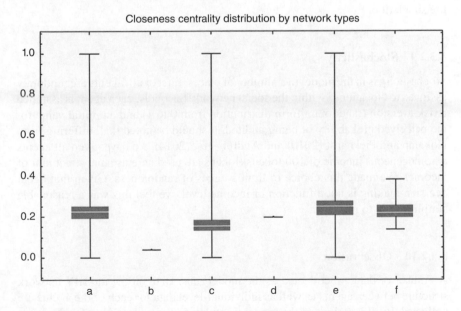

Figure 8.4 Closeness centrality distributions for the networks, labels are: (a) random network, (b) ring network, (c) small world network, (d) von Neumann network, (e) power law network, (f) preferential attachment network

8.3.2.6   Interaction

As this model focuses exclusively on tax paying behavior, agents have relatively simple interactions. These interactions include observation of the ratio of reported to actual income of their immediate network neighbors and whether or not a neighbor has been audited.

8.3.2.7   Collectives

There is a single collective in the model made up of all taxpayers. This collective is represented as a social network and is imposed exogenously at instantiation. Furthermore, it is static for the duration of the run.

8.3.2.8   Heterogeneity

Agents are heterogeneous in a number of ways. Each agent is provided with an income level and risk-aversion that is potentially unique. Moreover, each agent is in a unique location within the social network. This position will provide each of them with a unique view into the social dynamics going on within the population and unique set of experiences over the course of the simulation run creating the overall dynamic. These features were varied for each agent for each replicate of the simulation.

8.3.2.9   Stochasticity

Stochasticity is in the model in a number of places. First of all, agents are randomly assigned to "locations" within the social network. Secondly, each agent is assigned a risk-aversion (random uniform distribution from 0 to 1) and an initial value for the perceived probability of being audited. It should be noted that, differing from Hokamp and Pickhardt (2010) and Andrei *et al.* (2014), we have given all agents a homogeneous income of 100 tokens (tokens are used here as an abstract unit of income). We made this choice to limit sources of randomness. Given that agent decision-making is not a function of income level, we feel this was a reasonable simplification to make.

8.3.2.10   Observation

A number of data are collected as the model runs. Data on the model's network structure between agents as well as individual-level data for each of the agents are collected. In all iterations of the model, complete initialization items for the agent are recorded, including income, personality type (honest, dishonest, imitator), risk aversion, and the individual's initial probability of being audited. At every time step of the model, the agent's declared/actual income is also recorded. The entire model records the mean VMTR over the course of all time steps across the model, where VMTR is the voluntary mean tax rate, specifically: taxes paid divided by total income for each agent averaged over the population.

8.3.3   Details

8.3.3.1   Implementation Details

While the original Andrei *et al.* model was created in NetLogo 4.3 (Wilensky, 1999) and is reported in Andrei *et al.* (2014), this implementation of the model was created in Python using our own ABM framework. The framework and this model are freely available here: https://github.com/ceharvs/mabm. While Python is not as high performance as other compiled languages such as C++, it is highly accessible, easy to learn, and has a very strong user community with many packages and libraries that can be leveraged by users. This being the case, and because one did not already exist, we chose to create this framework in Python.

8.3.3.2   The Agent-Based Modeling Framework

We chose to create our own ABM framework for a number of reasons. The main reason was that distributed computing has the ability to facilitate very large-scale agent-based models by addressing the traditional scaling limitations. However, reliable and efficient communication strategies are needed to synchronize agent states across multiple processors. The protocol used to synchronize agent states across processors has a significant impact on the efficiency of the tool. While there are many management methods to synchronize activities across a processor, including conservative methods that perform a complete synchronization of all agent state information at every time step (Fujimoto, 2000), we chose an alternative approach that uses an event-driven technique to synchronize agents. This procedure only synchronizes changes to pertinent information in the model at each time step. This technique requires less information to be broadcast, which reduces the run time of the simulation while maintaining consistency in the model. This method works in this context due to the relative simplicity of our model and the fact that coordination among agents is "loose," meaning if agents occasionally receive information from a previous time step the overall system dynamics are not negatively impacted. These features allow us to make the trade-off between performance and perfect synchronicity.

Our massive-scale ABM framework was developed in Python using the Message Passing Interface, or MPI, capabilities to support distributed agents. The framework was designed using both of the aforementioned synchronization techniques: conservative and event driven. A sample rumor model was used in this experiment, where each agent had exactly two assigned neighbors that were distributed according to the probability of having a neighbor on a foreign processor. Trials were run with 20 million and 80 million agents with 4, 8, 16, and 32 processors. The completed timing analysis shows that the event-driven technique was significantly faster for this model than the conservative method. In addition to the overall speed increase, the event-driven technique also scales up in number of processors as well as number of agents per processor more efficiently than the conservative method. As agent activation regimes can have a significant impact on the dynamics of the model (Axtell, 2000a), one should be careful when choosing a synchronization method when moving from a "traditional" computational platform such as a laptop to a cluster computing system. Knowing that conservative synchronization may be needed to create the dynamics needed for a particular model (strong synchronization overriding execution efficiency), we included both in our framework.

8.3.3.3 Initialization

The process of initialization is always the same; however, depending on the input files used the resulting tax system may not be the same as any other. Upon initialization, the simulation creates the taxpayer network. Initially, this is simply a set of links and nodes for each type of network at each scale. Once the "skeleton" network is created, via functions available in python-igraph Python package (igraph .org), the nodes are given individual characteristics that then make the tax paying system. Further, this allows us to ingest the networks once and then run many other design points on this network without needing to recreate the basic network structure. Given the time necessary to generate the network, this is a more efficient method for running parameter sets against the same network structure. By randomly assigning taxpayer attributes to nodes within the network we are able to test the impact of having different types of agents in different network locations without the computational expense of destroying and recreating the network for each run of the model. It should be noted that this is a significant deviation from the initialization procedure used in Andrei *et al.* (2014) where the networks are recreated each time the simulation is instantiated. As agents are randomly created for each simulation run and randomly assigned to nodes in the network, akin to randomly distributing agents across a fixed topological landscape, we felt this was a reasonable approach to take given the trade-offs in performance we faced with network generation. Moreover, to further control randomness and isolate the impact of network structure, where appropriate, we held the ratio of nodes to edges as close to 1.2471 times the number of nodes, rounded to the nearest integer. This number was used as it allowed us to create the networks while holding some characteristics constant: in this case the ratio of nodes to edges. Clearly, this specific ratio does not work for all topologies, such as a ring or a von Neumann network as each node has a fixed degree of 2 and 4, respectively; rather, this heuristic was used for power law, preferential attachment, and random network topologies. Finally, it should be noted that the impact of holding the network structure fixed was not explicitly tested.

In the original model, the population of agents was made up of 441 total agents, 50 of which were designated as honest and 50 of which were designated as dishonest. As one of the points of the current study is to change the scale of the population this specific number will be replaced by a percentage. To correlate with the original work we will use 11.34% of the population to be designated as honest and an additional 11.34% of the population to be designated as dishonest.

8.3.3.4   Input Data

The model uses two input files. The first file specifies the network size and structure. This file specifies the tax paying population. For efficiency, we created the network specification separately from the simulation (as discussed above). The input file for the social networks was created with the igraph python package. The second file specifies the agent population and global parameters. These values will be discussed in more detail in our discussion on the design of experiments used here.

8.3.3.5   Submodels

There are three submodels used in this agent-based model: auditing, declaring income, and updating an agent's perceived likelihood of being audited.

Auditing Submodel
This submodel is quite simple. At each time step agents draw a random uniform number from 0 to 1. If this number is less than the objective audit probability, then the agent undergoes an audit. The agent then draws another random number, if this random number is less than the probability of apprehension then the audit is assumed to uncover any income not declared. If this occurs and there is undeclared income, the agent is severely penalized and then updates its perceived probability of audit (see further text). As the number of taxpayers in the simulation increases this becomes an unrealistic design element of the simulation. We are aware of no taxing authority that performs audits as a fixed proportion of the population they serve. We chose to implement this audit submodel to do as little disruption to the original formulations as possible. As discussed in the conclusion, we intend to add a budget/personnel constraint on the taxing authority in our future work.

Declaring Income Submodel
Honest agents always declare all of their income; therefore

$$X_{i,t} = I_{i,t} \tag{8.1}$$

where $X_{i,t}$ is agent i's declared income at time t, and $I_{i,t}$ is agent i's actual income at time t. Dishonest agents calculate a minimum income to declare based on experience and risk. The assumption for the following three equations is: if the agents feel they have a high likelihood of being audited they will declare all of their income even though they would prefer not to, if the agents feel there is very little risk of being audited then they will declare none of their income, finally, if the agents feels there is some risk they will declare some of their income. Following Hokamp and Pickhardt (2010), if

$$\alpha_s < \frac{\theta}{\theta + (\pi - \theta)e^{r\pi I_{i,t}}} \tag{8.2}$$

then $X_{i,t}$ will be 0. However, if

$$\alpha_s > \frac{\theta}{\pi} \tag{8.3}$$

then $X_{i,t}$ will be $I_{i,t}$ as the agents assume it is too risky to not declare all their income. In the above equations α_s is the *subjective* probability of apprehension, r is the agent's risk aversion, π is the penalty rate, and θ is the tax rate. However, if α_s falls within the interval defined by Eqs (8.2) and (8.3) then an agent's declared income becomes

$$X_{i,t} = I_{i,t} - \frac{ln\left(\frac{(1-\alpha_s)\theta}{\alpha_s(\pi-\theta)}\right)}{r\pi} \tag{8.4}$$

Once again, following Hokamp and Pickhardt (2010), imitating agents observe the behavior of their social network neighbors and base their declared income on the product of the ratio of actual to declared income of their neighbors and their own income

$$X_{i,t} = \frac{1}{v} \sum_{j=i-v}^{i-1} \frac{X_{j,t-1}}{I_{j,t-1}} I_{i,t} \tag{8.5}$$

where v is the number of link neighbors an agent possesses. Following Andrei *et al.*, if an agent is not apprehended and its $\alpha_{s,t} > \alpha$ (the objective probability of apprehension) then the agent decreases its α_s by 0.2. If an agent is apprehended, it adjusts its declared income and subjective probability of apprehension as follows

$$X_{i,t} = \theta(I_t - X_t)(1 + \pi I_t), \alpha_{s,t} = 1 \tag{8.6}$$

These dynamics were used to incorporate the human decision-making heuristic of availability (Tversky and Kahneman, 1973) and the fact that humans are loss adverse (Kahneman and Tversky, 1984).

Audit Likelihood Submodel
This submodel is also relatively simple. If an agent is audited it sets its subjective probability of audit to 1.0. Again, following Andrei *et al.*, if the agent is not audited in a given time step, then the agent decreases its subjective probability of audit by 0.2. The agent's subjective probability of audit is bounded by 0 and 1.

Table 8.1 The design of experiments that was executed

Factor	Values	Number of values
Objective probability of audit	0–0.1 in steps of 0.02	6
Number of taxpayers	100–1 million by powers of ten	6
Type of social network	None von Neumann Power law Preferential attachment Erdös–Rényi Ring Small worlds	7
Number of processors	2–32 by powers of 2	5
Total number of design points in the DOE:	–	1260

8.4 The Experiment

The model is run for 40 tax cycles. Each time step is a year and consists of a full "tax cycle." Meaning, agents may be audited and may be apprehended by the taxing authority, discuss what happened within their social network the previous year, decide what to do in the current tax year, and then file a tax return.

We ran a simple full factorial experimental design for this study. The factors that were varied included the objective probability of being audited, α, the structure of the social network, the number of taxpayers, and the number of processors on which the model was run. The experiment is summarized in Table 8.1. All other parameters are held constant, save agent risk aversion and subject audit probability, α_s, which, as discussed earlier, are drawn from uniform distributions.

8.5 Results

The simulation results are measured by average VMTR at the end of the 40 tax cycles. In general, while some differences are observed among the networks, the differences are not as dramatic as those reported by Andrei et al. (2014). Likely, this is a function of our implementation. As discussed earlier several implementation decisions were made that could cause differences in results. However, it is worth noting these differences as it may point to fragility of the results reported by Andrei et al. As shown in Figures 8.5 and 8.6, while there are few end state differences between networks of different scales and the basic outcome of the simulation, there are differences between networks of different types, which observation is consistent with the findings reported by Andrei et al. These differences hold as we increase the scale of the population of agents. The dramatic increase in

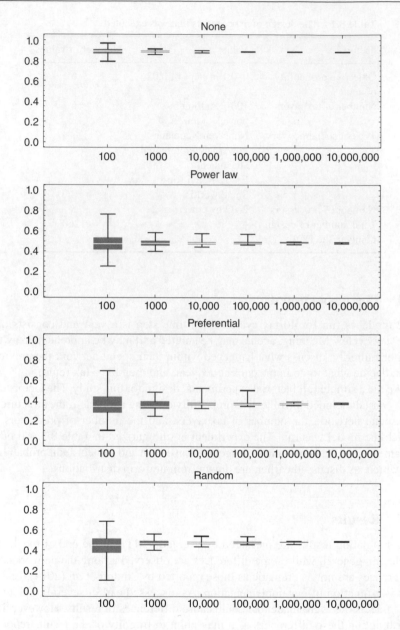

Figure 8.5   End of run distributions of VMTR by network type and size

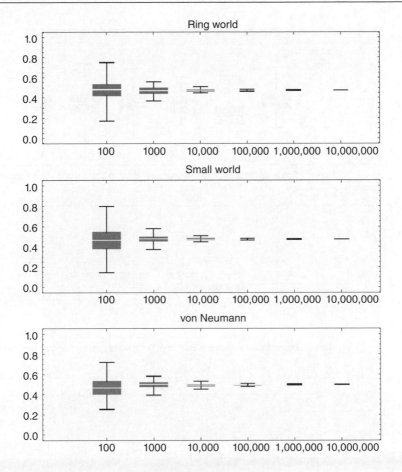

Figure 8.5 (*Continued*)

declared income observed in the no network case was caused by the imitator agents defaulting to honest behavior if they had no neighbors. However, as can be seen in the other cases, the mean level on income reported is approximately 40%. Thus, the taxing authority is faced with a very dramatic tax gap. Rather than receiving 25 tokens of tax from each individual the taxing authority is only receiving 10 tokens on average, producing a gap of 15 tokens.

8.5.1 Impact of Scale

In this case we did not see a noticeable change caused by changing the scale of the system, consistent with our intuition about the specifics of this model, as discussed earlier. What was observed, however, is a much stronger signal about

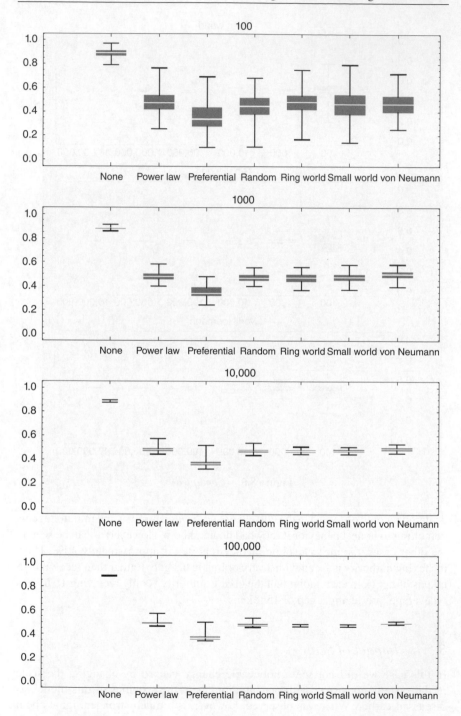

Figure 8.6 End of run distributions of VMTR by network size and type

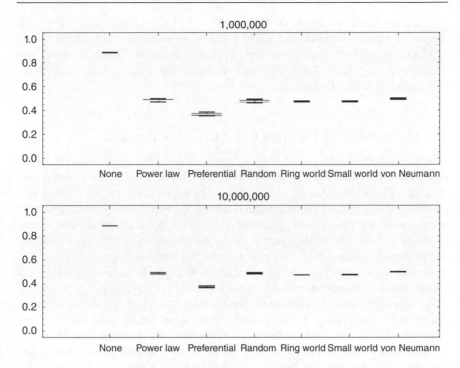

Figure 8.6   (*Continued*)

mean behavior, which is likely attributable to the central limit theorem. However, what is interesting is that while there was a similar collapse in the variance it was not as dramatic. We have observed this dynamics in other complex systems. As the system scales up there is a clearer and clearer signal regarding mean behavior but the variance in agent behavior continues to remain high, the so-called macro-level stability with micro-level dynamism. This sort of stability/dynamism can be seen in real-world systems such as employment in the United States. Although there is a relatively stable nationwide unemployment rate there is a great deal of dynamism at the employee level. Roughly five million individuals change jobs, two million leave the workforce, and two million join the workforce (Fallick and Fleischman, 2004). Another way to think about this dynamics is comparing it to a standing wave in a river. A standing wave is caused by water flowing downhill over an obstacle such as a rock. The wave remains stationary and collocated with the obstacle. Despite this stability, the water molecules that make up the wave are constantly changing. Just like nationwide employment, at one level we see stability, the wave, while witnessing micro-level dynamism. While being able to represent that both the micro-level dynamism and macro-level stability are not necessary for all analyses, it is important to be able to create analytics that can capture both when necessary.

One issue we do not see at the large scale is the impact of the unrealistic audit regime. If a fixed proportion of audits had a disproportionate impact, we would expect to see higher voluntary tax rates as scales increased. This is, however, not what we see (Figures 8.7 and 8.8), while variance does decrease, we do not see a systematic increase in taxes paid.

8.5.2   Distributing the Model on a Cluster Computer

Here we show that a massive-scale agent-based model of tax evasion can be created based upon previously published model formulations, specifically Hokamp and Pickhardt (2010) and Andrei *et al.* (2014). The typical reasons for moving to a parallel and distributed simulation architecture involve memory needs and how much computing power each agent will need, which is usually a function of how cognitively sophisticated the agent is. Given the simplicity of the agents involved in the simulation used here, one could argue that parallel execution is likely to be unnecessary. However, as shown further, even relatively simple agents at a US national scale consume more memory than most individual computers possess.

As discussed earlier, simulations are an important tool for social research. Furthermore, simulating social systems at the appropriate scale may be very important for a complete understanding of their likely dynamics. However, designing, building, and executing massive-scale agent-based models are nontrivial. When working with a simulation run across a set of computers that do not share a common memory, maintaining synchronization among the computing processors and designing agent communications so as to allow for social processes while not slowing the simulation down with relatively slow cross-processor messaging can be quite difficult.

Furthermore, agent activation can be tricky. As shown in Axtell (2000a), the specified agent activation regime can impact the results of a simulation significantly. Activation refers to how agents are "woken up" and do things. Although in the real world events happen simultaneously, it is difficult to do this en silico. Executing an agent-based model on a single processor necessitates waking up agents in a given order, one at a time. In order to simulate simultaneity under these conditions one typically activates agents in a new random order at each time step and "buffers" the agent states until after all agents have been activated. Buffering the state means the agents determine their new state but do not "reveal" it while other agents are still being activated. In this way, all agents are dealing with information from the same time step. There are other activation regimes that can be used, of course, and one should choose the method that is most appropriate for the system being simulated (Axtell, 2000a). As one moves to multiple processors, events can happen at the same time. However, now the simulation may become out of sync across the processor if, due to other processes running on the processors, execution time differs from one processor to another. For example, this could mean that one processor may be executing simulation time step 23 while

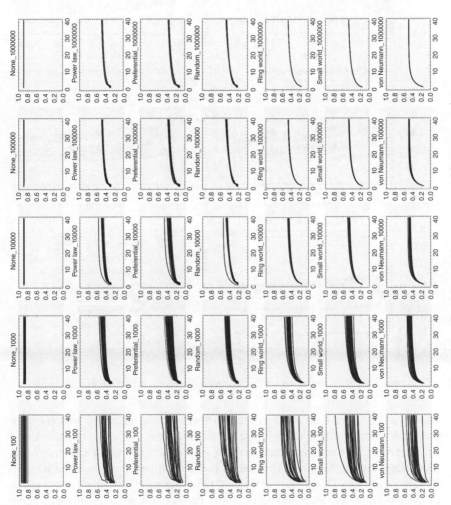

Figure 8.7 Mean ratio of declared income to actual income over time

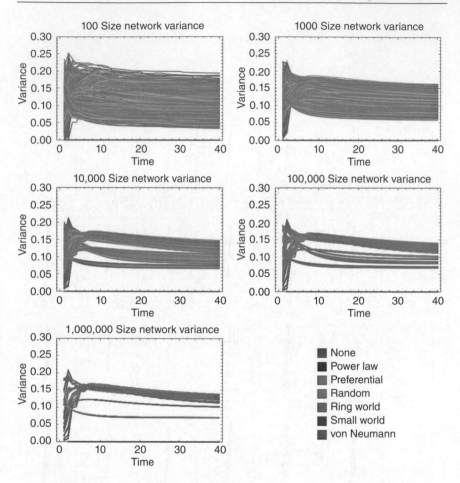

Figure 8.8   Variance in the ratio of declared income to actual income over time

other processors are executing time step 26 because an agent on the first processor had a particularly time-consuming set of calculations. Typically, this requires a central process to track the simulation progress on each processor and ensure that processors advance together. This means the simulation cannot move faster than its slowest process. Of course, the focus of the experiment described here was that of system scale, and so it should be noted that we did not explicitly test agent activation regimes in this study, rather we chose a single, logical one (discussed earlier) for this situation.

Another issue with distributed simulations is the communication among processes on different processors. This is typically very inefficient and can even negate any efficiency gain that is achieved by distributing a simulation to multiple processors. As can be seen in Figures 8.9 and 8.10, there is actually very

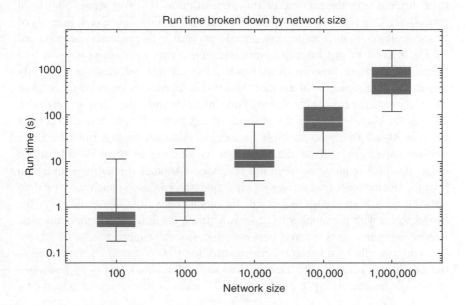

Figure 8.9 Simulation run time (in log scale) by scale, for all runs of the simulation

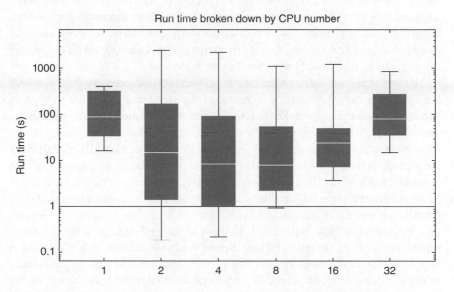

Figure 8.10 Simulation run time (in log scale) by number of processors, for all runs of the simulation

little difference in the runtime of this simulation as it is run across additional processors. This is because the agent level activity is quite simple and, in terms of time consumption, the speedup associated with additional processors is surpassed by that needed to synchronize communication across multiple processors and keeping everything in sync. Additionally, the network structure may heavily impact the execution time of the model, as it may increase or decrease the amount of cross processor messaging that will take place during a run. Certain networks, such as the ring, as well as the absence of any network have less connections between agents that span multiple processors. Simulations that run with these network structures will be the least time consuming as there is little to no communication required between the processors. Another simulation with a high level of interaction between agents on disparate processors, such as a random network with high connectivity, will be much more time consuming. As the model scales with the number of processors used, the amount of communication needed between agents to travel between processors also increases.

As an example of single run performance, for 10,000 agents on the preferential attachment network, the average time to run the simulation is 34.7 seconds on a single processor, which decreases to 10.7 seconds when run on 8 processors. This speedup is very small compared to the potential, since this model, for this size of a population, is already quick to complete. Looking at the same timing for simulations with one million agents, the run time decreases from 2360 seconds on 2 processors to just 827 seconds on 16 processors. The preferential attachment network is highly connected between processors. A better example of the advantages of parallelization is shown when we use the ring world network: the run time decreases from 1728.5 seconds on 2 processors to 226.4 seconds on 32 processors (aggregated performance is shown in Figure 8.10).

Memory usage may also necessitate distribution. In the present case, even with very simple agents the memory footprint of the simulation grows quite large as the agent population approaches national scales. A simulation with only 10 agents consumes 23 MB in memory, but as the size of the simulation increases to one million agents, the memory usage increases to roughly 2686 MB. Continually increasing the number of agents will scale this problem up to a point where it is unreasonable to run the simulation on a single machine and the problem needs to be distributed, especially on the scale of millions to billions of agents. While other languages are more efficient with their use of memory, such as C++, this problem cannot be fully mitigated; regardless of chosen language, scale will eventually necessitate distribution (given current hardware availability). Moreover, it is not unusual for the memory use of an agent-based model to scale super-linearly with the number of agents, especially when modeling a social network among the agents. Meaning, one can very quickly run into memory issues when scaling models up, especially network-focused models. In the case presented here, clearly we do not need to use a cluster computer due to memory demand, rather we could test and demonstrate the use of this agent-based modeling framework with a simple model already existent within the tax evasion analysis literature.

8.6 Conclusion

This chapter introduced a relatively efficient, Python-based, framework for creating massive-scale agent-based model which may be run on cluster computers. The model used as a tax evasion test case was based upon the work previously reported by Hokamp and Pickhardt (2010) and extended by Andrei *et al.* (2014). Our primary concern, here, is to test the importance of scale within this model of tax evasion. In this particular case, we found little sensitivity to scale, with VMTRs staying reasonably consistent. However, we continue to feel that this is a feature of complex social systems that warrants additional careful study. While our findings do differ from those reported by Andrei *et al.* important differences exist between our implementation and prior work. These differences exist for computational efficiency when using a cluster computer and to limit some sources of variation. We chose to reduce the sources of variation in this case to better isolate the potential impact of scale. In future work, we intend to extend the model to improve its verisimilitude as scale increases including, for example, a budget/personnel constraint on the taxing authority. Furthermore, we feel that careful analysis of the interrelationships among network structure, size, and simulation memory use is important. It would be useful for the ABM community, in general, to better understand the trade-offs involved in networks, agent population size, and agent verisimilitude. In conclusion, with modern software and hardware, researchers are now able to carefully explore the importance of scale when undertaking a study of tax evasion and other social complex systems.

References

Anderson, P.W. (1972) More is different. *Science*, **177** (4047), 393–396.

Andrei, A.L., Comer, K., and Koehler, M. (2014) An agent-based model of network effects on tax compliance and evasion. *Journal of Economic Psychology*, **40**, 119–133.

Axtell, R. (2000a) Effects of interaction topology and activation regime in several multi-agent systems. Brookings Institution working paper.

Axtell, R. (2000b) Why agents? On the varied motivations for agent computing in the social sciences. Brookings Institution Working Paper.

Axtell, R.L. and Epstein, J. (1999) *Coordination in Transient Social Networks: An Agent-Based Computational Model of the timing of Retirement*, Brookings Institute, Washington, DC.

Barkoczi, D. and Galesic, M. (2016) Social learning strategies modify the effect of network structure on group performance. arXiv preprint arXiv:1606.00753.

Bettencourt, L.M., Lobo, J., Helbing, D., Kühnert, C., and West, G.B. (2007) Growth, innovation, scaling, and the pace of life in cities. *Proceedings of the National Academy of Sciences of the United States of America*, **104** (17), 7301–7306.

Bloomquist, K. (2011) Tax compliance as an evolutionary coordination game: an agent-based approach. *Public Finance Review*, **39**, (1), 25–49.

Boccaletti, S., Latora, V., Moreno, Y., Chavez, M., and Hwang, D.-U. (2006) Complex networks: structure and dynamics. *Physics Reports*, **424** (4), 175–308.

Buss, S., Papadimitriou, C.H., and Tsitsiklis, J.N. (1990) On the predictability of coupled automata: an allegory about Chaos. *Complex Systems*, **5** (5), 788–793.

Clauset, A., Shalizi, C.R., and Newman, M.E. (2009) Power-law distributions in empirical data. *SIAM Review*, **51** (4), 661–703.

Dunbar, R.I. (1992) Neocortex size as a constraint on group size in primates. *Journal of Human Evolution*, **22** (6), 469–493.

Easley, D. and Kleinberg, J. (2010) *Networks, Crowds, and Markets: Reasoning About a Highly Connected World*, Cambridge University Press.

Epstein, J.M. (2006) *Generative Social Science: Studies in Agent-Based Computational Modeling*, Princeton University Press.

Fallick, B. and Fleischman, C.A. (2004) Employer-to-employer flows in the US labor market: the complete picture of gross worker flows. US Federal Reserve working paper.

Fujimoto, R.M. (2000) *Parallel and Distributed Simulation Systems*, vol. **300**, John Wiley & Sons, Inc., New York.

Gotts, N.M. and Polhill, J.G. (2010) Size matters: large-scale replications of experiments with FEAR-LUS. *Advances in Complex Systems*, **13** (4), 453–467.

Gulden, T., Koehler, M., Scott, S., and Henscheid, Z. (2015) Evidence for Allometric Scaling of Government Services in American Cities. *Computational Social Science Society of the Americas Annual Meeting*, Santa Fe, NM.

Hokamp, S. and Pickhardt, M. (2010) Income tax evasion in a society of heterogeneous agents–evidence from an agent-based model. *International Economic Journal*, **24** (4), 541–553.

Hölldobler, B. and Wilson, E.O. (2009) *The Superorganism: The Beauty, Elegance, and Strangeness of Insect Societies*, W. W. Norton & Company.

Jackson, M.O. (2008) *Social and Economic Networks*, Princeton University Press, Princeton, NJ.

Kahneman, D. and Tversky, A. (1984) Choices, values, and frames. *American Psychologist*, **39** (4), 341–350.

Korobow, A., Johnson, C., and Axtell, R. (2007) An agent–based model of tax compliance with social networks. *National Tax Journal*, **60** (3), 589–610.

Mantegna, R.N. and Stanley, H.E. (1999) *Introduction to Econophysics: Correlations and Complexity in Finance*, Cambridge University Press.

Müller, B., Bohn, F., Dreßler, G., Groeneveld, J., Klassert, C., Martin, R., Schlüter, M., Schulze, J., Weise, H., and Schwarz, N. (2013) Describing human decisions in agent-based models–ODD+ D, an extension of the ODD protocol. *Environmental Modelling and Software*, **48**, 37–48.

Nardin, L.G., Rosset, L.M., and Sichman, J.S. (2014) Scale and Topology Effects on Agent-Based Simulations: A Trust-Based Coalition. Interdisciplinary Applications of Agent-Based Simulation and Modeling, IGI Global.

Newman, M. (2010) *Networks: An Introduction*, Oxford University Press, Oxford.

Parker, J. and Epstein, J.M. (2011) A distributed platform for global-scale agent-based models of disease transmission. *ACM Transactions on Modeling and Computer Simulation (TOMACS)*, **22** (1), 2–27.

Schelling, T.C. (2006) *Micromotives and Macrobehavior*, W. W. Norton & Company.

Simon, H.A. (1972) Theories of bounded rationality. *Decision and Organization*, **1**, (1), 161–176.

Tversky, A. and Kahneman, D. (1973) On the psychology of prediction. *Psychological Review*, **80** (4), 237–251.

Watts, D.J. and Strogatz, S.H. (1998) Collective dynamics of "small-world" networks. *Nature*, **393** (6684), 440–442.

Weaver, W. (1948) Science and complexity. *American Scientist*, **36**, 536–544.

Weisstein, E.W. (2002) Cellular Automaton, http://mathworld.wolfram.com/CellularAutomaton.html (accessed 29 June 2017).

West, G.B. and Brown, J.H. (2005) The origin of allometric scaling laws in biology from genomes to ecosystems: towards a quantitative unifying theory of biological structure and organization. *Journal of Experimental Biology*, **208** (9), 1575–1592.

West, G.B., Brown, J.H., and Enquist, B.J. (1997) A general model for the origin of allometric scaling laws in biology. *Science*, **276** (5309), 122–126.

Wilensky, U. (1999) *"NetLogo", Center for Connected Learning and Computer-Based Modeling*, Northwestern University, Evanston, IL, http://ccl.northwestern.edu/netlogo/ (accessed 19 May 2017).

9

Agent-Based Simulations of Tax Evasion: Dynamics by Lapse of Time, Social Norms, Age Heterogeneity, Subjective Audit Probability, Public Goods Provision, and Pareto-Optimality

Sascha Hokamp and Andrés M. Cuervo Díaz

9.1 Introduction

"Essentially, all models are wrong, but some are useful!," a well-known claim attributed to George E. P. Box (Box and Draper, 1987, p. 424). Of course, all models necessarily are a brief representation of a real problem and, thus, simplify the complexity of the real world. In addition, concerning tax evasion and the shadow economy, another issue arises, that is, how to measure what needs to be hidden in the shadow.

The early attempts to measure the shadow economy (Gutmann, 1977; Feige, 1980; Tanzi, 1980; Klovland, 1984) date back to the equations on currency

Agent-Based Modeling of Tax Evasion: Theoretical Aspects and Computational Simulations,
First Edition. Edited by Sascha Hokamp, László Gulyás, Matthew Koehler, and Sanith Wijesinghe.
© 2018 John Wiley & Sons Ltd. Published 2018 by John Wiley & Sons Ltd.

demand by Cagan (1958) and the transaction approach by Feige (1979). Kirchgässner (2017, p. 99) identifies three approaches, which dominate the literature, (i) the "direct measurement" employing a survey method (Isachsen and Strøm, 1982; Feld and Larsen, 2005, 2012), (ii) the "indirect measurement" applying, a modified currency demand approach (Tanzi, 1980; Klovland, 1984), and, (iii) a model approach that originates from Weck (1983) and Frey and Weck-Hannemann (1984), well known as the (DYnamic) Multiple Indicators, MultIple Causes, (DY)MIMIC, approach. The latter is often used by Friedrich Schneider (with various co-authors, for example, Schneider, 2005, 2014; Schneider *et al.*, 2011, 2015; Buehn and Schneider, 2012; Schneider and Enste, 2013) and their estimates truly dominate the field. Feige (2016a) and Kirchgässner (2017) find severe problems to measure tax noncompliance and the shadow economy by the (DY)MIMIC approach: (i) confusion with definitions and measurements of the shadow economy, it seems that Friedrich Schneider and co-authors have "simply defined the entity [they] have measured" (Feige, 2016a, p. 26), (ii) "estimations produce only relative weights" (Kirchgässner, 2017, p. 103), (iii) conclusions are influenced by "arbitrary choices of indicators and normalizing coefficients," (Feige, 2016a, p. 26) and, (iv) impossibility "to draw statistically confirmed conclusions about causal relations in the real world […] from these estimates" (Kirchgässner, 2017, p. 103). Feige (2016b) and Schneider (2016) continue to deepen the discussion on flaws and strengths to measure the shadow economy. Hashimzade and Heady (2016) briefly summarize the dispute.

To overcome the severe problems in the measurement of tax compliance and the shadow economy, agent-based models (ABMs) might help via what–if studies. In this chapter, we focus on tax evasion at a micro level by agent-based computational simulations. We present novel insights into tax noncompliance driven by lapse of time, social norms, age heterogeneity, subjective audit probability, public goods provision, and Pareto-optimality within the framework presented by Hokamp and Pickhardt (2010) and Hokamp (2013, 2014). We reconsider the political cycle by Hokamp and Pickhardt (2010), the public goods provision cycle by Hokamp (2013, 2014) and, in addition, we provide new information on compliance at group level by the four behavioral types of taxpayers, i) neoclassical A-Types, ii) social interacting B-Types, iii) ethical C-Types, and iv) erratic D-Types. We find that neoclassical A-Type taxpayers play a key role within the model by pursuing the expected utility maximizing approach by Allingham and Sandmo (1972). Therefore, *ceteris paribus* we investigate the dynamics of tax evasion by neoclassical A-Types due to lapse of time, subjective audit probability, public goods provision, and age heterogeneity combined with social norms. The lapse of time effects strongly declines when raising the tax rate under a regime with a constant sanction rate and audit probability. Further, we build upon the calibration procedure of an econophysics ABM by Bazart *et al.* (2016) and calibrate our economic ABM with a tax declaration experiment by Alm *et al.* (2009). Finally, our sensitivity analysis provides numerical estimates that reveal the strong effect the penalty and tax rate have on the evasion by neoclassical A-Type and social interacting B-Type taxpayers. The chapter is structured as follows. Section 9.2 describes our ABM of tax evasion by a standard protocol. Section 9.3 discusses the scenarios and results while Section 9.4 provides the conclusion. Appendix 9A provides the political and public goods provision cycle by Hokamp and Pickhardt (2010) and Hokamp (2013, 2014).

9.2 The Agent-Based Tax Evasion Model

This section describes the agent-based model of tax evasion by[1] (Hokamp and Pickhardt, 2010; Hokamp, 2013, 2014) and our amendments according to the Overview, Design Concepts and Details (ODD) protocol by Grimm et al. (2006, 2010), including the extension to human decision-making by Müller et al. (2013), outlined in Chapter 1.

9.2.1 Overview of the Model

The agent-based tax evasion model by Hokamp and Pickhardt (2010) and Hokamp (2013, 2014) considers a heterogeneous set of taxpayers faced with a tax scheme explicitly modeling the environment consisting of a government and a tax authority. The model aims to analyze the tax compliance dynamics and the social interactions in a behaviorally heterogeneous population under lapse of time, social norm updating, age heterogeneity, adjustment of subjective audit probability, public goods provision, and Pareto-optimality. One out of the four behavioral archetypes of taxpayers rests on expected utility maximization by Allingham and Sandmo (1972); social norms evolve with the age of the taxpayers reflecting psychological findings by Kirchler (2007), Traxler (2010), and Traxler and Winter (2012), and social interactions are affected by the endorsement of Pareto-optimal allocations within closed groups or neighborhoods (Hokamp and Pickhardt, 2011; Pickhardt, 2012).

9.2.1.1 Purpose

The purpose of our ABM is to investigate, at the macro- and the micro level, the dynamics of tax noncompliance via thought experiments, which allows users, for example, researchers, tax practitioners, and policy makers, a usage of our approach for their own purposes, for example, to estimate the effects of tax reforms on the extent of tax evasion. Hokamp (2014) and Bazart et al. (2016) show how

---

[1] From the point of view discussed in Chapter 1, this ABM should be termed as a tax noncompliance model. However, in our chapter, we make use of the term "agent-based tax evasion model," since the main cause of not paying tax is evasion by intention and not noncompliance by randomness.

simulations of tax noncompliance add to tax evasion theories and experiments. The computational simulation of the tax evasion problem with the opportunity to adjust parameters through an ABM gives users the possibility to overcome real life experimental barriers and to study how taxpayers at the micro level affect the aggregated outcome at the macro level, for example, the tax revenue, of the system. The ABM provides an expansion of neoclassical-theory prediction, as it is able to account for the complex interactions in heterogeneous environments. Furthermore, our model allows to investigate the characteristics of interest in what-if studies. In particular, examination of tax evasion dynamics under public goods provision and testing the relevance of Pareto-optimal allocations become possible (Hokamp, 2014).

9.2.1.2   Entities, State Variables, and Scales

In the model, time steps reflect tax relevant periods or tax years and space is not modeled but we consider the environment, in which the tax compliance decision takes place. Our model includes a behaviorally heterogeneous population of N taxpayers[2], a tax authority, and a government, described in Hokamp (2014) as three kinds of agents. The government provides a public good to the population of taxpayers; the provision level depends on total revenues, gathered by the government, consisting of all paid tax and penalties minus the payment for tax reimbursement[3] and administrative expenses. The tax authority collects the tax voluntarily contributed by the taxpayers, and, in addition, it conducts at each time step a random audit process and charges the corresponding penalties and reimbursement from tax evasion or overpayment discovered, respectively. To run a tax scheme, the tax authority incurs administration expenses, which relate to the system efficiency. The tax authority is characterized by the effectiveness of the tax collection, ϵ_{TC}, the effectiveness of the audit process, ϵ_{AP}, and fixed costs, ϵ_{FC}. The government decides on the tax rate, θ, the penalty rate, π, the audit probability, α, and the tax complexity, γ while providing a public good with the efficiency, β. Hence, varying these parameters determines a political and public goods provision cycle (Hokamp and Pickhardt, 2010; Hokamp, 2013, 2014), which reflects the environment in which tax declarations, payments, and social interactions of taxpayers take place. A list of parameters and mathematical symbols is found in Table 9.1.

We define behavioral archetypes of taxpayers based on Hokamp and Pickhardt (2010), which divide a heterogeneous population into four kinds of taxpayers showing different behaviors. In particular, the taxpayers are classified as, (i) neoclassical A-Types, (ii) social interacting B-Types, and (iii) ethical C-Types and (iv) erratic D-Types.

---

[2] We drop the index i for an individual taxpayer whenever possible to simplify the equations.

[3] Some taxpayers mistakenly pay more income tax than they should because of the complexity of the system. A reimbursement is due when the tax authority becomes aware of the overpayment via an audit. For details see erratic D-Type taxpayers.

Table 9.1 Model parameters and mathematical symbols

	Government					Tax authority		
						Effectiveness of		
Model parameter	Tax rate	Penalty rate	Audit probability	Public goods efficiency	Tax complexity	Audit process	Tax collection	Fixed costs
Abbreviation	θ	π	α	β	γ	ϵ_{AP}	ϵ_{TC}	ϵ_{FC}

Neoclassical A-Types

This kind of taxpayers shows behavior patterns described in the neoclassical theory proposed by Allingham and Sandmo (1972). At each time step the decision, made by neoclassical A-Type taxpayers, is the one that maximizes their utility, $\mathcal{U}$, so that the definition of this function governs their behavior. Reflecting the utility of money, an exponential utility function on the after tax and penalties net income, $ATPNI$, is chosen[4],

$$\mathcal{U}(ATPNI) = 1 - e^{-\rho \ ATPNI} \tag{9.1}$$

which is sensitive to a subjective audit probability, income, tax, and penalty rate, but also incorporates lapse of time effects and a dependency on the risk parameter, ρ, which may change due to social norms. The $ATPNI$ incorporates the effects of public goods provision in the utility of the taxpayers via

$$ATPNI^{No \ Audit} = I - \theta X + \beta(\theta X(1 - \epsilon_{TC}) + PG_{-i}) \tag{9.2}$$

and

$$ATPNI^{Audit} = ATPNI^{No \ Audit} - (1 - \beta(1 - \epsilon_{AP}))(\pi(I - X) + BA) \tag{9.3}$$

where I represents the income earned in a time step; X is the decision variable of a taxpayer, how much income he voluntarily declares to the tax authority; PG_{-i} describes the contribution to a public good by all other taxpayers, except the taxpayer i under consideration; and BA denotes the penalties due to back audits for a lapse of time, $LOT > 0$,[5] described by Hokamp (2014). The $ATPNI$ is defined once for the case when the taxpayer is audited and once for no audit. In the version without an audit, the net income after tax and penalties is defined as the earned income minus the tax paid plus the benefits received from public goods. In the

---

[4] We assume risk aversion and, therefore, we exclude the case $\rho = 0$. Higher values of ρ indicate a higher preference for risk aversion. See Pavel (2015) for a discussion of risk aversion and risk-neutrality in the context of tax evasion.

[5] Without lapse of time effects, $LOT = 0$, we get $BA = 0$. For details on lapse of time and public goods provision see Section 9.2.1.3, which includes a definition of BA and PG_{-i} given in Eqs (9.10) and (9.11), respectively.

version when the taxpayer is audited, the charged penalty over evaded tax plus the change in public goods provision are taken into account.

Given Eqs (9.1)–(9.3) the neoclassical A-Type taxpayers apply the expected utility maximization procedure adopted by Allingham and Sandmo (1972). Hokamp (2014) presents the interval for the subjective audit probability, α_S, which allows for an inner solution, given by the lower bound

$$\alpha_S > \frac{1}{1 + \left(\frac{(1-\beta(1-\epsilon_{AP}))\pi}{(1-\beta(1-\epsilon_{TC}))\theta} - 1 \right) e^{\rho(1-\beta(1-\epsilon_{AP}))(\pi I + BA)}} \tag{9.4}$$

and the upper bound

$$\alpha_S < \frac{1}{1 + \left(\frac{(1-\beta(1-\epsilon_{AP}))\pi}{(1-\beta(1-\epsilon_{TC}))\theta} - 1 \right) e^{\rho(1-\beta(1-\epsilon_{AP}))BA}} \tag{9.5}$$

If the audit probability exceeds the upper limit in Eq. (9.5), the taxpayer becomes fully tax compliant, that is, $X = I$, and when α_S falls below the lower bound in Eq. (9.4), the taxpayer fully evades, that is, $X = 0$. For an audit probability in the range for an inner solution, the taxpayer voluntarily declares[6]

$$X = X^* = I + \frac{BA}{\pi} - \frac{\ln \left(\frac{(1-\alpha_S)(1-\beta(1-\epsilon_{TC}))\theta}{\alpha_S((1-\beta(1-\epsilon_{AP}))\pi - (1-\beta(1-\epsilon_{TC}))\theta)} \right)}{\rho\pi(1 - \beta(1 - \epsilon_{AP}))} \tag{9.6}$$

Social Interacting B-Types

These taxpayers check the behavior of their vicinity, v, which is of a parametrically specified size. For each taxpayer, the size of the social network, S, is equal. If taxpayer i is v-close to taxpayer j, the reverse may not be true. Further, v-close is a relation that is not transitive. Hence, we assume a randomized list of acquaintances (Gilbert and Troitzsch, 2005; Hokamp, 2013, 2014) employed by the taxpayers to estimate an average benefit. If the average reductions of income were less than the tax rate in the previous period, then the condition to evade tax[7] is fulfilled and they voluntarily announce

$$X_{i,t} = \frac{1}{v} \sum_{s \in S \subseteq \{1, \ldots, N\}; |S| = v} \frac{X_{s,t-1}}{I_{s,t-1}} I_{i,t} \tag{9.7}$$

declaring the same as the mean in their neighborhood. If the condition to evade is not fulfilled then they honestly declare their total income. Full tax compliance also applies as a default option, when a B-Type is born as well as in the initialization period of the model. Finally, if the tax evasion of a B-Type is uncovered, this

---

[6] The optimal income declaration includes a case without public goods provision and lapse of time effects according to Hokamp and Pickhardt (2010) shown in Chapter 8.

[7] For a mathematical description of the condition to evade tax see Hokamp (2014, p. 192, Eq. (24)).

subject mutates to a fully compliant taxpayer for a preassigned number of periods, τ. To put this differently, an uncovered B-Type mutates into a C-Type, an agent type, which will be described in the next paragraph.

Ethical C-Types

Such taxpayers are ethical or altruistically motivated. They stick to full declaration at all times; they are not affected by other taxpayers but affect the behavior of neoclassical A-Type and social interacting B-Type taxpayers, for example, via their contribution to the provision of public goods.

Erratic D-Types

These taxpayers try to follow a behavior, for instance, full tax compliance, but they randomly fail to accomplish their goal since they are faced with the tax complexity, γ, of the system. Their voluntarily declared income follows a normal distribution centered in the desired behavior[8]. They are not influenced by other taxpayers but they indirectly interact with A-Type and B-Type taxpayers via a public good.

Model Output: The Extent of Tax Evasion

At each tax relevant period the extent of tax evasion, ETE, of the whole group of taxpayers is calculated as

$$ETE_t = 1 - \frac{\sum_{i=1}^{N} X_{i,t}}{\sum_{i=1}^{N} I_{i,t}} \tag{9.8}$$

while the group behavior of the different kinds of taxpayers is expressed as

$$ETE_{t,K} = 1 - \frac{\sum_{i \in K} X_{i,t}}{\sum_{i \in K} I_{i,t}} \tag{9.9}$$

where $K \subseteq \{1, \ldots, N\}$ is the subset of taxpayers of the same behavioral archetype. In Section 9.3 we present graphical visualizations of these indicators to show their evolution over time. Further, social interacting B-Types employ ETE to estimate the extent of tax noncompliance in their social network.

9.2.1.3 Process Overview and Scheduling

At the beginning of each tax relevant period, the government and the tax authority inform the taxpayers about the tax environment, characterized by the parameters shown in Table 9.1. The taxpayers utilize this information to calculate (or optimize) their voluntarily declared income to the tax authority by applying the behavioral heuristics described in Section 9.2.1.2. After tax collection, the tax authority

---

[8] To avoid degenerated cases, we assume that erratic D-Type taxpayers are restricted to declare an income between zero and I/θ.

randomly conducts an audit, which might include back audits. The corresponding penalties and reimbursement are charged and the government may provide a public good. Finally, each agent grows one tax relevant period older and the taxpayers update their history of tax declarations and after 40 tax years the political and public goods cycle repeats. Further, the model allows to switch "on" and "off" the operation modes, (i) lapse of time, (ii) subjective audit probability, (iii) age heterogeneity, (iv) social norm updating, (v) public goods provision, and (vi) a check for Pareto-optimality.

Lapse of Time
Without considering lapse of time effects, $LOT = 0$, only the actual period might be subject to an audit by the tax authority. With lapse of time effects, $LOT > 0$, additional periods in the past might be audited. The whole population of taxpayers is affected by such a change in the audit procedure. However, lapse of time only affects the tax declaration by neoclassical A-Types and social interacting B-Types. A-Type taxpayers directly take into account that their penalties due to back audits, BA, may become greater and greater

$$BA = \sum_{k=t-LOT}^{t-1} \pi_k(I - X_k) \tag{9.10}$$

which shifts the interval for an inner solution towards zero and, at the same time, the optimal declaration toward the true income. B-Types are indirectly influenced by these amendments via the average extent of tax noncompliance in their social network.

Subjective Audit Probability
Typically, the tax declaration by neoclassical A-Types makes use of a (subjective) audit probability $\alpha_s = \alpha$ for the expected utility maximization proposed by Allingham and Sandmo (1972). However, after the experience of an audit, which uncovers a nonzero extent of tax evasion, the neoclassical A-Types endogenously adjust their subjective audit probability. Hence, they declare their true income in the following tax relevant period, since they set their subjective audit probability to unity, $\alpha_s = 1$. Thereafter, α_S is reduced in a stepwise fashion, by the updating parameter, δ, until the objective audit probability is reached. Hence, the subjective audit probability directly influences A-Types and indirectly affects B-Types via their social network.

Age Heterogeneity
The initial population of taxpayers is heterogeneous with respect to age. In particular, we assume that in the initial period the taxpayers are at an age between a minimum age, AGE_{MIN}, and a maximum age, AGE_{MAX}; initial age is uniformly distributed on $[AGE_{MIN}; AGE_{MAX}]$. Without "age heterogeneity" the taxpayers grow

one year older after a tax relevant period and they live forever. In addition, if "age heterogeneity" is activated, the taxpayers die when they have reached the maximum age. In this case, to allow for *ceteris paribus* conditions, the dead are replaced by newborn taxpayers with identical initial properties, except for age, which is set to the minimum age. Thus, age heterogeneity affects social interacting B-Types, since they are fully tax compliant in their first period of life. Further, age heterogeneity influences neoclassical A-Types iff $LOT > 0$.

Social Norm Updating

Social norms are updated via the risk parameter ρ. If taxpayers are getting older, we assume that they become more and more risk averse, which results in a higher extent of tax compliance. Such behavioral patterns are reflected by raising, ρ, for the elderly (Kirchler, 2007; Traxler, 2010; Nordblom and Žamac, 2012). Hence, social norm updating directly influences neoclassical A-Types via their utility function while social interacting B-Types are affected indirectly via their social network.[9]

Public Goods Provision

If this feature is activated, the government provides a public good, PG, which can be nonrivalrously consumed by the whole population of taxpayers

$$PG = -\mathcal{A} + \sum_{i=1}^{N}(\mathcal{T}_i + \mathcal{P}_i) \tag{9.11}$$

and which is financed by tax, $\mathcal{T}$ and penalties, $\mathcal{P}$, (consisting of fines due to tax cheating and reimbursement caused by tax overpayment) minus administrative costs, $\mathcal{A}$, described by Hokamp (2014). However, what really matters for expected utility maximization is the contribution to the public good by every other taxpayer, PG_{-i}.

$$PG_{-i} = -\epsilon_{FC} + (1 - \epsilon_{TC}) \sum_{j=1; j \neq i}^{N} \mathcal{T}_j + (1 - \epsilon_{AP}) \sum_{j=1; j \neq i}^{N} \mathcal{P}_j \tag{9.12}$$

In addition to the tax authority efficiency parameters, ϵ_{AP}, ϵ_{TC}, the optimal solution, shown in Eq. (9.6), of the neoclassical A-Types is influenced by the effectiveness of the public sector, β. Finally, public goods indirectly affect social interacting B-Types via their social network.

Check for Pareto-Optimality

The operation mode "Pareto-optimality" only makes sense when a public good is provided. Social interacting B-Types check for Pareto-optimality and, in case of detecting a Pareto-optimal allocation, do not change their voluntarily

---

[9] For details on the empirical background see Section 9.2.2.1.

declared income. Such a behavior reflects the efforts by B-Types to maintain a Pareto-optimal allocation. However, to identify Pareto-optimality of allocations, we make use of the procedure adopted by Hokamp and Pickhardt (2011) and Pickhardt (2012) which employs a maximum number of free-riders, N_{MAX}, tolerated by the Pareto-optimality concept. If more than N_{MAX} taxpayers do not pay their tax due, the allocation is not Pareto-optimal. Hence, the check for Pareto-optimality is a step function and directly influences social interacting B-Types as well as indirectly via their social network. Neoclassical A-Types are indirectly affected by the resulting change in the provision of public goods.

9.2.2   Design Concepts

9.2.2.1   Theoretical and Empirical Background

The model is able to capture a huge variety of scenarios because it allows for a wide selection of parameter values. For instance, in Sections 9.3.1–9.3.3 we examine an artificial political and public goods provision cycle by Hokamp and Pickhardt (2010) and Hokamp (2013, 2014). Targeting the tax declaration behavior of the taxpayers at the micro level, we employ theories as well as empirical findings. The neoclassical A-Types apply the theory of expected utility maximization proposed by Allingham and Sandmo (1972). Further, the investigation of lapse of time effects is motivated by changes to the German tax law. In tax year 2009, the relevant periods subject to an audit were changed from 5 to 10 years (Hokamp and Pickhardt, 2010), that is, $LOT = 5$ and $LOT = 10$.[10]

To adjust the subjective audit probability, so that it reflects psychological impacts (Kirchler, 2007), we assume an update of $\delta = -0.2$ per period in line with Hokamp and Pickhardt (2010). Furthermore, for the social interaction by the B-Types we follow Hokamp and Pickhardt (2010) and apply a mutation to full tax compliance, that is, $\tau = 4$, four tax years due to psychological reasons (Kirchler, 2007; Zaklan et al., 2008). Age heterogeneity and social norms date back to the findings by Kirchler (2007), Traxler (2010), and Nordblom and Žamac (2012), with the elderly showing a higher tax compliance than younger taxpayers. Following Hokamp (2014) we assume $AGE_{MIN} = 21$, $AGE_{MAX} = 60$ and an update of social norms with respect to a taxpayer's AGE is summarized in Table 9.2

For the simulations in Sections 9.3.1–9.3.3 we employ the political and public goods provision cycle by Hokamp and Pickhardt (2010) and Hokamp (2014). For instance, the efficiency of tax collection, ϵ_{TC}, is taken from Slemrod and Yitzhaki

---

[10] Pavel (2015) provides an introduction to lapse of time effects in tax compliance experiments. The scenarios are investigated with six, three, and one relevant period subject to an audit as well as no lapse of time effects. The tax compliance is observed to be higher when the tax authority is allowed to audit more periods in the past and a measurement of such lapse of time effects is provided.

Table 9.2 Social norm updating. Risk aversion changes with respect to age

Age	$21 \leq Age \leq 30$	$30 < Age \leq 40$	$40 < Age \leq 50$	$50 < Age \leq 60$
Risk aversion ρ	$[0; 0.25]$	$[0.25; 0.50]$	$[0.50; 0.75]$	$[0.75; 1]$

(2002). Thus, a public good is examined with a view to represent the standard theory and empirical findings on linear public goods experiments (e.g., by Ledyard, 1995; Zelmer, 2003; Croson, 2007) and we confirm the counter-intuitive result provided by Falkinger (1988, 1995) and Cowell (1992) that a higher provision level increases the extent of tax evasion. The check for Pareto-optimality originates from theoretical findings by Hokamp and Pickhardt (2011) and Pickhardt (2012).

Andreoni *et al.* (1998) estimate that about 7% of US taxpayers overpay their tax liabilities. In Sections 9.3.1 and 9.3.2 we use this empirical background for erratic D-Types to assume a share of about 15% in the population to account for unintended tax noncompliance (Hokamp and Pickhardt, 2010). In particular, we suppose that this group of taxpayers tries to behave like ethical C-Types, centering their declared income around perfect compliance, allowing for some overpayment, which results in a normal distribution with expectation $\mu = I$ and standard deviation $\sigma = \gamma I$. However, the tax declaration experiment conducted by Alm *et al.* (2009) does not allow for overpayment, so that we suppose as a best guess a normal distribution with expectation $\mu = 0.5I$ and standard deviation $\sigma = 0.5\gamma I$ for our calibration and sensitivity analysis in Section 9.3.4.

9.2.2.2 Individual Decision-Making

Spatial aspects do not play a role in our agent-based framework but the environment is explicitly modeled via a government and tax authority. The government provides public goods and the tax authority collects tax and penalties; bureaucratic activities do not involve any kind of individual decision-making. In contrast, the taxpayers follow four behavioral archetypes or individual decision heuristics (i) expected utility maximization, (ii) social interaction, (iii) ethical motivation, and (iv) erratic perception described in Section 9.2.1.2. The taxpayers adapt their tax declaration behavior to exogenously changing parameters (e.g., tax rate θ, and penalty π), but also may use a subjective audit probability (A-Types) or an average extent of tax noncompliance within their social network (B-Types), which are derived endogenously. Social norms are implemented via an increase of risk aversion, ρ, for senescent neoclassical A-Types. Hence, we model tax noncompliance as a social process, in which temporal aspects play an essential role in the declaration process via growing older and lapse of time effects. Erratic D-Types may err due to the complexity of the tax law. Further, uncertainty in the model is reflected by a random selection of the taxpayers subject to an audit.

9.2.2.3   Learning

The tax authority learns a taxpayer's true income and history via an audit. A neoclassical A-Type is able to learn the objective audit probability, iff "subjective audit probability" is activated, while a social interacting B-Type gathers information on the tax declaration behavior of his vicinity or social neighborhood. All other entities do not learn. While collective learning is not implemented in the model, the average extent of tax noncompliance learned by B-Types may be thought of as a type of "collective" learning.

9.2.2.4   Individual Sensing

The taxpayers correctly sense the information provided by the government and the tax authority. Nonetheless, erratic D-types may err in voluntarily declaring their income due to the complexity of the tax law, which reflects their misperception of the tax regime. If "subjective audit probability" is activated, the neoclassical A-Types misperceive the objective audit probability in some periods after an audit has discovered their cheating behavior on taxes. However, tax declarations are correctly perceived by the tax authority, but the government and the tax authority have no information on the true income besides the information that is provided by the taxpayers, voluntarily or via an audit. When considering the provision of a public good, the model also takes into account the costs for the tax collection and the audit process to uncover tax noncompliance.

9.2.2.5   Individual Prediction

The model is not forward looking and, therefore, does not include any individual prediction. However, the neoclassical A-Types use an expected utility maximization procedure and subjective audit probabilities that represent their perception of the future. The social interacting B-Types form beliefs of the tax compliance behavior in their neighborhoods. Further, iff "Pareto-optimality" is activated, B-Types do not change their contribution to the public good to maintain in the future the Pareto-optimal allocation observed. All decisions of the agents are based on present time and the tax relevant periods subject to back audits.

9.2.2.6   Interaction

Between the government and the tax authority on the one side and the taxpayers on the other side, the interaction is direct via tax enforcement variables (α, θ, and π), the audit process, tax payment, reimbursement and penalties, and the interplay takes place indirectly via public goods provision. The interaction among taxpayers is affected by the average rate of tax compliance in the social network for B-Types and public goods provision for A-Types. Hence, the communication channels between taxpayers are the level of public goods provision and the average extent of tax noncompliance.

9.2.2.7 Collective

Agents do not form a collective. However, the social network of B-Types may be considered as some kind of "collective." The size, v, of the social network, S, is a parameter of the model, such that the chosen number of acquaintances is randomly selected. Neighborhoods are constructed in such a way that if taxpayer i is in the social network of taxpayer j, the reverse might not be true. In other words, the relationships between taxpayers within a social network are basically one-directional.

9.2.2.8 Heterogeneity

The parameters, announced by the government and the tax authority, may change with respect to time under the exception or condition to repeat every 40 tax years. The taxpayers might be heterogeneous concerning their initial properties, (i) age, (ii) income, (iii) risk aversion, and (iv) list of acquaintances. In addition, heterogeneity may be caused by the behavioral archetype of the taxpayers, (i) neoclassical A-Type, (ii) social interacting B-Types, (iii) ethical C-Types, and (iv) erratic D-Types. Further, when time evolves, the heterogeneity spreads regarding the individual experience of tax declarations, audits, and payments of penalties and reimbursement.

9.2.2.9 Stochastic

Age, income, risk aversion, and the list of acquaintances are randomly assigned in the initial period. However, the model allows for switching "off" the random assignment and, instead, to specify, for example, risk aversion, which we set to $\rho = 0.01$ in Section 9.4. Further, the audits of taxpayers by the tax authority are random.

9.2.2.10 Observation

We collect data at each time step on the extent of tax compliance at the macro level and the behavioralgroup level described in Eqs (9.8) and (9.9), respectively. In particular, we are interested in the interaction among the taxpayers themselves within the tax declaration environment, and we study various scenarios as described in Section 9.3. We reaffirm the key findings by Hokamp (2014) that the dynamics by lapse of time and social norm updating have a particularly strong effect on the extent of tax evasion and we add an explanation at the behavioral group level. In addition, we visualize key aspects of the expected utility maximization approach adopted by Allingham and Sandmo (1972). Finally, in Section 9.3.4 we calibrate our model to the experimental tax compliance data collected by Alm *et al.* (2009) and provide novel insights into the influence of the tax rate, θ, the penalty, π, the audit probability, α, the tax complexity, γ, and the risk aversion, ρ, on tax noncompliance.

9.2.3 Details

9.2.3.1 Implementation Details

The model is implemented in RePast making use of the Java software packages. Hokamp (2013) presents computational simulations with different random seeds and finds that the fluctuations due to random effects are rather small. Hence, we restrict our simulations in Section 9.3 to random seed 11223344 to allow for an identification of Pareto-optimal allocations. Following the work by Hokamp (2014), we updated the code with a view to reconsider the extent of tax compliance at a behavioral group level. Code is available upon request. Coincidently, such a procedure simplifies an independent reproduction of our results.

9.2.3.2 Initialization

At the initial state of the artificial world no taxpayer evades tax and therefore his history of tax noncompliance does not reveal any kind of tax offence. The government and the tax authority have no costs and no revenues at the initial state. The initialization of the model may differ, for example, in the distribution of income and in the tax declaration by the taxpayers. For instance, a change in the composition of behavioral archetypes in the population of taxpayers might affect the audit process. To allow for *ceteris paribus* conditions it is possible to employ an identical random seed for generating the random numbers, which govern random effects such as the auditing process.

9.2.3.3 Input Data

The model does not utilize input data from external sources. All input data are specified within the model, for instance, the political and public goods cycle, as referred in Hokamp and Pickhardt (2010) and Hokamp (2013, 2014).

9.2.3.4 Submodels

There are no submodels.

9.3   Scenarios, Simulation Results, and Discussion

This section provides the scenarios under consideration, the resulting outcome and a brief discussion of our key findings. *Ceteris paribus* and jointly, we examine the tax evasion dynamics by six operation modes, (i) lapse of time, (ii) social norms, (iii) age heterogeneity, (iv) subjective audit probability, (v) public goods provision and (vi) Pareto-optimality described in Section 9.2.1.3. First, we replicate (Hokamp, 2014) with a view to add novel insights into the behavioral group level in Sections 9.3.1 and 9.3.2. Second, we employ a behaviorally homogeneous population of neoclassical A-Types to elucidate the expected utility maximization approach by Allingham and Sandmo (1972) in Section 9.3.3. Finally, in Section 9.3.4 we calibrate our model with the experimental tax compliance data provided by Alm *et al.* (2009) and present our sensitivity analysis on tax and penalty rate, audit probability, tax complexity, and risk aversion. Table 9.3 summarizes the scenarios and related specifications.

9.3.1   Age Heterogeneity and Social Norm Updating

Figures 9.1 and 9.2 elaborate Hokamp (2014, p. 195, Figures 1.1 and 1.2) by showing the effects of age heterogeneity and social norm updating under lapse of time and an adjustment of the subjective audit probability as described in Sections 9.2.1.3 and 9.2.2.1. We adopt the size of the population, $N = 2000$, and the behavioral type distribution: 50% A-Types, 35% B-Types, 0% C-Types, and 15% D-Types; and for income, I, and risk aversion, ρ, the uniform distribution on $[0; 100]$ and $]0; 1]$, respectively (from Hokamp, 2014), and, in addition, we employ the political cycle from Hokamp and Pickhardt (2010) outlined in the Appendix 9A. The simulations are done with the active mode "subjective audit probability." We present the extent of tax evasion for 80 tax relevant periods and the corresponding results at the behavioral group level for A- and B-Types. We do not show the graphs for C- and D-Types, since the former are fully tax compliant and the latter account for (unintended) tax noncompliance of less than 3%.

Figure 9.1(a) confirms that without lapse of time effects, $LOT = 0$, at the macro level no effect seems to be visible by age heterogeneity alone (Hokamp, 2014). However, the inspection of Figure 9.1(b) and (c) reveals that, contrary to A-Types, B-Types can be influenced by age heterogeneity alone (e.g., by -1.7% in tax relevant period 58). The B-Types become more tax compliant since newborn taxpayers declare their true income in their first tax relevant period. Social norms alone represent the lower bound since, after 30 tax relevant periods, each A-Type has reached the cohort with a risk aversion uniformly distributed on $[0.75; 1]$; the effect on tax evasion is smaller for B-Types, strongly influenced by the size of the social network, $v = 4$, than it is for A-Types. Further, jointly considering social norms and age heterogeneity allows taxpayers to update their social norms, to die and to be born; creating a scenario in which we find a strong impact on the A-Types.

Table 9.3 Overview: specification of agent-based simulations

Feature	Lapse of time	Tax relevant periods	Age heterogeneity	Social norm updating	Public goods	Pareto-optimality	Subjective audit probability	Figure
Scenario inspired by Hokamp (2014)								
Political and public goods cycle; Risk ρ uniform distributed [0;1]								
Distribution of behavioral types: A = 0.5; B = 0.35; C = 0; D = 0.15								
Reference 1;	LOT = 0	80	Off	Off	Off	Off	On	9.1–9.4
Reference 2;	LOT = 10	80	Off	Off	Off	Off	On	9.1–9.4
AH;	LOT = 0	80	On	Off	Off	Off	On	9.1
AH and SNU;	LOT = 0	80	On	On	Off	Off	On	9.1
SNU;	LOT = 0	80	Off	On	Off	Off	On	9.1
AH;	LOT = 10	80	On	Off	Off	Off	On	9.2
AH and SNU;	LOT = 10	80	On	On	Off	Off	On	9.2
SNU;	LOT = 10	80	Off	On	Off	Off	On	9.2
PGP;	LOT = 0	80	Off	Off	On	Off	On	9.3
… and AH and SNU;	LOT = 0	80	On	On	On	Off	On	9.3
… and PO (C = 0.15; D = 0);	LOT = 0	80	On	On	On	On	On	9.3
PGP;	LOT = 10	80	Off	Off	On	Off	On	9.4
… and AH and SNU	LOT = 10	80	On	On	On	Off	On	9.4
… and PO (C = 0.15; D = 0)	LOT = 10	80	On	On	On	On	On	9.4
Scenario inspired by Allingham and Sandmo (1972)								
Audit probability α = 0.01; Risk ρ uniform distributed [0;1]								
Distribution of behavioral types: A = 1; B = C = D = 0								
Allingham and Sandmo	LOT = 0	80	Off	Off	Off	Off	Off	9.5
Lapse of time effects	LOT = 5	80	Off	Off	Off	Off	Off	9.5
Lapse of time effects	LOT = 10	80	Off	Off	Off	Off	Off	9.5

Subjective audit probability	LOT = 0	80	Off	Off	Off	On	9.6
Public goods provision	LOT = 0	80	Off	On	Off	Off	9.6
Age heterogeneity and social norm updating	LOT = 0	80	On	Off	On	Off	9.6

Calibration to Alm et al. (2009) and Bazart et al. (2016)

Tax rate $\theta = 0.35$; Penalty rate $\pi = 0.525$; Audit probability $\alpha = 0.4$; Tax complexity $\gamma = 0.5$

Distribution of behavioral types: A = 0.26; B = 0; C = 0.39; D = 0.35

SAP On	LOT = 0	15	Off	Off	Off	On	9.7

Risk ρ uniform distributed [0;1]

Risk ($\rho = 0.01$)	LOT = 0	15	Off	Off	Off	On	9.7
Social interacting B-Types	LOT = 0	15	Off	Off	Off	On	9.7

Distribution of behavioral types: A = 0.21; B = 0.15; C = 0.34; D = 0.30

Sensitivity analysis

Distribution of behavioral types: A = 0.21; B = 0.15; C = 0.34; D = 0.30

	LOT = 0	15	Off	Off	Off	On	9.7

Scenarios: Tax rate $\theta = 0.385$; Penalty rate $\pi = 0.5775$; Audit probability $\alpha = 0.44$; Tax complexity $\gamma = 0.55$; Risk $\rho = 0.011$

Two scenarios are considered based on (i) Hokamp (2014) and (ii) Allingham and Sandmo (1972), and, in addition, a calibration to Alm et al. (2009) and Bazart et al. (2016) and a sensitivity analysis are presented. For each group of simulations the parameters of the model are specified, that is, tax rate θ, penalty rate π, audit probability α, tax complexity γ, risk ρ, and the distributions of the four behavioral types, (i) neoclassical A-Types, (ii) social interacting B-Types, (iii) ethical C-Types, and (iv) erratic D-Types. The political and public goods cycle is taken from Hokamp and Pickhardt (2010) and Hokamp (2014) provided in the Appendix 9A. Further, the specifications of the 28 simulation runs are presented with respect to seven features, which are, (i) "lapse of time" (LOT), (ii) "tax relevant periods," (iii) "age heterogeneity" (AH), (iv) "social norm updating" (SNU), (v) provision of pure "public goods" (PGP), (vi) a check for "Pareto-optimality" (PO), and (vii) an adjustment of the "subjective audit probability" (SAP). "on" marks activated features. "off" indicates options that are switched off. " ... " extends the preceding simulation, for example, with "Pareto-optimality" (PO).

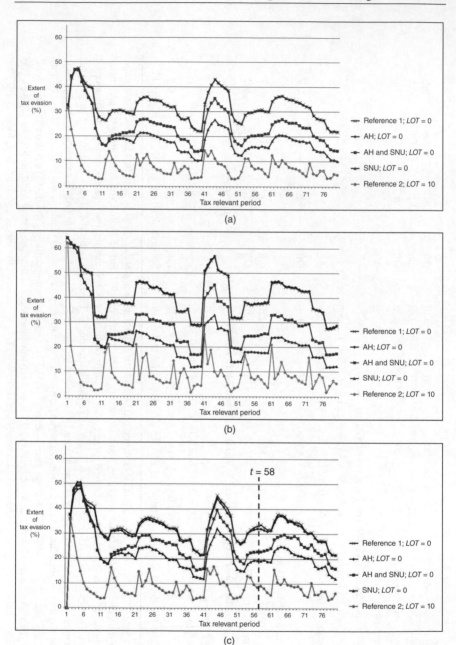

Figure 9.1 Scenario inspired by Hokamp (2014): age heterogeneity (AH) and social norm updating (SNU); lapse of time $LOT = 0$; (a) behaviorally heterogeneous society, (b) neoclassical A-Types, (c) social interacting B-Types. Panel (a) source: Hokamp (2014), reproduced with permission of Elsevier.

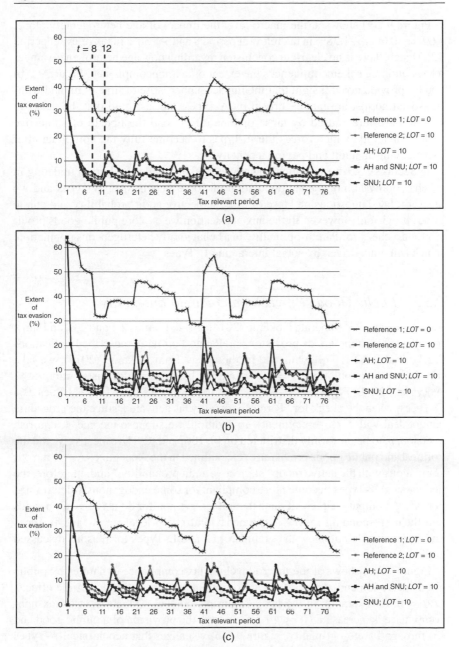

Figure 9.2   Scenario inspired by Hokamp (2014): Age heterogeneity (AH) and social norm updating (SNU); lapse of time *LOT* = 10; (a) behaviorally heterogeneous society, (b) neoclassical A-Types, (c) social interacting B-Types. Panel (a) source: Hokamp (2014), reproduced with permission of Elsevier.

Figure 9.2(a) shows at the macro level the impact of age heterogeneity alone, $LOT = 10$ (e.g., $+1.58\%$ in tax relevant period 8 and -1.38% in tax relevant period 12). Hence, there is no clear-cut conclusion regarding how age heterogeneity under a back auditing scheme influences the extent of tax noncompliance. Figure 9.2(b) and (c) provide novel insight that the impacts under lapse of time, with and without social norms, are mainly driven by A-Types. We also find that the peaks in Figure 9.2(b) are caused by lapse of time effects and the update of subjective audit probability, which forces the A-Types to become fully tax compliant after experiencing an audit and paying a penalty.

In Section 9.3.3 we investigate age heterogeneity and social norm updating in a behaviorally homogeneous population of 100% neoclassical A-Types and we answer related questions on lapse of time, subjective audit probability and public goods provision. However, in the next subsection we explore public goods provision and a check for Pareto-optimality in a behaviorally heterogeneous population, which is mainly driven by social interacting B-Types.

9.3.2   Public Goods Provision and Pareto-optimality

Figures 9.3 and 9.4 elaborate Hokamp (2014, p. 196, Figures 2.1 and 2.2) and show the effects by public goods provision and Pareto-optimality described in Sections 9.2.1.3 and 9.2.2.1. Following the methodology from Hokamp (2014), we take the size of the population, $N = 2000$, and the behavioral type distribution: 50% A-Types, 35% B-Types, 0% C-Types, and 15% D-Types (and 15% C-Types, 0% D-Types, when the check for "Pareto-optimality" is activated, since the procedure cannot deal with tax overpayment). In addition, we suppose income, I, and risk aversion, ρ, to be uniformly distributed on [0; 100] and [0; 1], respectively, and the political and public goods provision cycle outlined in the Appendix 9A. We ran our simulations with the active mode "subjective audit probability," and, therefore, the neoclassical A-Types become more compliant for some endogenously determined periods. As output we present the extent of tax evasion for 80 tax relevant periods and the corresponding results at the behavioral group level for A- and B-Types and, again, we do not show the graphs for C- and D-Types for reasons discussed earlier.

Figure 9.3(a) shows at the macro level the tax compliance dynamics by public goods provision and a check for Pareto-optimality without lapse of time effects, $LOT = 0$; for example, in the tax relevant periods 37–40 the extent of tax non-compliance is increased, (i) by $+1.5\%$ through the provision of a public good and (ii) through Pareto-optimality. Figure 9.3(b) visualizes that neoclassical A-Types account for an increase by $+2.3\%$ of tax evasion by the provision of a public good in period 37–40. Regarding Pareto-optimality, the effects on tax evasion are unclear, for instance, in period 19 we recognize a decrease while in period 38 there is an increase. Figure 9.3(c) presents the tax evasion dynamics by social interacting

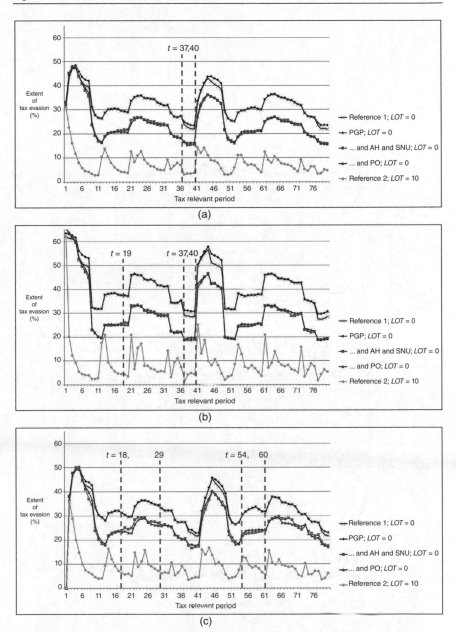

Figure 9.3 Scenario inspired by Hokamp (2014): public goods provision (PGP) and Pareto-optimality (PO); lapse of time $LOT = 0$; (a) behaviorally heterogeneous society, (b) neoclassical A-Types, (c) social interacting B-Types; " ... " extends the preceding scenario, for example, with age heterogeneity (AH) and social norm updating (SNU). Panel (a) source: Hokamp (2014), reproduced with permission of Elsevier.

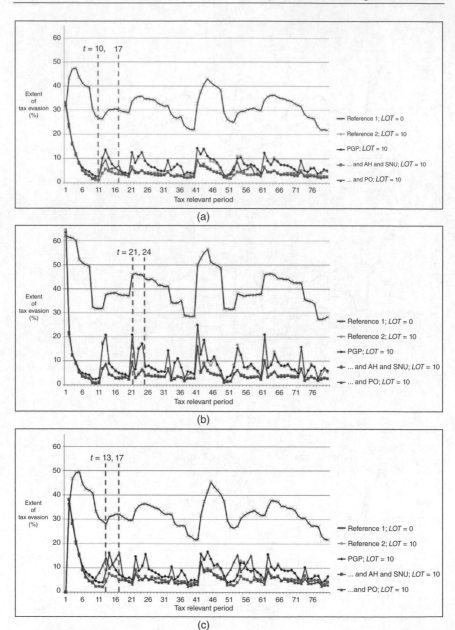

Figure 9.4 Scenario inspired by Hokamp (2014): public goods provision (PGP) and Pareto-optimality (PO); lapse of time *LOT* = 10; (a) behaviorally heterogeneous society, (b) neoclassical A-Types, (c) social interacting B-Types; " ... " extends the preceding scenario, for instance, with age heterogeneity (AH) and social norm updating (SNU). Panel (a) source: Hokamp (2014), reproduced with permission of Elsevier.

B-Types; for example, public goods provision and Pareto-optimality decrease tax evasion in periods 18–29 while they increase in periods 54–60.

Figure 9.4(a) shows at the macro level the extent of tax noncompliance by public goods provision and Pareto-optimality with lapse of time effects, $LOT = 10$. In particular, we confirm the existence of Pareto-optimal allocations in periods 10–17. Figure 9.4(c) shows two peaks in the extent of tax evasion by social interacting B-Types reaching 14.21 and 16.48% in period 13 and 17, respectively. Hence, under lapse of time effects, the check for Pareto-optimality leads to a higher extent of tax evasion than the reference case without public goods provision. The scenario with age heterogeneity and social norm updating provides a baseline of tax evasion under back auditing. Figure 9.4(b) represents the dynamics caused by neoclassical A-Types; among others, we recognize peaks in period 21 and 24, which are due to lapse of time effects and subjective audit probability updating. We investigate such effects in more detail in the next subsection.

9.3.3   The Allingham-and-Sandmo Approach Reconsidered

We recognize that neoclassical A-Type taxpayers particularly strongly influence the extent of tax evasion at the macro level. Within the behaviorally homogeneous population of $N = 2000$ neoclassical A-Types, each taxpayer is endowed with a risk parameter, ρ, uniformly distributed on $[0; 1]$. We assume a constant income, $I = 100$ Tokens, a constant objective audit probability, $\alpha = 0.01$ and vary the tax rate θ between 0.05 and 0.65 (in steps of 0.05). We apply a constant sanction rate, $\zeta = 150\%$, to account for the critique by Yitzhaki (1974). Hence, the penalty rate π ranges between 0.075 and 0.975. As output, we investigate for 80 tax relevant periods, the extent of tax evasion by A-Types with respect to, (i) lapse of time effects, (ii) subjective audit probability, (iii) public goods provision, and (iv) age heterogeneity combined with social norm updating.

9.3.3.1   Lapse of Time Effects

Figure 9.5 visualizes the dynamic effects by lapse of time. The standard approach by Allingham and Sandmo (1972) without lapse of time effects, that is, $LOT = 0$, is presented as a reference case in Figure 9.5(a). We find that tax compliance takes place even for a comparatively low audit probability. A tax rate of $\theta = 0.05$ allows for 95% tax evasion while a tax rate of 0.65 corresponds to 21%. The penalty rate and the uniformly distributed risk parameter ρ are the key factors that explain this outcome since the interval described in Eqs (9.4) and (9.5) depends on both. Hence, a higher penalty rate shifts the lower bound, Eq. (9.4), toward zero, such that full evaders start to declare a part of their true income. The upper bound, Eq. (9.5), does not change, because of selecting a constant sanction rate. In addition, the uniform distribution of risk aversion allows that a neoclassical A-Type

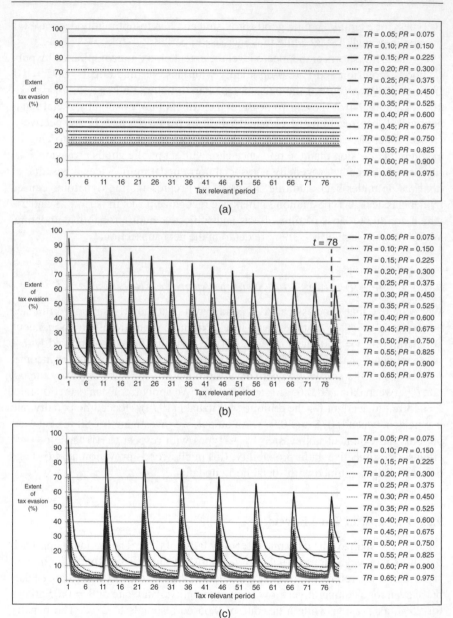

Figure 9.5 Scenario inspired by Allingham and Sandmo (1972): (a) reference without lapse of time effects, (b) lapse of time $LOT = 5$, (c) lapse of time $LOT = 10$

may fully evade, partially declare, or honestly announce his income. However, our results show that increasing the tax rate contributes to lowering the extent of tax evasion in an Allingham-and-Sandmo setting.

Figure 9.5(b) provides the extent of tax evasion with $LOT = 5$ tax relevant periods subject to back audits. The results show peaks in tax evasion in periods $1, 7, 13, \ldots, \ldots, 79$ that occur because (i) the first tax relevant period allows for a maximum extent of tax evasion, and (ii) periods drop out of the expected maximization procedure after six periods. Recall that in the model the actual period and, in addition, up to five periods can be subject to an audit by the tax authority. As a result, the upper level of tax evasion decreases with respect to time. A tax rate of 0.05 reaches 63% tax evasion in period 79 while a tax rate of 0.65 corresponds to 12%. The lower level is reached before the periods with a high extent of tax evasion drop out, that is, in periods $6, 12, 24, \ldots, \ldots, 78$. A tax rate of 0.05 corresponds to 25% tax evasion in period 78 while a tax rate of 0.65 allows for 4%. Thus, a back audit of 5 tax relevant periods significantly decreases tax evasion (e.g., 95 vs. 25 and 21 vs. 4%).

Figure 9.5(c) presents the tax evasion behavior when the actual period and 10 previous tax relevant periods, $LOT = 10$, are considered for the expected utility maximization procedure. Hence, periods $1, 12, 23, \ldots, 78$ are characterized by a peak in tax evasion. The upper level of tax evasion reaches 57% for a tax rate of 0.05 while the lower level is 17%. Considering a tax rate of 0.65 the upper level of tax evasion becomes 12% while the lower level is 2%. To conclude, the lapse of time of 10 tax relevant periods contributes to reducing the extent of tax evasion (e.g., 95 vs. 12 and 21 vs. 2%). The next subsection considers the dynamics by subjective audit probability, public goods provision, age heterogeneity, and social norm updating.

9.3.3.2 Subjective Audit Probability, Public Goods Provision, Age Heterogeneity, and Social Norm Updating

Figure 9.6 *ceteris paribus* shows the effects by an adjustment of the subjective audit probability, public goods provision (data by Hokamp, 2014, is provided in Tables 9A.1 and 9A.2 in the Appendix 9A), and age heterogeneity combined with social norm updating. Figure 9.6(a) elucidates the effects of adjusting the subjective audit probability described in Sections 9.2.1.3 and 9.2.2.1. After initializing our model, which needs five tax years, the extent of tax evasion fluctuates around a value significantly lower than without the adjustment of the subjective audit probability; for $\theta = 0.05$ the reduction of tax evasion is -3.2% and the tax rate 0.65 corresponds to -1%.

Figure 9.6(b) investigates the dynamics of public goods provision explained in Section 9.2.1.3. We show that public goods provision increases the extent of tax evasion; for $\theta = 0.05$ the maximum increase of tax evasion is $+2.78$ in periods 5–8 while the tax rate 0.65 corresponds to $+2.12$. Note that this relationship is

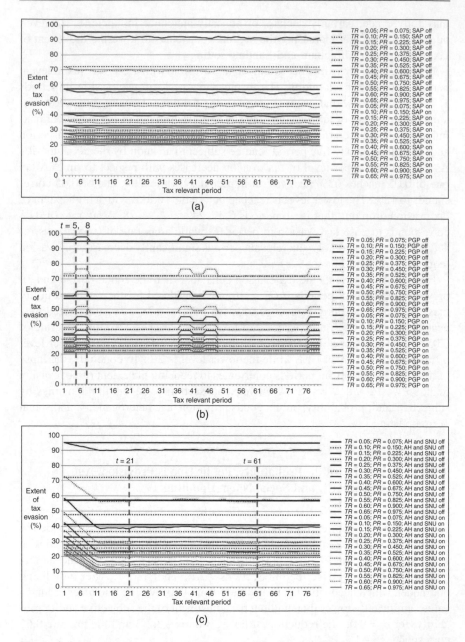

Figure 9.6 Scenario inspired by Allingham and Sandmo (1972): (a) update of subjective audit probability (SAP), (b) public goods provision (PGP), (c) age heterogeneity (AH), and social norm updating (SNU)

nonlinear. In behaviorally heterogeneous populations, the extent of tax evasion at the behavioral group level becomes even higher for neoclassical A-Types, for example, see Section 9.3.2.

Figure 9.6(c) jointly presents the extent of tax evasion by age heterogeneity and social norm updating outlined in Sections 9.2.1.3 and 9.2.2.1. After initialization of our model, which needs 11 tax relevant periods, the tax evasion fluctuates around a value significantly lower than without the inclusion of "age heterogeneity" and "social norm updating" effects; for $\theta = 0.05$ the reduction of tax evasion is -4.8%, the tax rates 0.15 and 0.65 correspond to -17.9 and -12%, respectively. Starting at time step 21 we identify a raise in the extent of tax evasion, for example, $+1.19$ for $\theta = 0.15$, and this phenomenon repeats starting at period 61, showing a cohort effect of social norm updating. However, in the first period the tax evasion is higher than the benchmark, since social norms influence the distribution of risk aversion (tax rates 0.05, 0.15, and 0.65 cause an increase of tax evasion in the initialization period by $+0.14$, $+1.12$, and $+0.24$, respectively). The next section calibrates our ABM to experimental data assuming a behaviorally heterogeneous population.

9.3.4   Calibration and Sensitivity Analysis

We calibrate our economic agent-based tax noncompliance model with experimental data provided by Alm et al. (2009). In each round of the experiment the participants in the tax declaration game earn an income of 60, 70, 80, 90, or 100 Tokens. The individuals in the study are faced with a tax rate of $\theta = 0.35$, a sanction rate of 150%, and an audit probability of $\alpha = 0.4$, which governs their decision on how much income to voluntarily declare to the tax authority. The population of 40 taxpayers is grouped into five social networks of eight group members. In addition, the taxpayers are informed on the average extent of tax noncompliance in their peer group, but they have no information on the individual outcome of the audit process; the environment, outlined ealier, does not change during the 15 rounds of the experiment. In particular, we make use of the experimental data from the session "official information (T2A)" (Alm et al., 2009) and we accordingly adjust the parameters of our ABM. We visualize the outcome of our calibration in Figure 9.7 and present our sensitivity analysis at the behavioral group level in Table 9.4.

Figure 9.7(a) shows the extent of tax noncompliance in the tax declaration game by Alm et al. (2009) as a benchmark. In addition, we present the corresponding results of the calibration by Bazart et al. (2016) to allow for a comparison. We recognize that the erratic D-Types are differently defined by Bazart et al. (2016), since Alm et al. (2009) do not consider tax overpayment. Hence, we recall our assumption that the tax declaration behavior by erratic D-Types is governed by a normal distribution with expectation $\mu = 0.5I$ and standard deviation $\sigma = 0.5\gamma I$ (see Section 9.2.2.1). Further, let the tax complexity be $\gamma = 0.5$.

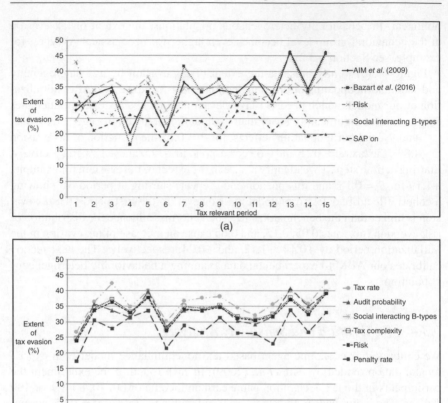

Figure 9.7 Calibration and sensitivity analysis: (a) calibration to Alm *et al.* (2009); (b) sensitivity analysis with respect to a 10% increase in tax rate, audit probability, tax complexity, risk, and penalty rate

Following Bazart *et al.* (2016, p. 138, Figure 3a) we take their behavioral type distribution: 26% A-Types, 0% B-Types, 39% C-Types and 35% D-Types. The active mode "subjective audit probability" described in Section 9.2.1.3 leads to a comparatively low level of tax noncompliance since neoclassical A-Types are highly tax compliant. Switching from a uniform distribution on]0; 1] to a constant risk aversion $\rho = 0.01$ increases the extent of tax evasion by A-Types and, thus, raises tax noncompliance in the whole population. To include imitating behavior patterns (Bazart *et al.*, 2016), we adjust the behavioral type distribution by increasing the share of social interacting B-Types at the expense of the three remaining archetypes to get a similar average extent of tax noncompliance as Alm *et al.* (2009), 33.6 versus 33.3%.

Hence, we conduct our sensitivity analysis under the behavioral type distribution: 21% A-Types, 15% B-Types, 34% C-Types, and 30% D-Types. Figure 9.7(b) visualizes the time evolution of tax compliance when investigating a 10% increase of the tax and penalty rate, the audit probability, the tax complexity, and risk aversion. Our results reveal that increasing the tax and penalty rate have the strongest impact on the extent of tax evasion, positively and negatively, respectively.[11]

Table 9.4 presents the extent of tax evasion averaged over 15 tax relevant periods. The tax declaration behavior by erratic D-Types differs when the behavioral

Table 9.4 Calibration and sensitivity analysis: average extent of tax evasion

Average extent of tax evasion	Behaviorally heterogeneous society	Neoclassical A-types	Social interactig B-types	Ethical C-types	Erratic D-types
Calibration					
Scenario inspired by Alm *et al.* (2009)	33.3				
SAP on Risk ρ uniform distributed [0;1]	23.5	3.8	0	0	50.5
Risk ($\rho = 0.01$)	29.9	52.2	0	0	50.5
Social interacting B-types	33.6	46.9	10.7	0	49.7
Sensitivity analysis					
Tax rate $\theta = 0.385$	+2.59	+6.96	+1.17	0	0
Penalty rate $\pi = 0.5775$	−5.79	−15.78	−1.54	0	0
Audit probability $\alpha = 0.44$	−0.16	−0.26	−0.90	0	0
Tax complexity $\gamma = 0.55$	+0.03	0	+0.07	0	+0.07
Risk $\rho = 0.011$	−0.57	−1.56	−0.15	0	0

"Calibration" shows the extent of tax evasion averaged over 15 tax relevant periods for a behaviorally heterogeneous society and the four groups of behavioral types according to the scenarios specified in Table 9.3. "Sensitivity Analysis" presents in absolute values the difference in the average extent of tax evasion compared to the benchmark "social interacting B-Types."

---

[11] On the technical side, we use the output for the tax relevant periods 41–55 to avoid initialization effects.

type distribution is changed, although we made use of the same random seed 11223344. The A-Types are strongly affected by the tax (+6.96) and penalty rate (−15.78), the effects by risk aversion (−1.56) and audit probability (−0.26) are orders of magnitudes smaller. The B-Types are influenced by all five parameters; again, the penalty rate (−1.54) causes the strongest impact. Surprisingly, tax complexity identically changes the tax declaration by B- and D-Types (+0.07). To summarize, the higher the penalty rate the stronger the effect on hampering tax evasion (−5.79) and the higher the tax rate the strongest the impact on the extent of tax evasion (+2.59). If we consider the aggregated influence of these two parameters, while keeping their ratio constant, then the net effect of increasing the tax rate will be a reduction of tax evasion, as the penalty rate dominates the effect.

9.4   Conclusions and Outlook

In this chapter, we introduced an agent-based tax evasion model based on Hokamp and Pickhardt (2010) and Hokamp (2013, 2014), which consists of a government, a tax authority, and four behavioral groups of taxpayers governed by, (i) neoclassical expected utility maximization, (ii) social interaction, (iii) ethical motivation, and (iv) erratic perception. We addressed the impact of these behavioral groups on the extent of tax noncompliance both at the macro level and the group level. In particular, we replicated the scenarios provided by Hokamp (2014), confirmed his findings and elucidated the extent of tax evasion at the micro level. We showed a strong impact by neoclassical A-Types, which provides the incentive to reconsider the Allingham-and-Sandmo approach. The model incorporated psychological findings, for example, by Kirchler (2007) and Traxler (2010), and we measured the resulting effects on social norms and the subjective audit probability. In addition, we provided figures for lapse of time effects, age heterogeneity, and social norm updating, which reaffirm the notion that tax evasion models should incorporate back audits and the time evolution of tax declarations (Hokamp, 2014). Finally, we calibrated our economic ABM with the tax declaration game devised by Alm *et al.* (2009) and conducted a sensitivity analysis. Our findings revealed that the tax rate positively influences the extent of tax evasion (+2.59%) and that the penalty rate negatively impacts tax evasion (−5.79%), when the parameters are increased by 10%. Table 9.4 summarized our results.

With respect to data, an extension of our economic ABM might be to include as input the income distribution at a state level, for example, see Llacer *et al.* (2013). Additional extensions might be to allow the taxpayers to accumulate wealth and to engage in a labor market. Further, more experiments are needed to estimate lapse of time effects, to elucidate the risk aversion of neoclassical A-Types and to identify the social network of interacting B-Types. However, these issues delineate a rich research agenda, which we leave for the future.

Acknowledgments

The work originates from the Ph.D. Thesis by Sascha Hokamp entitled "Income Tax Evasion and Public Goods Provision – Theoretical Aspects and Agent-Based Simulations" (Hokamp, 2013) and was financially supported by the Westfälische Wilhelms-Universität Münster, Germany, the Brandenburg University of Technology Cottbus, Germany, the Deutsche Bundesbank, the German Academic Exchange Service (DAAD), and the European Social Simulation Association (ESSA). Investigating the micro perspective on tax noncompliance under social information was partly supported by the Cluster of Excellence "Integrated Climate System Analysis and Prediction" (DFG EXC 177 CliSAP), Universität Hamburg, Germany. We would like to thank two reviewers for valuable comments. However, all errors remain ours.

References

Allingham, M.G. and Sandmo, A. (1972) Income tax evasion: a theoretical analysis. *Journal of Public Economics*, **1**, 323–338.

Alm, J., Jackson, B.R., and McKee, M. (2009) Getting the word out: enforcement information dissemination and compliance behavior. *Journal of Public Economics*, **93**, 392–402.

Andreoni, J., Erard, B., and Feinstein, J. (1998) Tax compliance. *Journal of Economic Literature*, **36** (2), 818–860.

Bazart, C., Bonein, A., Hokamp, S., and Seibold, G. (2016) Behavioural economics and tax evasion: calibrating an agent-based econophysics model with experimental tax compliance data. *Journal of Tax Administration*, **2** (1), 126–144.

Box, G.E.P. and Draper, N.R. (1987) *Empirical Model-Building and Response Surfaces*, John Wiley & Sons, Inc., New York.

Buehn, A. and Schneider, F. (2012) Shadow economies around the world: novel insights, accepted knowledge, and new estimates. *International Tax and Public Finance*, **19**, 139–171.

Cagan, P. (1958) The demand for currency relative to money supply. *Journal of Political Economy*, **66**, 303–329.

Cowell, F.A. (1992) Tax evasion and inequity. *Journal of Economic Psychology*, **13**, 521–543.

Croson, R.T.A. (2007) Theories of commitment, altruism, and reciprocity: evidence from linear public goods games. *Economic Inquiry*, **45** (2), 199–216.

Falkinger, J. (1988) Tax evasion and equity: a theoretical analysis. *Public Finance/Finances Publiques*, **43**, 388–395.

Falkinger, J. (1995) Tax evasion, consumption of public goods and fairness. *Journal of Economic Psychology*, **16**, 63–72.

Feld, L.P. and Larsen, C. (2005) *Black Activities in Germany in 2001 and 2004, A Comparison Based on Survey Data*, The Rockwool Foundation Research Unit, Copenhagen.

Feld, L.P. and Larsen, C. (2012) *Undeclared Work, Deterrence and Social Norms: The Case of Germany*, Springer-Verlag, Berlin/Heidelberg.

Feige, E.L. (1979) How big is the irregular economy. *Challenge*, **22** (1), 5–13.

Feige, E.L. (1980) Macroeconomics and the unobserved sector, in *Taxation: An International Perspective* (eds W. Block and M. Walker), The Fraser Institute, Vancouver, pp. 21–26.

Feige, E.L. (2016a) Reflections on the meaning and measurement of unobserved economies: what do we really know about the "shadow economy"? *Journal of Tax Administration*, **2** (1), 5–41.

Feige, E.L. (2016b) Professor Schneider's shadow economy: what do we really know? A rejoinder. *Journal of Tax Administration*, **2** (2), 93–107.

Frey, B.S. and Weck-Hannemann, H. (1984) The hidden economy as an unobserved variable. *European Economic Review*, **26**, 33–53.

Gilbert, N. and Troitzsch, K.G. (2005) *Simulation for the Social Scientist*, Open University Press.

Grimm, V. *et al.* (2006) A standard protocol for describing individual-based and agent-based models. *Ecological Modelling*, **198**, 115–126.

Grimm, V., Berger, U., DeAngelis, D.L., Polhill, J.G., Giske, J., and Railsback, S.F. (2010) The ODD protocol: a review and first update. *Ecological Modelling*, **221**, 2760–2768.

Gutmann, P.M. (1977) The subterranean economy. *Financial Analysts Journal*, **34** (1), 26–27.

Hashimzade, N. and Heady, C. (2016) Reflections on the meaning and measurement of unobserved economies: an editorial comment. *Journal of Tax Administration*, **2** (2), 108.

Hokamp, S. (2013) Income tax evasion and public goods provision: theoretical aspects and agent-based simulations. PhD thesis. Brandenburg University of Technology Cottbus, Germany.

Hokamp, S. (2014) Dynamics of tax evasion with back auditing, social norm updating and public goods provision: an agent-based simulation. *Journal of Economic Psychology*, **40**, 187–199.

Hokamp, S. and Pickhardt, M. (2010) Income tax evasion in a society of heterogeneous agents: evidence from an agent-based model. *International Economic Journal*, **24** (4), 541–553.

Hokamp, S. and Pickhardt, M. (2011) Pareto-optimality in Linear Public Goods Games. Center of Applied Economic Research Münster (CAWM) Discussion Paper No 45, University of Münster, Germany [download via https://www.wiwi.uni-muenster.de/cawm/sites/cawm/files/cawm/download/Diskussionspapiere/cawm_dp45.pdf] (accessed 15 November 2016).

Isachsen, A.J. and Strøm, S. (1982) The size and growth of the hidden economy in Norway. *Review of Income and Wealth*, **31**, 21–38.

Kirchgässner, G. (2017) On estimating the size of the shadow economy. *German Economic Review*, **18** (1), 99–111.

Kirchler, E. (2007) *The Economic Psychology of Tax Behavior*, Cambridge University Press, Cambridge.

Klovland, J.T. (1984) Tax evasion and the demand for currency in Norway and Sweden: is there a hidden relationship? *Scandinavian Journal of Economics*, **86**, 423–439.

Ledyard, J.O. (1995) Public goods: a survey of experimental research, in *The Handbook of Experimental Research*, Princeton University Press, Princeton, NJ, pp. 111–194.

Llacer, T., Miguel, F.J., Noguera, J.A., and Tapia, E. (2013) An agent-based model of tax compliance: an application to the Spanish case. *Advances in Complex Systems* **16** (04n05), 1–33.

Müller, B. *et al.* (2013) Describing human decisions in agent-based models: ODD + D, an extension of the ODD protocol. *Environmental Modelling & Software*, **48**, 37–48.

Nordblom, K. and Žamac, J. (2012) Endogenous norm formation over the life cycle – the case of tax morale. *Economic Analysis and Policy*, **42** (2), 153–170.

Pavel, R. (2015) Tax compliance dynamics: theoretical and experimental evidence. PhD thesis. Université de Montpellier, France.

Pickhardt, M. (2012) Pareto meets Olson – A Note on Pareto-optimality and Group Size in Linear Public Goods Games. *ORDO*, **63**(1), 195–202 [download via https://www.degruyter.com/view/j/ordo.2012.63.issue-1/ordo-2012-0115/ordo-2012-0115.xml].

Schneider, F. (2005) Shadow economies around the world: what do we really know? *European Journal of Political Economy*, **21**, 598–642.

Schneider, F. (2014) Work in the shadow: micro and macro results. *International Economic Journal*, **28** (3), 365–379.

Schneider, F. (2016) Comment on Feige's paper "reflections on the meaning and measurement of unobserved economies: what do we really know about the 'shadow economy'? *Journal of Tax Administration*, **2** (2), 82–92.

Schneider, F., Buehn, A., and Montenegro, C.E. (2011) Shadow economies all over the world: new estimates for 162 countries from 1999 to 2007, in *Handbook on the Shadow Economy* (ed. F. Schneider), Edward Elgar, Cheltenham.

Schneider, F. and Enste, D. (2013) *The Shadow Economy: An International Survey*, Cambridge University Press, New York.

Schneider, F., Linsbauer, K., and Heinemann, F. (2015) Religion and the shadow economy. *Kyklos*, **68** (1), 111–141.

Slemrod, J. and Yitzhaki, S. (2002) Tax avoidance, evasion and administration, in *Handbook of Public Economics*, North-Holland, Amsterdam, pp. 1425–1470.

Tanzi, V. (1980) The underground economy in the United States. Estimates and implications. *Banca Nazionale del Lavoro Quarterly Review*, **135**, 427–453.

Traxler, C. (2010) Social norms and conditional cooperative taxpayers. *European Journal of Political Economy*, **26**, 89–103.

Traxler, C. and Winter, J. (2012) Survey evidence on conditional norm enforcement. *European Journal of Political Economy*, **28**, 390–398.

Weck, H. (1983) *Schattenwirtschaft: Eine Möglichkeit zur Einschränkung der öffentlichen Verwaltung? Eine ökonomische Analyse*, Bern, Peter Lang.

Yitzhaki, S. (1974) A note on income tax evasion: a theoretical analysis. *Journal of Public Economics*, **3**, 201–202.

Zaklan, G., Lima, F.W.S., and Westerhoff, F. (2008) Controlling tax evasion fluctuations. *Physica A: Statistical Mechanics and its Applications*, **387** (23), 5857–5861.

Zelmer, J. (2003) Linear public goods experiments: a meta-analysis. *Experimental Economics*, **6**, 299–310.

Appendix 9A

We present the political cycle and the public goods provision cycle by Hokamp and Pickhardt (2010), Hokamp (2014) used for our simulations in Sections 9.3.1–9.3.3.

In addition to the public goods provision cycle, Table 9A.1 presents resulting values for the voluntary marginal per capita return

$$MPCR_{VOL} = (1 - \epsilon_{TC})\beta \tag{9A.1}$$

the forced Marginal Per Capita Return

$$MPCR_{FOR} = (1 - \epsilon_{AP})\beta \tag{9A.2}$$

and the maximum number of freeriders, N_{MAX}, tolerated by the Pareto-optimality concept (Hokamp and Pickhardt, 2011; Hokamp, 2014 and Pickhardt, 2012).

Table 9A.1 Political cycle

Parameter	Tax rate	Penalty rate	Tax complexity	Audit probability
Abbreviation	θ	π	γ	α
Tax relevant period				
1	0.20	0.30	0.10	0.01
5	0.20	0.30	0.10	**0.03**
9	0.20	**0.45**	0.10	0.03
13	**0.30**	0.45	0.10	0.03
17	0.30	0.45	**0.20**	0.03
21	**0.40**	0.45	0.20	0.03
25	0.40	0.45	0.20	**0.04**
29	0.40	0.45	0.20	**0.05**
33	0.40	**0.50**	0.20	0.05
37	**0.30**	0.50	0.20	0.05
40	0.30	0.50	0.20	0.05

Source: Reproduced from Hokamp and Pickhardt (2010, p. 544, Table 1) and Hokamp (2014, p. 197, Table A1). Changing Parameters are denoted in bold.

Table 9A.2 Public goods provision cycle

Parameter	Tax collection	Auditing process	Fixed costs	Public goods	Voluntary MPCR	Forced MPCR	Maximum number of free-riders
Abbreviation	ϵ_{TC}	ϵ_{AP}	ϵ_{FC}	β	$MPCR_{VOL}$	$MPCR_{FOR}$	N_{MAX}
Tax relevant period							
1	0.006	0.50	10,000	0.2000	0.198800	0.100000	5
5	0.006	**0.25**	10,000	0.2000	0.198800	0.150000	5
9	0.006	0.25	10,000	**0.0010**	0.000994	0.000750	1006
13	**0.100**	0.25	10,000	0.0010	0.000900	0.000750	1111
17	0.100	0.25	10,000	**0.0050**	0.004500	0.003750	222
21	0.100	0.25	**0**	0.0050	0.004500	0.003750	222
25	0.100	0.25	0	**0.0075**	0.006750	0.005625	148
29	0.100	0.25	0	**0.0100**	0.009000	0.007500	111
33	**0.006**	0.25	0	0.0100	0.009940	0.007500	100
37	0.006	0.25	0	**0.2000**	0.198800	0.150000	5
40	0.006	0.25	0	0.2000	0.198800	0.150000	5

Source: Reproduced from Hokamp (2014, p. 195, Table 3).
Four parameters are assigned to values, that are, the effectiveness of "tax collection," ϵ_{TC}, for example, Slemrod and Yitzhaki, (2002), argue for $\epsilon_{TC} = 0.1$ on p. 1426 and $\epsilon_{TC} = 0.006$ on p. 1449) the effectiveness of "auditing process," ϵ_{AP}; the "fixed costs" of tax authority, ϵ_{FC}; and the effectiveness of the "public goods" sector, β. Given these four figures the "marginal per capita return" of voluntary ($MPCR_{VOL}$) and forced ($MPCR_{FOR}$) contributions to public goods as well as the "maximum number of free-riders" (N_{MAX}) are derived. "Abbreviation" lists related mathematical symbols. "Tax relevant period" denotes the point in time when changes take place, which is every fourth period. Changing parameters are denoted in bold. Fixed costs and the maximum number of free-riders are denoted in tokens and number of agents, respectively. Remaining figures are provided in percent.

10

Modeling the Co-evolution of Tax Shelters and Audit Priorities*

Jacob Rosen, Geoffrey Warner, Erik Hemberg, H. Sanith Wijesinghe and Una-May O'Reilly

10.1 Introduction

It seems Benjamin Franklin was only half right – death may be no less certain today than it was in 1789, but taxes are far from inevitable. In the United States alone, the so-called tax gap, which is the difference between the tax owed and the amount paid on time in any given tax year, amounts to a whopping $450 billion in lost revenue, much of it accounted for by individual taxpayer noncompliance. There are, of course, many innocent reasons for noncompliance, including simple ignorance or error, but a significant portion is due to deliberate, and sometimes fraudulent, activity in the form of tax shelters.

An abusive tax shelter, or scheme, involves a sequence of transactions within a network of related business entities. These transactions are specifically designed to create and transfer artificial losses to designated beneficiaries, who then claim those losses in order to reduce their overall tax liability. The associated networks are usually composed of flow-through entities such as partnerships, trusts, and

* Select text, illustrations and figures for this chapter have been sourced from: Hemberg *et al.* (2016) © Springer Science+Business Media Dordrecht 2016. With permission of Springer.

Agent-Based Modeling of Tax Evasion: Theoretical Aspects and Computational Simulations, First Edition. Edited by Sascha Hokamp, László Gulyás, Matthew Koehler, and Sanith Wijesinghe. © 2018 John Wiley & Sons Ltd. Published 2018 by John Wiley & Sons Ltd.

S-corporations[1]. Such entities do not pay tax directly, but instead pass any net income or loss to their owners. Since partnerships can themselves own shares of other partnerships, it is possible to build multitiered ownership networks of enormous complexity, rendering manual audits prohibitively difficult.

The role that large, multitiered corporate structures play in the world of tax evasion has become evidently clear with the recent Panama Papers case (Boyd, 2016), that is being analyzed with complex graph database tools. Initial investigations here indicate the extensive use of off-shore shell corporations to hide income and assets that might otherwise be taxed by their home jurisdictions. An example in point is the eighteenth century English manor bought by the American film producer Stanley Kubrick that was found to be owned by three shell holding companies. These were in turn held by trusts for his children and grandchildren further obscuring tax ownership relationships (Goodman, 2016).

Much of the existing literature on tax evasion is concerned with the economic and psychological determinants of individual taxpayer compliance (Allingham and Sandmo, 1972; Andrei *et al.*, 2013; Bloomquist, 2011). Even highly sophisticated studies of heterogeneous and dynamic interactions between taxpayers and auditors assume a single, hypothetical tax evasion method (Balsa *et al.*, 2006). While these models lend insight into the factors influencing agent behavior, they do not explain the structure of evasion schemes. Our own focus is rather different. Rather than try to model the *decision* to evade, we seek to understand the actual *structure* of schemes, and in particular, how that structure evolves in response to changing enforcement priorities. Hence, the fundamental unit of investigation is not about the intention to evade – but how to evade. Such knowledge would help guide audit policy toward more efficient discovery of emerging schemes.

The most sophisticated schemes are generally constructed in such a way that their constituent transactions satisfy the letter of the law. When considered as a whole, however, and in the context of the surrounding ownership network, it sometimes occurs that a particular *sequence* of transactions is deemed illegal by the courts. In such cases, the risk associated with the original scheme spurs the invention of novel variants. Usually, these variants exploit the same underlying mechanism, but do so in ways that skirt the new enforcement regime.

We note here that a distinction must be made between illegal tax evasion and legal tax avoidance schemes. The nuanced differences between these two types are discussed in Chapter 1. Our own work focuses on the illegal exploitation of a particular subset of the US tax code, corresponding to the availability and accessibility of data to the authors. The specific rules we focus on surround the adjustment of a quantity called "basis." Basis is just the set point from which gains or losses on an asset are computed. Usually, the basis of an asset is just the cost of acquiring it, but there is provision for adjustment of basis under various circumstances. These basis adjustment rules are intended to allow businesses to account for things like capital

---

[1] The "S" in S-corporations corresponds to "small." These corporations are restricted by law to a maximum of 100 shareholders.

expenditures, liabilities, and depreciation, but it is sometimes possible to subvert these intentions by arranging transactions whose net effect, under the rules, is to inflate the basis of certain assets artificially. Applications of this methodology to other countries should first consider the local tax laws pertaining to basis.

In what follows, we provide a description of our methodology using the ODD+D protocol (Müller *et al.*, 2013). We emphasize here the high-level concepts and features to help drive intuition about our approach. For finer grain details please refer to the Appendix in Hemberg *et al.* (2016).

10.2 Overview

It is our hypothesis that every known scheme can ultimately be represented according to the following taxonomy: a set of assets; a set of tax entities (including partnerships, individual taxpayers, and trusts); a directed network indicating the initial ownership structure connecting entities to other entities and assets; and finally, a sequence of pairwise transactions in which assets are exchanged between entities. Since our goal is to discover the likely forms of emerging schemes, and since most new schemes are constructed from older versions that use the same underlying mechanism, one usually has a good notion as to the first three components of the taxonomy. The problem of discovering new candidate schemes thus reduces to the problem of searching the space of possible pairwise transactions in the "neighborhood" of an existing one.

The "best" schemes are those that afford the largest reduction in tax liability with the smallest possible risk. Viewed in this way, we are faced with a classic optimization problem on the space of pairwise transactions. This space has a natural syntactic structure that lends itself to search using a variant of genetic algorithms (GAs) known as grammatical evolution (GE) (Warner *et al.*, 2014). Our team has built a functioning, end-to-end GE codebase that has already proved capable of finding transaction sequences corresponding to two known schemes, namely Installment-sale Bogus Optional Basis (IBOB) and Distressed Asset Debt (DAD). We will describe our efforts using the IBOB scheme as an example in the upcoming sections. For further details about the DAD scheme please refer to Rosen *et al.* (2015). Note that GA/GEs are but one class of algorithms that can be used for co-evolutionary optimization (de Jong *et al.*, 2007). Further investigation is required to assess the relative merits of other techniques.

In the most basic sense, genetic algorithms are procedures for managing populations of candidate solutions to a particular problem. Typically, the first step of this procedure is to evaluate the fitness of each candidate solution. In the next step, one selects some subset of the fittest members of this population, introduces a degree of random variation into these solutions by some means (usually an analog of mutation and crossover from biology), and then repeats the process on the new populations generated thereby. Note, the use of biological concepts in the context of tax compliance has been previously studied (Torgler, 2014), but mostly in determining the incentive to evade. Here, we leverage the study of incentives and propagation in Darwinian evolution to evaluate how the logistics of a tax system can be used to reduce one's tax liability. In theory, this leads to better and better candidate solutions after many repetitions.

All GA's require a procedure for mapping candidate solutions into a data structure amenable to random variation. GE provides just such a structure by affording a clean separation between a space of "genotypes" (namely, integer lists) and a corresponding space of "phenotypes" (or candidate solutions). GE uses context-free grammars to map elements of the genotype space into candidate solutions. For our purposes, candidate solutions consist of sequences of transactions that can be simulated on a network of asset and entity objects. Such simulations require a method for allocating gain and loss recursively within these networks.

In parallel to simulating the population of candidate tax evasion schemes, we also model a population of auditors (represented as an audit score sheet) to mimic the enforcement actions of the tax authorities. As we shall see in the simulation results, co-evolutionary behavior can be observed as the tax evasion schemes adapt to different audit priorities and vice versa. This is akin to biological predator–prey type population interactions where there is a need for constant adaptation of one species with another to ensure survival. Co-evolution here can occur across a continuum of scales, from the microcellular level through to the species level, depending on the specific trait that would provide an advantage. This observation, embodied in the Red Queen hypothesis (Van Valen, 1973), has also been compared to an evolutionary arms race. In the context of tax evasion, the "arms" correspond to "strategies" for either evasion or detection. The evolutionary biologist Richard Dawkins defined such evolutionary strategies as "memes" (Dawkins, 1989). An example here is the so-called Son of BOSS tax shelter that emerged in the mid-1990s after its immediate predecessor, a scheme referred to as "shorting against the box," was rendered defunct by changes in the tax code (Wright, 2013).

We model audit policy as a collection of flagged transaction subsequences that have each been assigned a score between 0 and 1. These scores are constrained to sum to unity, and can therefore be thought of as representing the relative probability that the associated transaction will be detected and result in the assessment of a penalty. By incorporating these scores into our objective function, we are able to "prune out" undesirable transaction subsequences, and thereby guide the search toward those schemes that are less likely to be detected under a particular enforcement regime. These are, of course, precisely the schemes most likely to emerge under that regime.

We discuss these various components in greater detail in subsequent sections. We then illustrate the principles of our search methodology by referring to experiments on the IBOB scheme.

10.3   Design Concepts

The overall design concept of our system, Simulating Tax Evasion and Law through Heuristics (STEALTH), is illustrated in Figure 10.1. As discussed in Section 2, it is possible to formulate the process of constructing potential tax evasion schemes more abstractly as a search problem in a combinatorial space of transaction sequences. This space contains all the possible pairwise transactions that can occur within an initial ownership network of assets and entities. The search terminates when solutions are found whose "fitness" exceeds some predefined threshold. Hence, the outermost layer of our computational approach is basically a genetic algorithm that manages a population of candidate solutions corresponding to potential evasion schemes.

The basic structure of the algorithm is as follows. First, a random initial population of "chromosomes" is generated. In our case, these chromosomes consist of simple integer lists. By themselves, these lists are meaningless. They are useful only as a kind of seed for growing candidate schemes – in effect, they index points in the search space (though it is possible for two distinct chromosomes to produce the same transaction sequence). Because chromosomes are just integer lists, it is straightforward to introduce random variation by the action of genetic operators

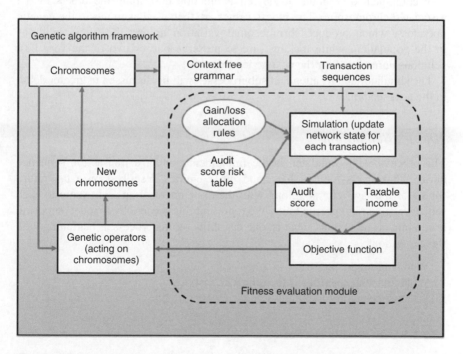

Figure 10.1   Design framework for our genetic algorithm. The framework for the evolution of audit scorecards is entirely analogous.

corresponding to mutation and crossover (see Warner *et al.*, 2014 for more details). This variation is required in order to allow the algorithm to explore different points in the search space, and thereby discover variants of existing schemes.

Of course, in order to simulate transactions within a network, there must first be a method for converting chromosomes into candidate transaction sequences. This takes place in the genotype-to-phenotype converter module, which uses a context-free grammar to transform each integer list into a transaction sequence. The details of this process can be found in Warner *et al.* (2014). Once these trans-action sequences have been generated, they are passed to the simulator module. The simulator parses through each transaction sequence, makes any state changes required to all involved entities and assets, and propagates any gains or losses incurred throughout the ownership network.

The simulator requires a module capable of implementing rules governing the allocation of gain and loss in complex partnership structures. The details of this module are described in depth in Rosen (2015). In addition, the simulator requires a system for representing auditing policy. This system must be capable of scoring the relative likelihoods that different transaction sequences incur audits resulting in the assessment of penalties. The output of the simulator consists of a taxable income and an audit score per each chromosome. These quantities are then combined to compute an objective function describing the fitness of the associated chromosome. The fitness values are then used as inputs to the genetic operators, which produce chromosomal variation around the fittest solutions in the population, while making sure to preserve one or two of the very best solutions unmodified for the next generation.

The simulator and its attendant subprocesses will be discussed in greater detail in the next section.

10.3.1   Simulation

The process being simulated is the implementation of a transaction sequence between taxpayers, partnerships, and other entities described by a tax return (or set of relevant returns), along with the assignment of an audit score to each sequence, given its specific traits. We represent this process by first decomposing it into inputs and outputs. That is, one can think of this process as beginning with three questions:

1. Who is filing the return?
2. What financial activities is the return describing?
3. What is the auditor looking for?

10.3.1.1   Ownership Network

Because partnerships are composed of multiple partners, the financial documents that a partnership files reference several taxpayers and potentially other

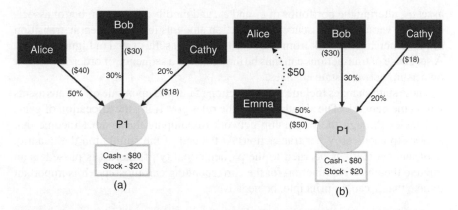

Figure 10.2 Example of ownership network and transactions between partners. (a) Example of a network of interconnected entities. The number in parenthesis denotes each owner's basis (usually just the purchase price of their original contribution). P1 corresponds to the partnership entity created by the partners Alice, Bob, and Cathy. (b) Example of a transaction between entities. The dotted line denotes the transfer of assets. Emma purchases Alice's 50% share of P1 in exchange for $50, and the outside basis of the share is increased to $50

partnerships, corporations, or trusts. Therefore, we must represent this related collection of entities as shown in Figure 10.2(a). Each entity is represented by a node, each of which holds a *portfolio* of assets. The edges on the graph show ownership linkages between an entity and a partnership, displaying both the Fair Market Value (*FMV*) and *outside basis* of the corresponding shares, which are themselves treated as assets. Thus, the *state* of the network can be fully described as a set of entities, with each entity owning a set of assets, and each asset having an FMV, adjusted basis, and other contextually relevant information.

Note that, in the example featured in Figure 10.2(a), the sum of outside bases is less than the total value of the assets contributed. This apparent mismatch illustrates a very common situation that can arise in partnerships like P1; namely, the values of assets are not static, and may change after they are purchased. The outside basis remains the same (except under certain special situations), but the asset values are constantly changing. In this particular example, the assets have accrued value since they were purchased.

10.3.1.2 Transaction Sequences and Taxable Income

Given that tax returns are essentially a description of financial activity, we can represent the information that is required to be filed on a tax return as a *transaction sequence*. A single *transaction* in this context consists of a set of two entities e_1 and e_2, and two respective assets a_1 and a_2. Entity e_1 gives e_2 asset a_1 in exchange for asset a_2, altering the portfolios of e_1 and e_2, and modifying any number of associated state variables. In essence, a transaction amounts to a very specific transition of the ownership network from one state to another as illustrated in Figure 10.2(b). A sequence of transactions can thus be thought of as a sequence of order-dependent ownership network state changes.

These state changes include the production of income upon sale of various assets within the network. The simulator uses the rules governing the allocation of gains and losses through the ownership network to compute the taxable income that accrues to each entity per transaction. At the end of the simulation the quantity X of income that has accrued to the particular entity of interest is passed as an input to the objective function for the corresponding chromosome. It is important to note that X can, in principle, be negative.

10.3.1.3   Audit Score Sheet

Almost by definition, an auditor scans through a stack of tax returns, looking for a piece of observable information, or a joint occurrence of several pieces of observable information, that would indicate a high likelihood of suspicious activity. Thus we represent a hypothetical auditor as an *audit score sheet*, with the first column being a set of *observable* events, as well as their combined joint occurrences. That is, if we suppose that there are n individually observable events, then an audit score sheet would contain $2^n - 1$ rows.

The second column on the audit score sheet is a corresponding list of weights

$$\{w_i\}_{i=1}^{2^n-1} : w_i \geq 0 \quad \text{and} \quad \sum_{i=1}^{2^n-1} w_i = 1$$

that are referred to as *audit points*. Each point indicates the amount by which the overall *audit likelihood* score should be incremented each time that the corresponding event (or joint occurrence of events) is observed (see Table 10.1).

At the start of simulation, the audit score is 0. As the simulation proceeds through each transaction k, the score is incremented by the weight w_i associated with the corresponding observables $i = i(k)$. At the end of a simulation an audit score $s = \sum_k w_{i(k)}$ is passed to the objective function.

10.3.1.4   Summary

Our simulation methodology establishes a representation of (i) the ownership network: the unit being investigated, (ii) the transaction sequence: the activity of the unit under investigation, and (iii) the audit score sheet: the behavior of the investigator. These three elements, when combined with the partnership tax calculator, form the simulation shown in Figure 10.3. The simulator produces both a measure of taxable income for all relevant entities and a score capturing the likelihood of audit.

Table 10.1 Example of an
audit score sheet with three
individually observable
events, resulting in seven
total rows and corresponding
audit points

Observable	Points
1	w_1
2	w_2
3	w_3
$1 \cup 2$	w_4
$1 \cup 3$	w_5
$2 \cup 3$	w_6
$1 \cup 2 \cup 3$	w_7

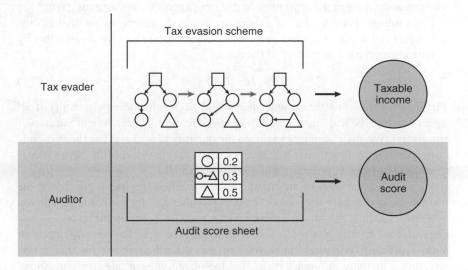

Figure 10.3 STEALTH tax ecosystem simulator.

10.3.2 *Optimization*

Once a method has been established for simulating transaction sequences (and the corresponding auditing process), we can begin establishing a means to search over potential financial activity and auditing policies to optimize certain objectives. In terms of the concepts defined in Section 10.3.1:

- What is an effective transaction sequence, given an initial ownership network and audit score sheet?
- What is an effective audit score sheet, given an initial ownership network and transaction sequence?

In this section, we address the question of what defines an "effective" evasion scheme or audit policy within our framework. We also briefly discuss our ability to model the co-evolutionary dynamics between evaders and auditors. This enables us to study the effect of changes in enforcement policy on the likely forms of emerging schemes.

10.3.2.1 Objective Functions

To recap, the simulation described in Section 10.3.1 produces two outputs per chromosome, given an initial ownership network; namely, the taxable income X of the entity of interest, and an audit score s associated with that entity's financial activity. These quantities can be combined to form "fitness functions" whose maximization defines the objectives of different agents in the tax ecosystem.

From the perspective of a tax evader, an effective scheme is one that affords the largest reduction of taxable income for the smallest risk. We capture this by stipulating a fitness function U of the form

$$U = (I - X)(1 - s) \tag{10.1}$$

The quantity I is just the income that would accrue to the evader entity if all assets to which he had any ownership rights were immediately sold at the start of simulation. We shall sometimes refer to the quantity $G = (I - X)$ as the "tax gap" corresponding to a particular scheme. It is clear that the goal of the tax evader is to make G as large as possible, while keeping s small.

By contrast, the goal of any particular enforcement regime is to choose the weights w_i that tend to keep U as small as possible for any particular population of schemes. Thus, the fitness function of an audit scorecard with respect to any particular scheme is just $-U$. The evolutionary process for the population of audit scorecards is entirely analogous to that described above for transaction sequences, the only difference being that the phenotypes in question correspond to different values of the weights w_i.

10.3.2.2 Co-evolution

Auditors and evaders exist in a continuous state of mutual co-adaptation, with evaders adjusting their patterns of activity to accommodate perceived changes in enforcement policy, and auditors altering their priorities to target the most successful schemes. The core genetic algorithm discussed above can be adapted to capture this predator–prey dynamic by initializing two populations

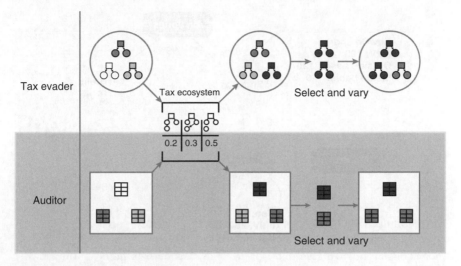

Figure 10.4 Concurrent optimization of high likelihood audit scores and low-risk task schemes.

simultaneously: one a collection of candidate schemes, and the other a population of audit scorecards, each with different distributions of weights w_i.

During the simulation phase, an individual is selected from each of these populations, and the corresponding fitness functions are evaluated. Specifically, each individual in the population shown on the top of Figure 10.4 represents both an initial ownership network and an associated transaction sequence. In every generation, each individual from that population selects a subset of the audit score sheet population shown at the bottom of the figure to evaluate against. The process is then repeated for the audit score sheets being compared against the ownership network–transaction sequence pair. A more detailed account can be found in Rosen (2015).

10.4 Details

We describe here the IBOB real estate tax scheme that is used as a case study for our experiments and outline the grammar from which it can be generated. The genetic algorithm parameter settings are also listed.

10.4.1 IBOB

The Installment Sale Bogus Optional Basis Transaction (IBOB) is one instance of a wider class of tax evasion strategies that rely on a mechanism known as

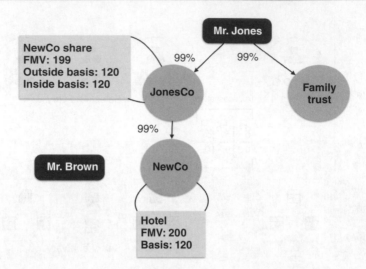

Figure 10.5 The steps in the IBOB abusive tax avoidance scheme. The basis of an asset is artificially stepped up and tax is avoided by using "pass-through" entities. IBOB step 1.

"basis-shifting." Such schemes are designed to decrease, defer, or eliminate any taxable income or capital gain incurred from the sale of an asset, usually an item of real estate. This is often accomplished by artificially increasing the asset's basis, a quantity equivalent to its purchase price under most circumstances.

The initial state for the IBOB scheme is shown in Figure 10.5. The subsequent transaction steps are shown in Figure 10.6(a) and (b). Suppose Jones, a real estate developer, owns a house he originally bought for $120 (i.e., the house has a *basis* of $120), which he now wishes to sell to another individual named Brown for $200. In order to avoid incurring $80 in capital gains upon sale of the house, Jones forms two partnerships, JonesCo and FamilyTrust, making sure to retain majority ownership (99%) of each. He then arranges for JonesCo to create, and retain majority ownership of, yet another partnership called NewCo. Jones then contributes the house through JonesCo to NewCo, an action which does not alter the basis of the house.

Next, Jones orchestrates a transaction wherein FamilyTrust purchases JonesCo's share in NewCo for an annuity (Figure 10.6(a)). Provided JonesCo has made an election under Section 754 of the Tax Code,[2] this action increases the basis of the house without incurring any current tax liability. Finally, NewCo sells the house to Brown (Figure 10.6(b)) for seemingly no capital gain because the basis of the house was increased by the previous transaction.

---

[2] Section 754 provides a provision to allow a partnership to adjust the basis of property when there is either a distribution of the property (Section 734(b)), or a transfer of partnership interest (Section 743(b)). Additional details can be found in Cornell (2016).

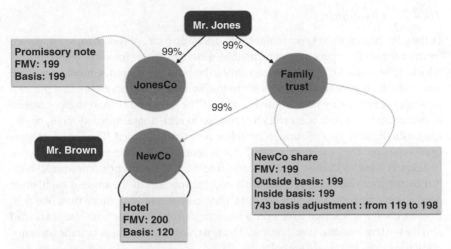

(a) JonesCo sells its share of NewCo to Family trust in exchange for an annuity

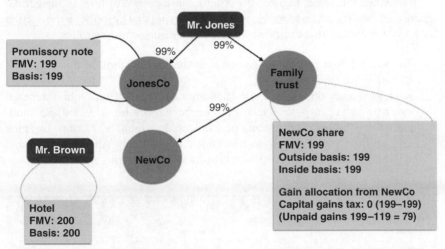

(b) Brown purchases the house from NewCo

Figure 10.6 The steps in the IBOB abusive tax avoidance scheme. The basis of an asset is artificially stepped up and tax is avoided by using "pass-through" entities. (a) IBOB step 2, (b) IBOB step 3.

Thus, rather than paying capital gains on the $80 from the cash sale to Brown, Jones has to pay tax on the gain incurred from the sale of JonesCo's interest in NewCo. But because it was purchased with installment payment, the tax is only paid as the payments are received, a deferral which is advantageous to Jones. Furthermore, some manifestations of this scheme involve FamilyTrust defaulting on the annuity payments, which means that no tax is ever paid.

10.4.2 *Grammar*

In the GE version of the genetic algorithm utilized in our approach, the compressed form of the search space is represented by a Backus–Naur form (BNF) grammar, which defines the language that describes the possible output sentences. a BNF grammar has terminal symbols, nonterminal symbols, a start symbol and production rules for rewriting nonterminal symbols. The grammar is used in a generative approach and the production rules are applied to each nonterminal symbol, beginning with the start symbol, until a complete program is formed. The list of integers (genotype) rewrites the start symbol into a sentence. An integer from the list of integers is used to choose a production rule from the current nonterminal symbol by taking the current integer input and the modulo of the current number of production choices. Each time a production from a rule with more than one production choice is selected to rewrite a nonterminal symbol, the next integer is read and the system traverses the genome. The rewriting is complete when the sentence comprises only terminal symbols.

To illustrate this logic, Figure 10.7 provides an example of how an integer list (genotype) can be converted to a sentence (phenotype) that describes a transaction between two entities that exchange assets. The parsing occurs as follows:

1. We pick the first rule in the grammar as the start symbol, in this case (1) <transactions>.
2. Next, expand the left most nonterminal symbol in our sentence <transactions>. We take the current integer input 3 and the modulo of the number of production choices 2, which is 1, thus we pick <transaction> the production choice at position 1 (the indexing starts at 0) and rewrite the <transactions> with <transaction>.

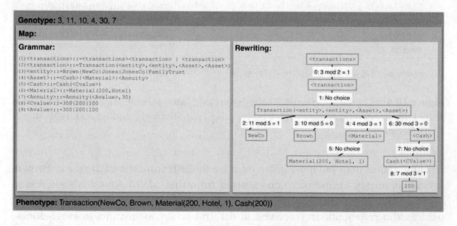

Figure 10.7 Example of mapping a list of integers (genotype) into a list of transactions (phenotype) by using grammatical evolution.

3. Again expand the left most nonterminal symbol <transaction>. There is only one production choice here, so it is rewritten to Transaction (<entity>, <entity>, <Asset>, <Asset>).

4. Again expand the left most nonterminal symbol <entity>. We take the current integer input 11 and the modulo of the number of production choices 5, which is 1, thus we pick NewCo. The sentence is now Transaction(NewCo, <entity>, <Asset>, <Asset>).

5. The left most nonterminal symbol is again <entity>. We take the current integer input 10 and the modulo of the number of production choices 5, which is 0, thus we pick Brown. The sentence is now Transaction(NewCo, Brown, <Asset>, <Asset>).

6. The left most nonterminal symbol is now <Asset>. We take the current integer input 4 and the modulo of the number of production choices 3, which is 1, thus we pick <Material>. The sentence is now Transaction(NewCo, Brown, <Material>, <Asset>).

7. The left most nonterminal symbol is now <Material>. There are no choices for <Material> so we rewrite it with Material(200, Hotel, 1). The sentence is now Transaction(NewCo, Brown, Material(200, Hotel, 1), <Asset>).

8. The left most nonterminal symbol is again <Asset>. We take the current integer input 30 and the modulo of the number of production choices 3, which is 0, thus we pick <Cash>. The sentence is now Transaction(NewCo, Brown, Material(200, Hotel, 1), <Cash>).

9. The left most nonterminal symbol is now <Material>. There are no choices for <Cash> so we rewrite it with Cash(<Cvalue>). The sentence is now Transaction(NewCo, Brown, Material(200, Hotel, 1), Cash(<CValue>).

10. The left most nonterminal symbol is Cash(<CValue>. We take the current integer input 7 and the modulo of the number of production choices 3, which is 1, thus we pick 200. The sentence is now Transaction(NewCo, Brown, Material(200, Hotel, 1), Cash(200)).

11. There are no more nonterminal symbols left to rewrite and our string rewriting is done.

For the purpose of detecting IBOB, we initialize a network with two tax filers, Mr. Jones and Mr. Brown, and three partnerships, JonesCo, NewCo, and FamilyTrust. These entities have portfolios of assets that include Cash, an Annuity, a Hotel, and various partnership shares. The assets can have different fair market values. The BNF grammar used to define the transaction space among these entities is shown in Figure 10.8.

The first recursive rule in the grammar shows that the search space (language) is bounded only by the length of the input (genome) used to map integers to transactions. We also note that the search space can be increased and customized by altering the structure of the grammar.

```
<transactions>::=<transactions><transaction>|<transaction>
<transaction>::=Transaction(<entity>,<entity>,<Asset>,<Asset>)
<entity>::=Brown|NewCo|Jones|JonesCo|FamilyTrust
<Asset>::=<Cash>|<Material>|<Annuity>|<PartnershipAsset>
<Cash>::=Cash(<Cvalue>)
<Material>::=Material(200,Hotel,1)
<Annuity>::=Annuity(<Avalue>,30)
<Cvalue>::=300|200|100
<Avalue>::=300|200|100
<Pname>::=NewCo|JonesCo|FamilyTrust
```

Figure 10.8   IBOB BNF grammar for STEALTH.

10.4.3   Parameters

For the experiments considered here, audit scores are determined as the sum of four audit points between 0 and 1. The initial average distribution of audit scores in the population is shown in Table 10.2. Note a "Single Link" audit observable here refers to transactions that occur between entities that are connected through a single direct ownership relationship, such as between Jones and JonesCo as shown in Figure 10.5. "Double Link" corresponds to transactions that occur between entities that have an ownership relationship 2 hops apart, such as between Jones and NewCo as shown in Figure 10.6. The value of each audit point can be thought of as the relative importance of the associated behavior to the auditor (the reasons for the numerical values of the three audit score sheets are explained in Section 10.5).

We ran 100 independent iterations of the co-evolutionary GA for 100 generations each with tax scheme and audit score populations of size 100. We chose half of the tax scheme population for evaluating the fitness of the solution in the other audit score population and vice versa. The parameters that govern the GA simulation are enumerated in Table 10.3.

The GA parameters in Table 10.3 were chosen through trial and error. Since our focus is on understanding the dynamics between taxpayers and auditors,

Table 10.2   Initial average distribution of audit points for each experiment

Audit observable	LimitedAudit	EffectiveAudit	CoEvolution
iBOB	0	0.25	0.25
Annuity	0.33	0.25	0.25
Single link	0.33	0.25	0.25
Double link	0.33	0.25	0.25

Table 10.3 Parameters for IBOB experiments

Parameter	Description	Value
Mutation rate	Probability of integer change in individual	0.1
Crossover rate	Probability of combining two individual integer strings	0.7
Tournament size	Number of competitors when determining most fit individuals	2
Number chosen	Fraction of other population that each individual is tested against	0.5
Population size	Number of individuals in each population	100
Generations	Number of times populations are evaluated	100

parameters were not calibrated to achieve an optimal result for either side. Thus the specific parameters need only demonstrate the co-evolutionary relationship between the two agent types, which is easily obtained by the above parameters. We discuss next the experiments based on these initializations.

10.5 Experiments

We conducted a number of experiments to verify that the co-evolutionary dynamics of our simulation are consistent and sufficient enough to find existing tax schemes. To first order, we expect that when schemes evolve to accommodate existing audit priorities, the audit points will themselves evolve to detect the new evasion behavior. This should result in either convergent or oscillatory dynamics.

10.5.1 Experiment LimitedAudit: Audit Observables That Do Not Detect IBOB

For this experiment, we purposefully left out audit points that can detect IBOB. Thus, once IBOB is found it will propagate through the tax scheme population on the merit of being the best tax scheme that is unauditable.

Figure 10.9 shows plots of tax scheme and audit score sheet fitness. Figure 10.9(a) shows the fitnesses of both the tax schemes and the audit scores for the best individual from each generation for a single iteration. Note that while IBOB was found in only 34% of the iterations here, we expect most iterations to converge to IBOB as the number of generations increases. The average audit point values displayed in Figure 10.9(b) show the variation in the points over time. Note here that the audit points are unchanged and in equilibrium once IBOB has been found. The average audit points of the best solutions for all the iterations are shown in Figure 10.9(c), where we see that transactions that exchange a material asset for an annuity are assigned a higher audit point.

Material-annuity transactions have a significantly higher audit point because they occur more frequently than the other two schemes. That is, any transaction in which the Hotel is exchanged for an annuity mitigates all of the taxable gain

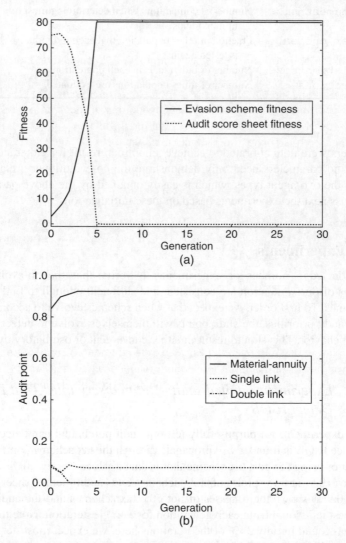

Figure 10.9 Evolution of IBOB in STEALTH experiment 1 (LimitedAudit). (a) Best fitness for one run (b) Distribution of audit points for one run (c) Distribution of audit points averaged over runs. Source: Hemberg *et al.* (2016). Reproduced with permission of Springer.

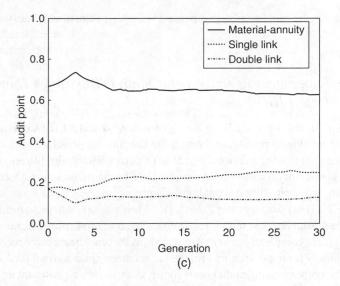

Figure 10.9   (*continued*)

on the ultimate sale of the hotel because annuities are nontaxable. Furthermore, a double link transaction requires that a material-annuity transaction takes place because Jones has to purchase the Hotel from NewCo with an annuity. Thus, the likelihood of a tax scheme involving a material-annuity transaction is higher than the likelihood of a single- or double-linked scheme. This results in a higher average audit point assigned to material-annuity transactions because it is the most common way to mitigate taxable gain in our example.

A clear pattern emerges when IBOB is evolved: initially, the pool of tax schemes gravitates toward a network of transactions that contains suspicious activity, which the audit scores are able to detect. Only after the audit scores evolve to reduce the fitness of such schemes does IBOB become dominant.

Two distinct metastable states emerge when the basic IBOB is not found. The most common is when a suspicious scheme is evolved in an early generation, which the audit scores can effectively detect early on, causing the scheme fitness to converge toward its minimum and the audit score fitness to converge to its maximum. Alternatively, the pools of both tax schemes and audit scores oscillate with respect to each other for the duration of the run, implying a process of suspicious schemes emerging and audit scores evolving to detect them, causing another suspicious scheme to become dominant. Many runs show oscillations or long-lived transients, as these show the kind of predator–prey dynamics we expect, and illustrate that the search can sometimes get stuck in a "metastable state."

Ultimately, the only stable configuration is one in which IBOB dominates the population – any "oscillations," whatever the intervals between transient peaks, and whatever the number of those peaks (1, 2, 3, ...) must eventually give way to IBOB.

10.5.2   Experiment EffectiveAudit: Audit Observables That Can Detect IBOB

In this experiment, we include an audit point that can detect IBOB. Thus, IBOB should not be able to propagate through the tax scheme population. Because the audit score sheets were previously unable to detect IBOB, the fitness of the tax schemes would only oscillate until a single IBOB scheme was introduced into the population, at which point it would quickly propagate.

Figure 10.10(a) displays the fitnesses of both the tax schemes and the audit score sheets from the best individual from each generation from 1 iteration. Since the audit points completely cover all transactions that can create large recognizable loss, the fitness is always minimal for the tax schemes and maximal for audit score sheets. The corresponding audit points for the iteration are all constant as shown in Figure 10.10(b). Furthermore, Figure 10.10(c) shows that the average audit points of the best individuals over all the runs correspond to the expected values at the initial state.

We conclude that the observed co-evolutionary dynamics are consistent with the expectations for this example.

10.5.3   Experiment CoEvolution: Sustained Oscillatory Dynamics Of Fitness Values

Our goal with this set of experiments was to generate sustained oscillatory dynamics, since we have shown in previous experiments that oscillations in tax scheme fitness are possible for a short amount of time before converging to equilibrium. This is a necessary step because a primary assumption underlying our model is that tax schemes and audit scores sheets are engaged in a perpetual co-evolutionary process in which no global attractor exists. Because the audit score sheets were unable to detect IBOB in experiment LimitedAudit, the fitness of the tax schemes would only oscillate until a single IBOB scheme was introduced into the population, at which point it would quickly propagate. At the same time, simply allowing the audit score sheets to detect IBOB would result in rapidly convergent dynamics, as demonstrated in experiment EffectiveAudit.

To generate sustaining oscillations, we augment the audit score sheets to assign the lowest audit point a value of 0, so that there will always be at least one scheme that is not detectable by the auditor. Our hypothesis is that once the population of audit score sheets begins to converge, a tax scheme will evolve that utilizes the type of behavior that is currently not detectable by the majority of audit score sheets.

The effective tax scheme will propagate within its population until the audit score sheets gradually evolve to detect the now dominant behavior.

Figure 10.11(a) displays the fitnesses of both the tax schemes and the audit scores from the best individual from each generation during a single iteration. In this scenario, since the audit points cannot completely cover all the transactions

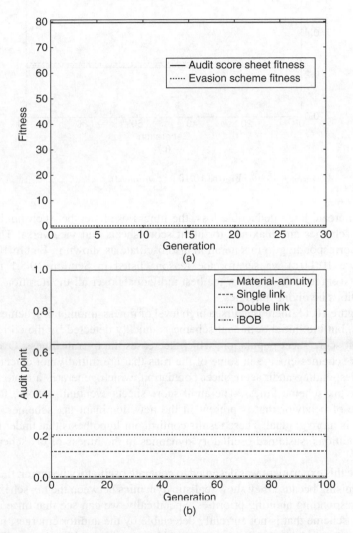

Figure 10.10   Evolution of IBOB in STEALTH experiment 2 (EffectiveAudit). (a) Best fitness for one run (b) Distribution of audit points for one run (c) Distribution of audit points averaged over runs. Source: Hemberg *et al.* (2016). Reproduced with permission of Springer.

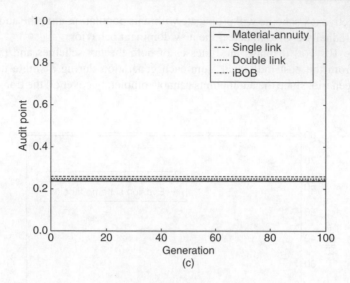

Figure 10.10   (*continued*)

that can create large deductible loss, the fitness oscillates between minimal for the tax schemes and maximal for audit score sheets and vice versa. The audit points corresponding to this iteration also oscillate as shown in Figure 10.11(b). In Figure 10.11(c) we see for the reasons listed in Section 10.5.1 that the highest average audit points of the best individuals over all the iterations are for transactions involving annuities.

In Figure 10.11, there is at first a high level of fitness among tax schemes across all runs, but the initial dominant scheme is quickly detected by the corresponding audit score sheet population, which decreases the overall fitness. Over time, new tax schemes emerge in some of the runs that are initially not detectable by the corresponding audit score sheet population, which generates a rapid upward surge in tax scheme fitness. The audit score sheets eventually evolve to detect the type of behavior that is present in the new dominant tax schemes, but the process is more gradual. These results confirm our hypothesis that under the correct conditions, sustained oscillatory dynamics in the fitness of tax schemes are possible.

While this experiment was designed to generate oscillatory behavior, the results are promising because they show realistic dynamics between the tax schemes and the corresponding auditing priorities. Specifically, we can see that once a single new tax scheme that is not currently detectable by the auditor emerges, it propagates throughout the population very quickly, as evidenced by the steep upward slope in the average scheme fitness plot. Conversely, the audit score sheets take a longer time to adapt to the new tax scheme.

10.6   Discussion

We have demonstrated an end-to-end methodology that automates the discovery of tax evasion schemes given a set of assets, tax entities, auditing policies, and quantitative risk-reward scoring measures. A core capability developed here was the representation of tax schemes as a sequence of asset exchanges between

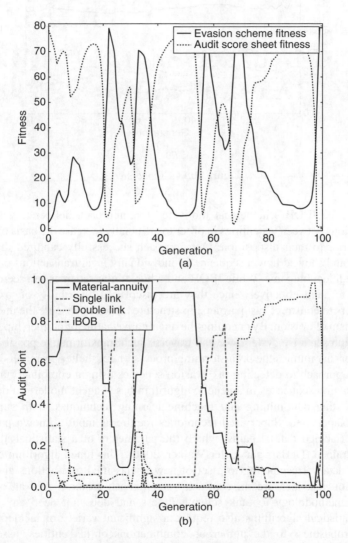

Figure 10.11 Evolution of IBOB in STEALTH experiment 3 (CoEvolution). Source: Hemberg *et al.* (2016). Reproduced with permission of Springer.

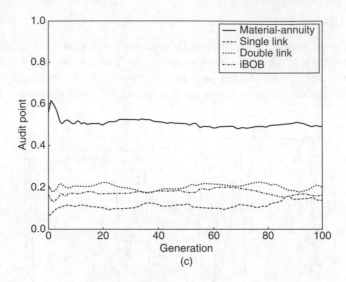

Figure 10.11   (*continued*)

entities. The IBOB scheme that motivated this approach deliberately arranges transactions between tax entities in order to artificially increase the basis of certain assets prior to their disposition. Our approach successfully captured the IBOB scheme and showed how a sequence of individually legal transactions can result in a fraudulent outcome. While IBOB and other similar schemes can seem simple enough to audit effectively once they are discovered, the ability of tax shelter promoters to construct vast partnership structures tends to obscure the underlying mechanism of evasion. By encoding the quantitative nature of partnership taxation into a representative system, we can traverse such transformative possibilities in a manner that mimics the brainstorming process of tax shelter promoters.

This approach to detecting tax evasion schemes using a combinatorial search of transaction sequences in a genetic algorithm-based agent model is a departure from existing data-mining and machine learning techniques, both supervised and nonsupervised. Supervised techniques require as input a known pattern of evasion (labeled data), against which the parameters of a statistical algorithm can be trained (DeBarr and Eyler-Walker, 2006). This tuned algorithm can then process data corresponding to a set of new unclassified transactions and return an evasion risk measure. In contrast, our approach requires only an objective scoring methodology to rank scheme fitness and does not need any training data. Statistical algorithms also require a significant variety of labeled data to ensure robustness to the numerous combinations of tax entities, assets, and transactions that may occur. This variety is often difficult to achieve. An even larger concern is that of "overfitting," where the algorithm may not be able to detect schemes with minor variations from those used for training purposes. Alternatively, nonsupervised machine learning techniques do not require *a priori* known patterns of evasion. While this may appear favorable, categorizing evasion is based here on a measure of anomaly from a baseline. This approach can suffer from high false positive rates when a suitable normative state cannot be defined.

We propose our methodology for tax scheme detection not as a replacement, but as a complement to existing data-mining techniques. In particular, we envision an iterative process that begins with manual brainstorming by subject matter experts (SMEs) to help identify combinations of entities, assets, sections of the tax code, and audit points of specific interest. Insights from past examinations and/or expectations about areas of future evasion can help guide this brainstorm. Transaction sequences can then be randomly initialized using identified asset and entity elements.

Alternatively, transaction sequences specific to a known fraudulent scheme can be set as the starting point. The subsequent schemes evolved by the search process should be examined by SMEs to validate their potential practical effect. If required, the audit plan can be updated to guide the search process further. Once a new scheme type is verified for efficacy, data-mining techniques can be employed to test for their existence. Moreover, these new scheme variants can be incorporated into existing training data sets to help calibrate the underlying statistical algorithms. This process can be repeated by integrating observations from the data into the initial conditions for the subsequent search.

An additional use-case for this methodology is to conduct "what–if" scenarios. The relative merits of different audit plan criteria can be compared in terms of the relative fitness of the different tax schemes that emerge. Similarly, the impact of different tax policy changes could also be examined. The goal with all these efforts is to help anticipate and scope the downstream impact of an enforcement policy prior to its deployment. As before, this methodology can help complement existing manual table-top strategy gaming and red-team/blue-team exercises designed for this purpose (Mirkovic *et al.*, 2008).

While we have demonstrated significant success with this approach, there are certain limitations that hinder broader applicability. The specific sections of the tax code we considered were reducible to first-order logic statements. For example, the calculation of an asset's inside/outside basis follows a complicated, yet fairly consistent and unambiguous, set of rules. In general, however, the tax code consists of many definitions, relationships, and exceptions that require semantic parsing, a process that can be ambiguous and inconsistent. Issues of interpretation are often only resolved in a court of law. As a workaround, expert rules can be specified in the audit plan to help direct the search toward or away from certain outcomes. Advances in natural language processing techniques (Moens *et al.*, 2007) could also be considered to supplement first-order logic calculations and allow for more complex scenarios to be explored.

References

Allingham, M.G. and Sandmo, A. (1972) Income tax evasion: a theoretical analysis. *Journal of Public Economics*, **1**, 323–338.

Andrei, A.L., Comer, K., and Koehler, M. (2013) An agent-based model of network effects on tax compliance and evasion. *Journal of Economic Psychology*, **40**, 119–133.

Balsa, J., Antunes, L., Respicio, A., and Coelho, H. (2006) Autonomous Inspectors in Tax Compliance Simulation na.

Bloomquist, K. (2011) Tax compliance as an evolutionary coordination game: an agent-based approach. *Public Finance Review*, **39**, 25–49.

Boyd, R. (2016) The Panama Papers Graph Database is Now Available for Download, http://tinyurl .com/gplpqz4 (accessed 29 July 2016).

Cornell University Legal Information Institute (2016) 26 U. S. Code 754 - Manner of Electing Optional Adjustment to Basis of Partnership Property, https://www.law.cornell.edu/uscode/text/26/ 754 (accessed 29 August 2016).

Dawkins, R. (1989) *The Selfish Gene*, 2nd edn, Oxford University Press.

DeBarr, D. and Eyler-Walker, Z. (2006) Closing the gap: automated screening of tax returns to identify egregious tax shelters. *ACM SIGKDD Explorations*, **8** (1), 11–16.

de Jong, E.D., Stanley, K.O., and Wiegand, R.P. (2007) Introductory Tutorial on Coevolution. Proceedings of the 9th Annual Conference Companion on Genetic and Evolutionary Computation, ACM, pp. 3133–3157.

Goodman, L.M. (2016) Panama Papers: What Stanley Kubrick Can Teach You About Tax Shelters, http://tinyurl.com/goppdyv (accessed 23 April 2016).

Hemberg, E., Rosen, J., Warner, G., Wijesinghe, S., and O'Reilly, U.M. (2016) Detecting tax evasion: a co-evolutionary approach. *Artificial Intelligence and Law*, **24** (2), 149–182.

Khwaja, M. and Iyer, I. (2014) Revenue potential, tax space and tax gap: a comparative analysis. World Bank Policy Research Working Paper, (6868).

Mirkovic, J., Reiher, P., Papadopoulos, C., Hussain, A., Shepard, M., Berg, M., and Jung, R. (2008) Testing a collaborative DDoS defense in a red team/blue team exercise. *IEEE Transactions Computers*, **57** (8), 1098–1112.

Moens, M.F., Boiy, E., Palau, R., and Reed, C. (2007) Automatic Detection of Arguments in Legal Texts. *Proceedings of the 11th International Conference on Artificial Intelligence and Law*, ACM, pp. 225–230.

Müller, B. *et al.* (2013) Describing human decisions in agent-based models-ODD+ D, an extension of the ODD protocol. *Environmental Modelling & Software*, **48**, 37–48.

Rosen, J. (2015) *Computer Aided Tax Avoidance Policy Analysis*, Massachusetts Institute of Technology, Cambridge, MA.

Rosen, J., Hemberg, E., Warner, G., Wijesinghe, S., and O'Reilly, U.M. (2015) Computer Aided Tax Evasion Policy Analysis: Directed Search Using Autonomous Agents. *Proceedings of the 2015 International Conference on Autonomous Agents and Multiagent Systems*, International Foundation for Autonomous Agents and Multiagent Systems, pp. 1825–1826.

Torgler, B. (2014) Can tax compliance research profit from biology? (No. 2014-08), Center for Research in Economics, Management and the Arts (CREMA).

Van Valen, L. (1973) A new evolutionary law. *Evolutionary Theory*, **1**, 1–30.

Warner, G., Wijesinghe, S., Marques, U., Badar, O., Rosen, J., Hemberg, E., and O'Reilly, U.M. (2014) Modeling tax evasion with genetic algorithms. *Economics of Governance*, **16** (2), 165–178.

Wright, D. (2013) Financial alchemy: how tax shelter promoters use financial products to bedevil the IRS (and how the IRS helps them). *Ariz. State Law J.*, **45**, 611.

11

From Spins to Agents: An Econophysics Approach to Tax Evasion

Götz Seibold

11.1 Introduction

Econophysics is quite a recent development in physics, which started in the beginning of the 1990s and, as the name suggests, deals with the application of concepts derived from physics to problems in the field of finance and economics. Predominantly, these concepts are adopted from statistical physics, in particular from the modern theory of phase transitions and comprise notions such as scaling and criticality, which are used to describe financial or economic data (cf. e.g., Stanley *et al.*, 1996; Lux and Marchesi, 1999; Liu *et al.*, 1999; Yamasaki *et al.*, 2005). Meanwhile, several books and review papers not only provide a good introduction (see e.g., Mantegna and Stanley, 2000; McCauley, 2004; Chakraborti *et al.*, 2011a; Chen and Li, 2012) but also include critical reflections on the development of this new field as in Stanley *et al.* (2001) and Gallegati *et al.* (2006). In statistical physics one derives the phenomenological laws of classical thermodynamics from the microscopic mechanical description of the constituting individual particles. In this spirit a lot of research in econophysics is based on agent-based models (Chakraborti *et al.*, 2011b) that go beyond the traditional economic description of a "representative agent" due to the incorporation of interactions between agents. In econophysics models these interactions then can have a correspondence in physics *and* economy as, for example, in kinetic exchange models

Agent-Based Modeling of Tax Evasion: Theoretical Aspects and Computational Simulations,
First Edition. Edited by Sascha Hokamp, László Gulyás, Matthew Koehler, and Sanith Wijesinghe.
© 2018 John Wiley & Sons Ltd. Published 2018 by John Wiley & Sons Ltd.

(Patriarca *et al.*, 2004) where agents exchange money in pairs between themselves. It can be shown (Patriarca *et al.*, 2004) that such models bear a close analogy to the kinetic theory of gases with agents representing particles and trades corresponding to the collision of particles.

The investigations and formalisms presented in this chapter deal with an econophysics-based multiagent theory of tax evasion within the Ising model. Originally developed for the description of ferromagnetism (Ising, 1925) the Ising model nowadays is also frequently used to investigate the dynamics of social and financial systems (for an overview see e.g., Castellano *et al.*, 2009; Galam, 2012; Sen and Chakrabarti, 2013; Sornette, 2014). One of the first applications in this context is attributed to Föllmer (1974) who investigated what he called an "Ising economy," that is, an egalitarian economy with two goods and two exclusive preferences. A review on more recent applications can be found in Sornette (2014). In Section 11.2 we will introduce the basic notions of the Ising model and in the following Section 11.2.3 will also illustrate a numerical approach for its solution based on the Monte-Carlo method. Further on, in Section 11.3, we discuss its application to tax evasion including the seminal papers by Zaklan *et al.* (2008, 2009) and in Section 11.4 the heterogeneous agent extension by Pickhardt and Seibold (2014) and Seibold and Pickhardt (2013). On the basis of discrete choice models the subsequent Section 11.4 connects the physical variables entering the Ising model to quantities that are more familiar in the economics community. In particular, we show that in this framework the Ising model is just a simple way to construct the utility function for interacting "binary" decision makers. Finally, in Section 11.6, we summarize the key issues of the econophysics approach to tax evasion and discuss further perspectives and possible developments.

11.2   The Ising Model

11.2.1   Purpose

The Ising model was originally proposed by Wilhelm Lenz as part of a Ph.D. problem to his student Ernst Ising and published in the Zeitschrift für Physik in 1925 (Ising, 1925). At that time the aim was to understand the magnetic properties of matter, in particular, the question why ferromagnetism breaks down in iron, cobalt, or nickel above a certain temperature T_c (so-called Curie temperature).

11.2.2   Entities, State Variables, and Scales

The model of Lenz and Ising is based on the assumption that the material hosts a collection of individual elementary magnets at positions $\mathbf{r}_i$. These magnets are also called magnetic moments and are mathematically characterized by a vector $\mathbf{S}_i \equiv \mathbf{S}(\mathbf{r}_i)$, which points from the elementary magnet's south pole to its north pole.

Uhlenbeck and Goudsmit (1925) proposed that the electron itself possesses a magnetic moment, the spin, which nowadays is known to dominate the ferromagnetic properties in the above mentioned materials. For this reason "spin" has become also a standard name for the vector $\mathbf{S}_i$ in the Ising model. Ising further assumed that interactions with the crystal restrict the orientation of the spin to a preferred direction, say z, so that only the component S_i^z is relevant. This component can take two values $S^z(\mathbf{r}_i) = \pm 1$ where the magnitude is only fixed for convenience and does not play a role in our further considerations. In order to generate a collective magnetic state the spins at locations $\mathbf{r}_i$ and $\mathbf{r}_j$ can interact via an exchange constant J_{ij}. Often, these interactions are short distance so that only the exchange J between nearest neighbors matters. The corresponding energy E of the Ising model reads as

$$E = -J \sum_{\langle ij \rangle} S_i^z S_j^z \qquad (11.1)$$

where the minus sign has been introduced for convenience and the brackets indicate summation over nearest neighbors. It should be noted that a different value for the magnitude of S^z would only imply a rescaling of the energy value. If we neglect for the moment the influence of temperature and are just interested in the state which minimizes the energy equation (11.1) then it becomes clear that the spin configuration strongly depends on the sign of the exchange constant J. For positive $J > 0$, Figure 11.1 shows the two ground state spin configurations on

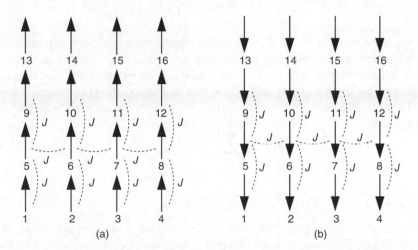

Figure 11.1   Ground state spin configurations of the Ising model on a 4×4 square lattice with nearest-neighbor interaction $J > 0$. At temperature $T = 0$ Eq. (11.1) yields two degenerate states where either all spins point in the "upward" (a) or "downward" (b) direction. Some of the nearest neighbor interactions J are indicated by dashed curvy lines. In addition, one often implements periodic boundary conditions where also sites of the left and right (upper and lower) boundary are connected by J, for example, sites 5 and 8 or sites 2 and 14

a 4×4 square lattice, both ferromagnetic and having the same energy.[1] Clearly, the ferromagnetic alignment is favored because each pair of neighboring spins, either both spin up or spin down, gives a contribution $-J$ to the energy. On the other hand, for $J < 0$ the ground state would be an anti-ferromagnet where spins alternately point in the up and down direction. In the following discussions we restrict to the $J > 0$ case.

While apparently the evaluation of the minimum energy spin configuration of Eq. (11.1) is rather trivial the situation becomes more involved at finite temperature. In this case, one considers the Ising model in connection with a heat reservoir of temperature T so that both systems can exchange portions of heat energy. This heat energy can in turn disorder the spins so that above a certain critical temperature T_c the ferromagnetic state breaks down. Formally, this problem has to be solved with methods of statistical mechanics (see e.g., Huang, 1987) and Ising in his 1925 thesis was only able to give a solution for a one-dimensional chain of spins. However, in this case the (disappointing) finding was that a phase transition toward a ferromagnetic state exists only at $T = 0$, that is, for an infinitesimal finite temperature the spins become immediately disordered. It was only in 1944 that the Ising model was solved for a two-dimensional square lattice by Onsager (1944). In this case, the phase transition occurs at a finite temperature

$$T_c = \frac{2J}{k \ln \left(1 + \sqrt{2}\right)}$$

where $k \approx 1.38 \times 10^{-23}$ J/K denotes the Boltzmann constant.[2] No analytical solution has been found until today for three-dimensional, cubics, lattices and so on.

A frequent extension of the Ising model, which we will also investigate in the following application to tax evasion, concerns the incorporation of a magnetic field B. In this case, even for the two-dimensional square lattice, no analytical solution is available.

$$E = -J \sum_{\langle ij \rangle} S_i^z S_j^z - B \sum_i S_i^z \tag{11.2}$$

11.2.3   Process Overview and Scheduling

The availability of analytical solutions for special systems only requires numerical methods in order to solve more complex Ising models as the ones we will use in our applications to tax evasion. For a given spin configuration $\{S^z\}_i$ we can compute the energy E_i from Eq. (11.1). At finite temperature T the corresponding

---

[1] The number of bonds of the 4×4 square lattice is $N_b = 24$ for open, and $N_b = 32$ for periodic boundary conditions corresponding to energies $E = -24J$ and $E = -32J$, respectively.

[2] J/K is the ratio between physical units for work (Joule) and temperature (Kelvin).

probability for this state is given by the Boltzmann weight

$$p(\{S^z\}_i) \equiv p(E_i) = \frac{1}{Z} e^{-E_i/kT} \tag{11.3}$$

and the normalization constant is called the partition function with

$$Z = \sum_i e^{-E_i/kT} \tag{11.4}$$

Note that the summation in Eq. (11.4) is over all possible spin configurations. Since each spin can take two orientations one has 2^N different realizations for a system with N spins and it is clear that an exact evaluation of the sum would only be possible for small lattice sizes, even numerically. In order to handle this problem one makes use of the so-called Monte-Carlo Metropolis algorithms that are based on the principle of detailed balance and have been introduced by Metropolis *et al.* (1953). This principle corresponds to the necessary condition (but not sufficient condition) for a system in equilibrium

$$p(E_i)p(E_i \rightarrow E_j) = p(E_j)p(E_j \rightarrow E_i) \tag{11.5}$$

or
$$\frac{p(E_i \rightarrow E_j)}{p(E_j \rightarrow E_i)} = \frac{p(E_j)}{p(E_i)} = e^{-(E_j-E_i)/kT} \tag{11.6}$$

which states that the probability of the system for having realized spin configuration $\{S^z\}_i$ and transitioning to configuration $\{S^z\}_j$ is the same as having realized $\{S^z\}_j$ and transitioning to $\{S^z\}_i$. The idea is now to approach the equilibrium state by a series of transitions $\{S^z\}_i \rightarrow \{S^z\}_j$ where two successive configurations only differ by a single spin flip. In order to implement this procedure one needs a probability to evaluate whether the spin flip is realized and which satisfies detailed balance. One popular implementation in this regard is the Glauber probability

$$p(\{S^z\}_i \rightarrow \{S^z\}_j) = \frac{1}{1 + \exp\{(E_j - E_i)/kT\}} \tag{11.7}$$

which satisfies Eq. (11.5) and where the energies E_j and E_i only differ by the contribution of a single spin flip.

The numerical implementation of this algorithm is then straightforward:

1. Start with an initial (e.g., random) spin configuration.
2. Choose a spin of the system at random.
3. Calculate the energy difference ΔE between the configuration where this spin would be flipped and the energy of the old configuration.
4. Generate a random number r with $0 \leq r \leq 1$.
5. If $p = 1/(1 + e^{\Delta E/kT}) > r$ accept the spin flip.
6. Start from (1) until the predefined number of steps is reached.

Results presented in this chapter have been obtained by a slight variation of this procedure; namely, instead of selecting spins at random [item (2)] we go through the lattice in a regular fashion and associate one complete sweep through the lattice as one (tax relevant) time period (see next section). More detailed discussions about Monte-Carlo schemes and practical implementations can be found for example, in Binder and Heermann (2010) and Krauth (2006).

11.3   Application to Tax Evasion

In a series of papers (Zaklan *et al.*, 2008, 2009; Lima and Zaklan, 2008) Zaklan and collaborators have suggested the application of the Ising model Glauber dynamics to the problem of tax evasion. According to their proposal a spin $S^z = +1$ is identified with a honest tax payer whereas $S^z = -1$ represents a cheater. Spins on a lattice thus represent a society of agents, which can either be compliant or noncompliant with regard to tax evasion. Although at first glance this seems to be a severe simplification, data from tax compliance laboratory experiments (Alm and McKee, 2006; Alm *et al.*, 2009; Bloomquist, 2011; Bazart and Bonein, 2014) and also data from the IRS National Research Program (2001) for small business filers (Bloomquist, 2011) support a bimodal distribution of the declared income which is peaked at zero and full income, respectively.

This feature is demonstrated in Figure 11.2, which in the main panel (full gray-shaded bins) reports the frequency of declarations obtained in the tax compliance experiment of Bazart and Bonein (2014). This experiment was designed as a pure declaration game, excluding redistribution of tax through the provision of public goods, and aimed to investigate reciprocal relationships in tax compliance decisions under various types of inequities. In each round (period) participants received an income of 100 points and were subject to certain fiscal policy parameters (e.g., tax audit, penalty rates, etc.; for details see Bazart and Bonein, 2014). Data shown in Figure 11.2 were obtained from the so-called horizontal inequity treatment where, at the end of each tax relevant period, participants learned the average income reported by the other members of their group. Important in the present context is that the declarations X_i^{data} of most participants are close to either $X_i^{data} = 100$ or $X_i^{data} = 0$, which results in the bimodal distribution shown with the full bins in the main panel of Figure 11.2. In order to map this distribution to Ising data X_i^{ising} one can introduce a threshold X_c and assign all declarations with $X_i^{data} \leq X_c$ to $X_i^{ising} = 0$ and all declarations with $X_i^{data} > X_c$ to $X_i^{ising} = 100$. In the Ising model, a participant with $X_i^{ising} = 0$ is then represented by a spin $S^z = -1$ (i.e., a noncompliant agent) and a participant with $X_i^{ising} = 100$ corresponds to a compliant agent with $S^z = +1$. Fixing the threshold to $X_c = 52$ yields the distribution shown by open bins in Figure 11.2 and one finds that the average of the experimental and Ising data is almost the same in each

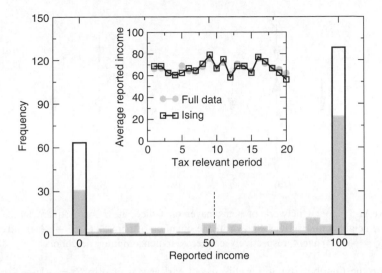

Figure 11.2   Main panel: The range of reported income ($0 \leq X_i^{data} \leq 100$) is divided into 21 bins and the height of each bin corresponds to the frequency of declarations within the corresponding X_i^{data}-interval. The Ising distribution (open bins) is obtained by assigning all declarations with $X_i^{data} \leq X_c$ to $X_i^{ising} = 0$ and all declarations with $X_i^{data} > X_c$ to $X_i^{ising} = 100$. Inset: Upon setting $X_c = 52$ (indicated by the dashed line) the Ising data give an excellent approximation to the average reported income as obtained from the experimental data (Bazart and Bonein, 2014) for all periods of the experiment. Source: Figure reproduced from Bazart *et al.* 2016 by courtesy of the Journal of Tax Administration.

period of the experiment as shown in the inset to Figure 11.2. This agreement also supports the validity of an Ising description of tax evasion behavior.

The horizontal inequity setting in the experiments of Bazart and Bonein (2014) aimed to study the question whether the evasion decision of one taxpayer will be conditional on those of all other taxpayers (see also Schnellenbach, 2010, and references therein). In fact, it was shown that some taxpayers did change their declaration decisions in the next period to get closer to the average reported income of other group members, that is, taxpayers tend to "copy" the predominant compliance behavior of those in their social network. In order to model this kind of behavior one has to take the interaction between agents into account, which in the Eq. (11.1) is due to the exchange constant J. As illustrated in Figure 11.3 this interaction causes an agent to copy the average behavior of its "social network," which for the simple structure depicted in Figure 11.3 consists of the four nearest neighbors but can be generalized in a straightforward manner to more complex networks.

In Figure 11.3(a) a honest tax payer ($S_i^z = +1$) interacts with noncompliant neighbors and, for the moment, we ignore the influence of an external field B.

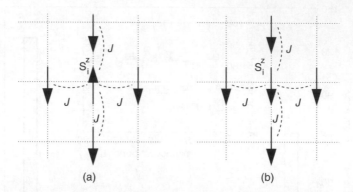

Figure 11.3 Social network of a tax payer on lattice site i for a square lattice with nearest-neighbor interactions. Shown are the situations of a honest ($S_i^z = +1$, (a)) and cheating ($S_i^z = -1$, (b)) agent, respectively, connected to noncompliant neighbors

Within the Glauber dynamics, discussed earlier, a spin flip $S_i^z = +1 \rightarrow S_i^z = -1$ would then lower the energy by $\Delta E = -8J$ ($-2J$ per bond) so that the associated transition probability Eq. (11.7) is large at low temperature (i.e., for $kT < J$). On the other hand, if the agent has the same evading behavior than its neighbors (Figure 11.3(b)) the energy change for the spin flip transition $S_i^z = -1 \rightarrow S_i^z = +1$ is positive ($\Delta E = +8J$) corresponding to a small transition probability at low temperature. In other words, the central agent tends to conform to the behavior of its neighbors where the adaptation increases with decreasing temperature. The latter parameter therefore is interpreted in Zaklan *et al.* (2008, 2009), Lima and Zaklan (2008) as a "social temperature parameter." In Section 11.5 we will show that in the framework of discrete choice models this parameter can also be related to the fluctuations of unobserved utility. In their numerical modeling of tax evasion dynamics, Zaklan *et al.* associated a tax relevant period with an adjustment of all lattice spins based on the Glauber probability Eq. (11.7). Moreover, they introduced the probability of an efficient audit which means that the detection of tax evasion (i.e., $S_i^z = -1$) forces the agent to stay honest over the following h time periods.

Figure 11.4 shows examples for the tax evasion dynamics obtained for selected parameter values on a 1000×1000 square lattice with nearest-neighbor interactions. Since only the ratio J/kT enters into the Glauber dynamics Eq. (11.7), the individual temperatures are specified with respect to J/k. Figure 11.4(a) and (b) display the extent of tax evasion (i.e., the fraction of -1-spins) as a function of time for small "social temperature" $T = 2$ and for small and large audit probabilities $\alpha = 0.05$ and $\alpha = 0.8$, respectively. The initial condition is full compliance for all agents.[3] Due to the small T-value between the economics and econophysics

---

[3] Note that the long-term behavior does not depend on the initial condition as is demonstrated by the dashed line in Figure 11.4(a). Here, all agents are set to noncompliance at the initial time step.

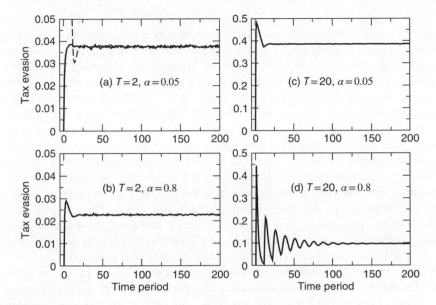

Figure 11.4   Tax evasion dynamics on a 1000×1000 square lattice with nearest-neighbor interactions J. (a, b) Results for agents with a social temperature of $T = 2$ and audit probabilities $\alpha = 0.05$ and $\alpha = 0.8$, respectively. (c, d) The corresponding results for a social temperature of $T = 20$. The scale of temperature is set by J/k. Tax audits enforce compliance over $h = 10$ time steps. Initial condition is full compliance for all agents ($S_i^z = +1$) except for (a) where the dashed line corresponds to an initial condition with full evasion ($S_i^z = -1$)

based approach only a small number of agents become noncompliant in the first time step and because the audit probability is also small (Figure 11.4(a)), tax evasion p_{te} increases continuously up to some saturation value $p_{te} \approx 0.038$. At this value the system of agents has reached equilibrium between the influence of the neighborhood to become a noncompliant taxpayer and the tax audits which enforce compliance over $h = 10$ time periods. Upon increasing the audit probability to $\alpha = 0.8$ (Figure 11.4(b)) the number of tax evaders reaches its maximum already after about three time steps and then starts to decrease due to the high audit probability. After 10 time steps, initially detected tax evaders may again switch from compliant to noncompliant behavior, which leads to the observed small increase of p_{te} before it reaches its saturation value of $p_{te} \approx 0.022$. Note that the "social temperature" $T = 2$ is below the ordering temperature $T_c = 2.269$ of the two-dimensional Ising model. For zero audit probability, the long-term tax evasion would therefore also converge to a small value ($p_{te} \approx 0.045$), corresponding to the fraction of "minority spins" at this particular temperature. The effect of a finite audit probability is, therefore, just a further reduction from this small value.

In contrast, much above the ordering temperature the zero audit equilibrium state is reached for an equal number of compliant and noncompliant tax payers, that is, $p_{te} = 0.5$. Finite audit probabilities then lead to a further reduction of this equilibrium value. This situation is depicted in Figure 11.4(c) and (d), which display the extent of tax evasion for a system build up from agents who decide mainly autonomously, due to the large "social temperature" $T = 20$ (i.e., much larger than the exchange energy). As a consequence, already after two time steps approximately half of the agents become noncompliant, which in turn induces a reduction of p_{te} due to the auditing. It is remarkable that even for small audit probability $\alpha = 0.05$ (Figure 11.4(c)) p_{te} decreases by $\approx 20\%$ before it saturates at $p_{te} \approx 0.39$ after more than 10 time steps. Upon invoking a larger audit probability, $\alpha = 0.8$, the number of autonomous agents decreases rapidly after the first initial time steps and consequently p_{te} decreases to almost zero within the first 10 time steps. Half the fraction of those who have been detected at the first time step will then select the possibility of noncompliance again, which leads to the oscillatory behavior of p_{te}. A stable situation is only reached at $\gtrsim 150$ time steps at $p_{te} \approx 0.1$.

A more detailed discussion of related results can be found in Zaklan *et al.* (2008, 2009), and Lima and Zaklan (2008). In particular, the influence of a homogeneous magnetic field B, interpreted as the influence of mass media, has been investigated by Lima and Zaklan (2008). In this context, a positive $B > 0$ would enhance the confidence in tax authorities whereas $B < 0$ would imply a reduction of tax payers trust in governmental institutions. It has also been shown that the results are not sensitive to the specific lattice geometry. This issue has been analyzed in Zaklan *et al.* (2008, 2009), Lima and Zaklan (2008) where, in addition, alternative lattice structures such as the scale-free Barabási-Albert network or the Voronoi–Delaunay network have been considered. In addition, Lima (2010) investigates Erdös–Rényi random graphs and finds that the results for these alternative lattices do not differ significantly from those obtained with a square lattice. More recently, a generalization of the model with regard to the incorporation of nonequilibrium dynamics has been proposed in Lima (2010, 2012a,b).

Extensions of the model concern also the implementation of different audit schemes. The original proposal (Zaklan *et al.*, 2009, 2008) with enforcement of compliance over a fixed number of time steps upon detection of tax evasion has been investigated in a randomized variant in Lima and Zaklan (2008). Moreover, also lapse of time effects have been studied (Seibold and Pickhardt, 2013), that is, the situation where a detected agent is screened over several years in the past by the (tax) authorities (i.e., backaudit).

11.4 Heterogeneous Agents

In the original Zaklan model (Zaklan *et al.*, 2008, 2009), each agent is either compliant or noncompliant and transitions between both behaviors are possible. However, the behavioral dynamics is governed by the Glauber probabilities (Eq. (11.7)) which, besides the contribution of the social network, depends on the "global parameters" temperature and magnetic field only. Therefore, agents in a social network consisting of the same number of cheating as well as honest tax payers have the same Glauber probabilities, which means that they are not endogenously different. However, most economic agent-based models assume behaviorally different agent types as, for instance, in the work of Hokamp and Pickhardt (2010) where four behavioral types have been introduced. First, they consider maximizing A-Types that show rational and risk-adverse behavior by maximizing their expected utility in agreement with Allingham and Sandmo (1972). Second, the model incorporates imitating B-Type agents that copy the behavior of the majority of their social network; that is, their decision on being compliant or noncompliant in the tax relevant period t_n is dependent on the average compliance of agents in their social network at the previous period t_{n-1}. The third type of agents are "ethical" ones which, motivated by behavioral norms, are mostly compliant. Finally, random D-Types constitute the fourth type of agents. Due to the complexity of the tax law these agents may make unintended mistakes with respect to declaring their true income. Therefore, D-Types randomly switch between being compliant and noncompliant.

How can we implement these behaviorally different agents into the Ising model description of tax evasion discussed in Section 11.3. Concerning the A-Type agents it is obviously not possible to have risk-adverse behavior in our ecophysics model. Within the approach of Allingham and Sandmo (1972) this would imply a concave utility function which allows for an "inner solution," that is, the possibility that agents declare part of their income. However, the variables in the Ising model can take only two values, ± 1 corresponding to full- or zero-income declaration. In the Allingham–Sandmo model, such agents are risk-neutral with an utility function which is linear in the after tax and penalty net income. This yields full (zero) declaration if the audit probability α is larger (smaller) than the ratio between the tax rates of declared and undeclared income. In practice this means that for real values of the various parameters risk-neutral A-Type agents mostly declare zero income. In the Ising model we can implement such behavior by coupling the spin that represents the A-Type to a local *negative* magnetic field $B_i^a < 0$, that is, the energy functional equation (11.1) is supplemented by a term

$$E^a = - \sum_{i \in \{a\}} B_i^a S_i^z \tag{11.8}$$

where the summation runs over all sites that are occupied by an A-Type. Inspection of the Glauber probability equation (11.7) reveals that the magnetic field has only a significant influence when it is (i) larger than the contribution from neighbors (i.e., $|B_i^a| \gg J$) and (ii) when it is not overcome by large temperature fluctuations. In order to locally satisfy the last criterion we couple each site i to a heat bath with temperature T_i, that is, in contrast to the original Zaklan model (Zaklan et al., 2008, 2009) temperature is not a global parameter but is locally defined at each site.[4] For A-Types this means that in addition $|B_i^a| \gg T_i^a$ should be satisfied in order to obtain large Glauber probabilities for $S_i^z = -1$ states on the A-Type sites. In the same way, C-Type agents can be modeled by the coupling to local magnetic fields $B_i^c > 0$ via

$$E^c = - \sum_{i \in \{c\}} B_i^c S_i^z \tag{11.9}$$

where again the relations $|B_i^c| \gg T_i^c$ and $|B_i^c| \gg J$ should be satisfied in order to have the dominant influence on the Glauber probabilities given by B_i^c. The positive sign of the field (as opposed to the negative A-Type field B_i^a) lowers the energy contribution Eq. (11.9) for $S_i^z = +1$ states.

For imitating B-Type agents the Glauber probability should be maximized for spin states close to the average spin of their social network. For this purpose we model these agents by a small local temperature $T_i^b \ll J$ so that the influence of the interaction energy $\sim J$ is dominant. Also, random D-Types are characterized by the local temperature, which has to satisfy $T_i^d \gg J$. In this case, one sees from Eq. (11.7) that the Glauber probabilities are close to $p \approx 0.5$, that is, the random agent is strongly fluctuating between compliant and noncompliant behavior.

Figure 11.5 illustrates the Ising model design for heterogeneous agents for a 4×4 lattice with nearest-neighbor couplings J. Every site is coupled to an individual heat bath with temperature T_i, represented by the boxes. Moreover, the spins on A-Type (C-Type) sites are coupled to a negative (positive) magnetic field (as visualized by the large gray arrow). The parameter ranges for the individual agent types are summarized in Table 11.1.

Clearly, the various agent types are textbook cases for the individual behavior of tax payers. In principle, one can parametrize an agent at lattice site i by an arbitrary combination for the field and temperature values B_i, T_i so that besides the "pure" types, also agents with an intermediate behavior (e.g., between A- and B-Types) are possible. In fact, this is what one finds when the parameters are calibrated with tax compliance laboratory experiments (Bazart *et al.*, 2016).

Figure 11.6 shows the Glauber probabilities (Eq. (11.7)) for a compliant tax payer switching to noncompliance in the (*magnetic field, temperature*) parameter space. Figure 11.6(a) (Figure 11.6(b)) reports the situation that all neighbors are noncompliant (compliant) and the "pure" agent-types are indicated in the panels. For example, a compliant A-Type (which can be the result of an audit) has always probability $p \approx 1$ to switch to noncompliance, irrespective of the behavior of neighbors. It is also apparent from Figure 11.6 that only the probability of B-Types strongly depends on the state of the nearest neighbors. In any case, it is possible to change the behavior of agents continuously by changing the field and (or) temperature parameters and therefore to have agents with an intermediate behavior between the "pure" types.

---

[4] Here the local temperature is a fixed "external" parameter, which differs from the case where a temperature gradient is applied to the system boundaries and local T_is are then obtained as a result of heat flow as, for example, in Harris and Grant (1988) and Agliari *et al.* (2010).

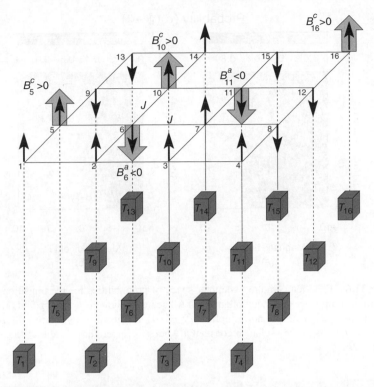

Figure 11.5 Ising model with heterogeneous agent types. Each site is coupled to a heat bath with temperature T_i as indicated by boxes. Moreover, A-Type agents are coupled to a *negative* magnetic field (e.g., site 6) whereas C-Type agents are coupled to a *positive* magnetic field (e.g., site 10). Nearest neighbors are coupled by the exchange constant J

Table 11.1 Summary of parameter ranges that specify the various agent types in the heterogeneous Ising model

Type	Parameter specification
a	$|B_i^a < 0| \gg T_i^a$; $|B_i^a < 0| \gg J$
b	$T_i^b \ll J$; $|B_i^b| \approx 0$
c	$|B_i^c > 0| \gg T_i^a$; $|B_i^a > 0| \gg J$
d	$T_i^b \ll J$; $|B_i^b| \approx 0$

Figure 11.7 reports the tax evasion dynamics (i.e., the fraction of $S_i^z = -1$ agents) for different compositions of the heterogeneous tax payer society. For simplicity we restrict to "pure" agent-types with the corresponding parameter values given in the caption to Figure 11.7. Initial conditions at time step $t = 0$

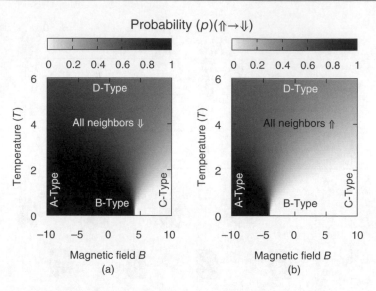

Figure 11.6 Field-temperature parameter space of probabilities for the transition of an agent from compliance to noncompliance. (a) All nearest neighbors are compliant ($S_i^z = 1$), (b) All nearest neighbors are noncompliant ($S_i^z = -1$) as illustrated in Figure 11.3. B-field values are in units of the exchange constant J whereas temperature T values are given in units of J/k

are such that $S_i^z = 1$ for C-Types and $S_i^z = -1$ for A-, B-, and D-Type agents. As shown in Figure 11.4 the equilibrium value at large times does not depend on these initial conditions as these are only relevant on a short time scale after $t = 0$. For this reason, the initial fraction of tax evasion in Figure 11.7 is always one minus the fraction of C-Types. Panel I.) of Figure 11.7 shows the evasion dynamics for 20% B- and D-Types, while the respective fraction of noncompliant A-Types and fully compliant C-Types varies by 20%. This also induces a variation in the long-term tax evasion of about $\sim 22\%$, which is slightly larger because of the contribution of B-Types that copy the behavior of A- and C-Types when they are nearest neighbors. In panel II.) the percentage of A- and C-Types is fixed to 20% whereas the relative fraction of B- and D-Types is varied. Note that an ensemble of B-Types is predominantly compliant (cf. Figure 11.4(a, b)) whereas the tax evasion of random D-Types approaches 0.5 for low audit rates (cf. Figure 11.4(c)). Therefore, a shift of the relative fraction from B- to D-Types, as in panel II.), increases tax evasion, where the increment decreases with decreasing percentage of B-Types. This is due to the decreasing amount of B-Type clusters that contribute to compliance so that for smaller fractions of B-Types these are predominantly connected to A- and C-Types. The corresponding imitating effect compensates and therefore tax evasion increases by a lesser degree when the fraction of B-Types is reduced from 20% to 0% as compared to the reduction

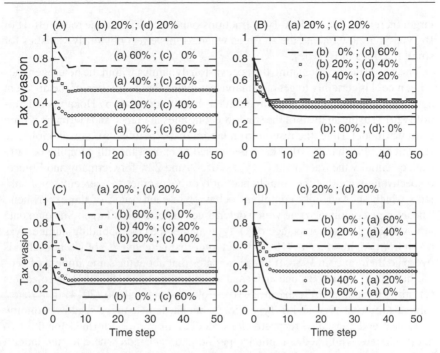

Figure 11.7 Tax evasion dynamics for the multi-agent Ising model. In each panel the percentage of two agent-types is fixed to 20% as indicated in the heading. The fraction of the other two agent-types is then varied in steps of 20%. System size: 1000×1000 square lattice with nearest-neighbor coupling J. Enforced compliance period after audit: $h = 4$ time steps. Agents are specified by the following parameters: A-Types ($T_i^a = 1$, $B_i^a = -20$), B-Types ($T_i^b = 1, B_i^b = 0$), C-Types ($T_i^c = 1, B_i^c = 20$), D-Types ($T_i^d = 20, B_i^d = 0$). Temperature values are in units of J/k and field values in units of J

from 60% to 40% (with a corresponding increase in the fraction of D-Types). The situation of fixed percentages ($= 20\%$) for A- and D-Types is illustrated in panel III.). For an additional fraction of 60% for B-Types, the long-term tax evasion is ~ 0.55, that is, about 0.25 points larger than what one expects from the contribution of A- (~ 0.2) and D- (~ 0.1) types. Clearly, this "excess" evasion comes from the imitating B-Types, which, due to the interaction with noncompliant A-Types, contribute to tax evasion. Decreasing the fraction of B-Types with a similar increase of C-Types obviously enhances compliance because now the still significant fraction of B-Types (cf. e.g., the result for 40% B-Types) can also imitate the compliant C-Type behavior. This effect vanishes for a small percentage of B-Types so that in this limit panel III.) only shows a weak variation of the corresponding long-term evasion. Finally, panel IV.) shows the result for fixed C- and D-Type fraction while the percentage of A- and C-Types is changed. Here the increase of A-Types naturally increases evasion with a larger increment for small A-Type fractions panel III.).This is due to the effect of B-Types that now imitate the A-Types whereas this effect obviously vanishes for small fractions of B-Types.

The key issue for calculating the tax evasion dynamics within the heterogeneous Ising model is certainly to get reasonable estimates for the fraction of the different agent types. A first step in this direction has been undertaken by Hokamp and Seibold (2014) in the analysis of the shadow economies in France and Germany within a similar model. Here, the results from laboratory experiments on tax compliance (Bazart and Pickhardt, 2011) have been used to extract the full compliance ratio (i.e., essentially the fraction of C-Types) of 5% and 20% for Germany and France, respectively. Note that the compliance ratio only specifies the percentage of subjects which showed fully compliant behavior in each round of the experiment. The average compliance rate which determines tax evasion is larger. Since no data are available for the percentage of D-Type agents the corresponding number was taken from Andreoni *et al.* (1998) which reports that about 7% of US households overpaid their taxes in 1988. If one anticipates that about the same amount of people underpays their taxes one arrives at a percentage of $\sim$ 15% of D-Type agents (cf. also Hokamp and Pickhardt, 2010). Hokamp and Seibold (2014) then determine the fraction of A- and B-Type agents that participate in the shadow economy in France and Germany, respectively. Calculations were performed for the two audit schemes with fixed compliance period and backaudit where for the latter a backaudit period of $b_p = 5$ time steps has been considered. The latter is compatible with the limitation period of five years in 2006/2007 where the experiments by Bazart and Pickhardt (2011) have been conducted. Data on the shadow economies for France and Germany were adopted from Buehn and Schneider (2012), which for the two countries reports 15% and 16% of the official GDP, respectively, averaged over the period 1999–2007. These numbers are supported by the data of Elgin and Öztunali (2012), which report 16.53% for OECD EU-countries in the period from 2001–2009. As a result, the analysis revealed a larger share of A-Type agents for Germany, where the absolute magnitude and the difference to France depends on the considered enforcement scheme. For example, the back audit mechanism results in a fraction of $\sim$ 16% A-Types for France and $\sim$ 11% A-Types for Germany. The missing fraction then corresponds to the B-Types, which yields 69% (Germany) and 49% (France). Note that a more direct estimate for the percentage of agent types has been performed recently in Bazart *et al.* (2016) where the field and temperature parameter has been determined for individual participants in a tax evasion laboratory experiment.

11.5   Relation to Binary Choice Model

There is a close relationship between the Glauber probabilities in Eq. (11.7) and the probabilities which can be derived within the logit discrete choice model. The development of the latter was awarded with the 2000 Nobel prize in economics to Daniel L. McFadden "for his development of theory and methods for analyzing discrete choice." This connection therefore allows to link the dynamics of the behavioral dynamics obtained from utility-maximizing behavior and therefore provides a link between econophysics and classical economics description of tax evasion.

Consider an agent, in the present context called decision maker, which can choose between the two alternatives of declaring zero or full income. Taking again an ensemble of heterogeneous agents, the decision maker n obtains a certain level of utility (or profit) from each alternative, which we denote by $\mathcal{U}_{nj}$ with $j = 0, 1$ corresponding to zero and full income declaration, respectively. Further on we assume that the individual's utility from choosing the alternative j takes the form $\mathcal{U}_{nj} = \mathcal{V}_{nj} + \mathcal{I}_{nj}$, where $\mathcal{V}_{nj}$ is common to all individuals with observed attributes j (i.e., declaration of full or zero income), while $\mathcal{I}_{nj}$ is particular to the drawn individual. Both utility terms are deterministic, the first reflecting "representative" (and observable) tastes in the population, and the second reflecting (unobservable) idiosyncratic taste variations. The probability P_{nj} that decision maker n chooses alternative j is then given by

$$
\begin{aligned}
P_{nj} &= \mathrm{Prob}(\mathcal{U}_{nj} > \mathcal{U}_{ni}) \quad \text{for } i \neq j \\
&= \mathrm{Prob}(\mathcal{V}_{nj} - \mathcal{V}_{ni} + \mathcal{I}_{nj} > \mathcal{I}_{ni}) \quad \text{for } i \neq j
\end{aligned}
\tag{11.10}
$$

and the probabilities are normalized, $P_{n0} + P_{n1} = 1$.

The basic assumption of the logit model is that the unobserved contributions, $\mathcal{I}_{nj}$, are independent with regard to agent n, and alternative $j = 0, 1$ and are specified by the Gumbel or double exponential probability density

$$
p(\mathcal{I}_{nj}) = e^{-\mathcal{I}_{nj}} e^{e^{-\mathcal{I}_{nj}}}.
\tag{11.11}
$$

The probabilities equation (11.10) can therefore be expressed via the corresponding cumulative distribution

$$
h(\mathcal{I}_{nj}) = e^{e^{-\mathcal{I}_{nj}}}
\tag{11.12}
$$

weighted by the density of the unknown $\mathcal{I}_{nj}$. One finds for the probability of zero income declaration for agent n

$$
\begin{aligned}
P_{n0} &= \int d\mathcal{I}_{n0} p(\mathcal{I}_{n0}) h(\mathcal{V}_{n0} - \mathcal{V}_{n1} + \mathcal{I}_{n0}) \\
&= \int d\mathcal{I}_{n0} e^{-\mathcal{I}_{n0}} e^{e^{-\mathcal{I}_{n0}}} e^{e^{-\mathcal{V}_{n0} + \mathcal{V}_{n1} - \mathcal{I}_{n0}}} \\
&= \frac{1}{1 + e^{-(\mathcal{V}_{n0} - \mathcal{V}_{n1})}}
\end{aligned}
\tag{11.13}
$$

where the integration can be easily performed by substituting $x = e^{-\mathcal{I}_{n0}}$. Similarly, the probability for full income declaration is obtained as

$$P_{n1} = \frac{1}{1 + e^{-(\mathcal{V}_{n1} - \mathcal{V}_{n0})}} \tag{11.14}$$

so that the condition $P_{n1} + P_{n0} = 1$ is satisfied. Thus the decision probability on declaring zero or full income depends on the difference between the corresponding observable parts of the utility function $\mathcal{V}_{n0} - \mathcal{V}_{n1}$. In case $\mathcal{V}_{n0} - \mathcal{V}_{n1} \gg 0$, which yields $P_{n0} \approx 1$, the agent will most probably declare zero income while full income declaration is most probable for $\mathcal{V}_{n0} - \mathcal{V}_{n1} \ll 0$. From Eq. (11.10) it turns out that the decision probabilities are solely determined by utility differences so that the utility functions themselves are only specified up to a constant. Moreover, since the agent chooses the alternative with the largest utility the overall scale of the utility function does not influence on the decision. One possibility (Train, 2002) to normalize the scale of utility is to fix the variance of the unobserved contributions $\mathcal{I}_{nj}$, that is, to scale the utility by a factor $1/\sigma_n$ such that $\mathrm{Var}(\mathcal{I}_{nj}/\sigma_n) = \mathrm{Var}(\mathcal{I}_{nj})/\sigma_n^2 = 1$. Note that we consider here the possibility that the variance σ_n (or the "range of taste variations") is different among agents. For each agent the rescaled utility functions $\widetilde{\mathcal{U}}_{nj} = \mathcal{V}_{nj}/\sigma_n + \widetilde{\mathcal{I}}_{nj}$ has then the property $\mathrm{Var}(\widetilde{\mathcal{I}}_{nj}) = 1$ and one finds for the corresponding probabilities of zero and full income declaration

$$P_{n0} = \frac{1}{1 + e^{-(\mathcal{V}_{n0} - \mathcal{V}_{n1})/\sigma_n}} \tag{11.15}$$

$$P_{n1} = \frac{1}{1 + e^{-(\mathcal{V}_{n1} - \mathcal{V}_{n0})/\sigma_n}} \tag{11.16}$$

For each agent the probabilities are therefore determined by the ratio between the observable utility parts and the standard deviation of the unobservable contribution. For independent (isolated) agents the difference between Eqs (11.13) and (11.14) and Eqs (11.15) and (11.16) is irrelevant since the relative order of probabilities is preserved under the rescaling. However, it matters if we consider interacting agents where the utility function of agent n may also depend on the decisions of the neighbors in the network. The construction of such an utility function for interacting agents is facilitated by noting the similarity of Eqs (11.15) and (11.16) with the Glauber probability equation (11.7), which provides a correspondence between the difference in observable utility divided by the standard deviation of unobserved utility and the energy difference in the Ising model divided by (local) temperature

$$\frac{\mathcal{V}_{n0} - \mathcal{V}_{n1}}{\sigma_n} \leftrightarrow \frac{E(S_n = 1) - E(S_n = -1)}{T_n} \tag{11.17}$$

In particular, this also suggests a correspondence between the observable part of the utility function and the (negative) energy in the Ising model which is reasonable since a state with lower energy (i.e., more negative) has larger probability in the Monte-Carlo dynamics and therefore larger utility from the perspective of a discrete choice model. Moreover, the correspondence (Eq.(11.17)) offers an interpretation of temperature in terms of the standard deviation of the unobservable utility part. The latter refers to the spread in the nonmeasurable taste or attitude so that according to our previous classification noncompliant (A-Types), copying (B-Types), and compliant (C-Types) agents are characterized by a small σ_n corresponding to their fixed moral attitude towards tax evasion (A- and C-Types) or their fixed copying behavior (B-Types). On the other hand, random D-Types display large variations of attitude toward evasion and thus can be described by a large σ_n. Thus the standard deviation of the unobserved utility σ_n specifies the agents in the same way than the temperature parameter T_n in the Ising model which was empirically fixed based on the functional form of the Glauber transition probabilities equation (11.7).

11.6   Summary and Outlook

Among the agent-based models for tax evasion discussed in this book, the Ising model-based approach may be considered as a "maverick" since it is not derived from an economic perspective but adopted from physics so that its validity in the present context at a first glance seems rather ill-founded. One issue concerns the fact that agents following the Ising model approach can only declare zero or full income. In this regard we have seen in Section 11.3 that a variety of data supports a bimodal distribution in the reported income compatible with an Ising-type behavior of taxpayers. In principle, the restriction of "two-state" Ising agents can be resolved by generalizing the description in terms of the so-called Potts model (Potts, 1952; Wu, 1982) where each spin (agent) can take q orientations and the limit $q = 2$ would then correspond to the standard Ising model. In this way it would be possible to discretize the taxable income into q bins and also to specify the reported income within these q values. This would also allow to attribute to each agent an after tax and income whereas the present Ising model formulation only provides information about compliance decisions. A first step in this direction has been recently undertaken by Crokidakis (2014).

We have seen in Section 11.5 that one can derive a correspondence between the observable part of the utility function and the (negative) energy in the Ising model, which thus provides a link between the economics-based approach and the econophysics-based approach. This allows to study the differences between both descriptions with regard to the implementation of the dynamics and also the direct (nonmarket based) interaction among agents. Concerning the dynamical aspect, agent-based models from the economics domain are often based on a generalization of the Allingham–Sandmo theory (Allingham and Sandmo, 1972) where either all or a subgroup of agents display rational behavior (cf. e.g., Hokamp and Pickhardt, 2010 and references therein) so that there exists an utility function with is inhomogeneous with respect to the parameters specifying net income, perceived audit probability, and so on. The dynamics is then implemented by requiring at each time step t_n (i.e., at each tax relevant period) the stationarity of the utility function. This means that rational agents decide instantaneously on their declared income $X_i(t_n)$ that only depends on the parameter values $\{p(t_n)\}$ in the same period. On the other hand, in the econophysics approach, the Ising model dynamics is governed by Glauber transition probabilities equation (11.7) [or equivalently Eqs (11.15) and (11.16)], which therefore corresponds to a Markov process.[5] This means that also in this case the dynamics is "memoryless" with the difference that the state of the system at time step t_{n+1} is solely determined by the state at time step t_n. As a consequence the "econophysics dynamics" does not necessarily correspond to the saddle point of the utility function and also explores the fluctuating environment. Concerning the interactions among agents, the econophysics Ising model description includes these as nonlinearities of the spin variables in the energy functional (or utility function). The dynamics of imitating taxpayers is then obtained within the same Glauber formalism as for (non)compliant agents, which only differ by their parameter values (i.e., local temperature and field). On the other hand, economic agent-based models often use distinct dynamics for imitating agents as compared to rational ones. For example, Hokamp and Pickhardt (2010) evaluate the ratio of declared $[X_i(t_n)]$ and full $[W_i(t_n)]$ income of imitating B-Types at time step t_n from the average ratio of the social network at the previous time step t_{n-1}, that is, contrary to rational agents the dynamics of B-Types is not instantaneous. Summarizing, the appealing feature of the Ising model-based econophysics approach to tax evasion is the coherent description of agent types within a minimal parameter set. The calibration of tax compliance experiments (Bazart et al., 2016) even reveals that it is possible to characterize the decision probability of participants with a temperature and field parameter, which offers a new perspective for the quantitative evaluation of such investigations. As discussed earlier a future generalization of the formalism can be achieved on the basis of the Potts model that allows for a more quantitative description of pay-offs and therefore of a more realistic implementation of audit schemes.

References

Agliari, E., Casartelli, M., and Vezzani, A. (2010) Microscopic energy flows in disordered Ising spin systems. *Journal of Statistical Mechanics: Theory and Experiment*, P10021-1–13.

Allingham, M.G. and Sandmo, A. (1972) Income tax evasion: a theoretical analysis. *Journal of Public Economics*, **1**, 323–338.

Alm, J. and McKee, M. (2006) Audit certainty, audit productivity, and taxpayer compliance. *National Tax Journal*, **59**, 801–816.

Alm, J., Deskins, J., and McKee, M. (2009) Do individuals comply on income not reported by their employer? *Public Finance Review*, **37**, 120–141.

---

[5] This only holds in the absence of audits which enforce the compliance of agents over several time steps and therefore violate detailed balance.

Andreoni, J., Erard, B., and Feinstein, J. (1998) Tax compliance. *Journal of Economic Literature*, **36**, 818–860.

Bazart, C. and Bonein, A. (2014) Reciprocal relationships in tax compliance decisions. *Journal of Economic Psychology*, **40**, 83–102.

Bazart, C., Bonein, A., Hokamp, S., and Seibold, G. (2016) Behavioral economics, tax evasion and the shadow economy – calibrating an agent-based econophysics model with experimental tax compliance data. *Journal of Tax Administration*, **2**, 126–144.

Bazart, C. and Pickhardt, M. (2011) Fighting income tax evasion with positive rewards. *Public Finance Review*, **39**, 124–149.

Binder, K. and Heermann, D.W. (2010) *Monte Carlo Simulation in Statistical Physics*, Springer-Verlag, Heidelberg, Dordrecht, London, New York.

Bloomquist, K. (2011) Tax compliance as an evolutionary coordination game. *Public Finance Review*, **39**, 25–49.

Buehn, A. and Schneider, F. (2012) Shadow economies around the world: novel insights, accepted knowledge, and new estimates. *International Tax and Public Finance*, **19**, 139–171.

Castellano, C., Fortunato, S., and Loreto, V. (2009) Statistical physics of social dynamics. *Reviews of Modern Physics*, **81**, 591–646.

Chakraborti, A., Toke, I.M., Patriarca, M., and Abergel, F. (2011a) Econophysics review: I. Empirical facts. *Quantitative Finance*, **11**, 991–1012.

Chakraborti, A., Toke, I.M., Patriarca, M., and Abergel, F. (2011b) Econophysics review: II. Agent-based models. *Quantitative Finance*, **11**, 1013–1041.

Chen, S.-H. and Li, S.-P. (2012) Econophysics: bridges over a turbulent current. *International Review of Finance Analysis*, **23**, 1–10.

Crokidakis, N. (2014) A three-state kinetic agent-based model to analyze tax evasion dynamics. *Physica A*, **414**, 321–328.

Elgin, C. and Öztunali, O. (2012) Shadow economies around the world: model based estimates. Working Papers 2012/05, Bogazici University, Department of Economics.

Föllmer, H. (1974) Random economies with many interacting agents. *Journal of Mathematical Economics*, **1**, 51–62.

Galam, S. (2012) *Sociophysics*, Springer-Verlag, Berlin.

Gallegati, M., Keen, S., Lux, T., and Ormerod, P. (2006) Worrying trends in econophysics. *Physica A*, **370**, 1–6.

Harris, R. and Grant, M. (1988) Thermal conductivity of a kinetic Ising model. *Physical Review B*, **38**, 9323–9326.

Huang, K. (1987) *Statistical Mechanics*, John Wiley & Sons, Inc., New York.

Hokamp, S. and Pickhardt, M. (2010) Income tax evasion in a society of heterogeneous agents - evidence from an agent-based model. *International Economic Journal*, **24**, 541–553.

Hokamp, S. and Seibold, G. (2014) How much rationality tolerates the shadow economy? An agent-based econophysics approach, in *Advances in Social Simulation - Proceedings of the 9th Conference of the European Social Simulation Association* (eds B. Kaminski and G. Koloch), Springer-Verlag, Berlin/Heidelberg, pp. 119–128.

Ising, E. (1925) Beitrag zur Theorie des Ferromagnetismus. *Zeitschrift fur Physik*, **31**, 253–258.

Krauth, W. (2006) *Statistical Mechanics; Algorithms and Computations*, Oxford University Press.

Lima, F.W.S. (2010) Analysing and controlling the tax evasion dynamics via majority-vote model. *Journal of Physics: Conference Series*, **246**, 012033, 12 pp.

Lima, F.W.S. (2012a) Three-state majority-vote model on square lattice. *Physica A*, **391**, 1753–1758.

Lima, F.W.S. (2012b) Tax evasion dynamics and Zaklan model on opinion-dependent network. *International Journal of Modern Physics C*, **23**, 1250047–11.

Lima, F.W.S. and Zaklan, G. (2008) A multi-agent-based approach to tax morale. *International Journal of Modern Physics C*, **19**, 1797–1808.

Liu, Y., Gopikrishnan, P., Cizeau, P. *et al.* (1999) Statistical properties of the volatility of price fluctuations. *Physical Review E*, **60**, 1390–1400.

Lux, T. and Marchesi, M. (1999) Scaling and criticality in a stochastic multi-agent model of a financial market. *Nature*, **397**, 498–500.

McCauley, J. (2004) *Dynamics of Markets, Econophysics and Finance*, Cambridge University Press.

Mantegna, R.N. and Stanley, H.E. (2000) *An Introduction to Econophysics - Correlations and Complexity in Finance*, Cambridge University Press.

Metropolis, N., Rosenbluth, A.W., Rosenbluth, M.N. *et al.* (1953) Equation of state calculations by fast computing machines. *Journal of Chemical Physics*, **21**, 1087–1092.

Onsager, L. (1944) Crystal statistics. I. A two-dimensional model with an order-disorder transition. *Physical Review B*, **65**, 117–149.

Patriarca, M., Chakraborti, A., and Kaski, K. (2004) Statistical model with a standard Γ distribution. *Physical Review E*, **70**, 016104-1–5.

Pickhardt, M. and Seibold, G. (2014) Income tax evasion dynamics: evidence from an agent-based econophysics model. *Journal of Economic Psychology*, **40**, 147–160.

Potts, R.B. (1952) Some generalized order-disorder transformations. *Mathematical Proceedings of the Cambridge Philosophical Society*, **48**, 106–109.

Schnellenbach, J. (2010) Vertical and horizontal reciprocity in a theory of taxpayer compliance, in *Developing Alternative Frameworks for Explaining Tax Compliance* (eds J. Alm, J. Martinez-Vasquez, and B. Torgler), Routledge International Studies in Money and Banking, Routledge, pp. 56–73.

Seibold, G. and Pickhardt, M. (2013) Lapse of time effects on tax evasion in an agent-based econophysics model. *Physica A*, **392**, 2079–2087.

Sen, P. and Chakrabarti, B.K. (2013) *Sociophysics: An Introduction*, Oxford University Press.

Sornette, D. (2014) Physics and financial economics (1776–2014): puzzles, Ising and agent-based models. *Reports on Progress in Physics*, **77**, 062001-1–28.

Stanley, M.H.R., Amaral, L.A.N., Buldyrev, S.V. *et al.* (1996) Scaling behaviour in the growth of companies. *Nature*, **379**, 804–806.

Stanley, H.E., Amaral, L.A.N., Gabaix, X. *et al.* (2001) Similarities and differences between physics and economics. *Physica A*, **299**, 1–15.

Train, K.E. (2002) *Discrete Choice Methods with Simulation*, Cambridge University Press.

Uhlenbeck, G.E. and Goudsmit, S. (1925) Ersetzung der Hypothese vom unmechanischen Zwang durch eine Forderung bezüglich des inneren Verhaltens jedes einzelnen Elektrons. *Naturwissenschaften*, **47**, 953–954.

Wu, F.Y. (1982) The Potts model. *Reviews of Modern Physics*, **54**, 235–268.

Yamasaki, K., Muchnik, L., Havlin, S. *et al.* (2005) Scaling and memory in volatility return intervals in financial markets. *Proceedings of the National Academy of Sciences of the United States of America*, **102**, 9424–9428.

Zaklan, G., Lima, F.W.S., and Westerhoff, F. (2008) Controlling tax evasion fluctuations. *Physica A*, **387**, 5857–5861.

Zaklan, G., Westerhoff, F., and Stauffer, D. (2009) Analysing tax evasion dynamics via the Ising model. *Journal of Economic Interaction and Coordination*, **4**, 1–14.

Index

Agent-Based Modeling of Tax Evasion: Theoretical Aspects and Computational Simulations,
First Edition. Edited by Sascha Hokamp, László Gulyás, Matthew Koehler, and Sanith Wijesinghe.
© 2018 John Wiley & Sons Ltd. Published 2018 by John Wiley & Sons Ltd.